Acoustical Factors Affecting Hearing Aid Performance

Acoustical Factors Affecting Hearing Aid Performance is a volume in the **PERSPECTIVES IN AUDIOLOGY SERIES**—Lyle L. Lloyd, series editor. Other volumes in the series include:

Publisher's Note
Perspectives in Audiology is a carefully planned series of clinically oriented and basic science textbooks. The series is enriched by contributions from leading specialists in audiology and allied disciplines. Because technical language and terminology in these disciplines are constantly being refined and sometimes vary, this series has been edited as far as possible for consistency of style in conformity with current majority usage as set forth by the American Speech-Language-Hearing Association, the *Publication Manual of the American Psychological Association,* and the University of Chicago's *A Manual of Style.* University Park Press and the series editors and authors welcome readers' comments about individual volumes in the series or the series concept as a whole in the interest of making **Perspectives in Audiology** as useful as possible to students, teachers, clinicians, and scientists.

A Volume in the Perspectives in Audiology Series

ACOUSTICAL FACTORS AFFECTING HEARING AID PERFORMANCE

Edited by

Gerald A. Studebaker, Ph.D.
Distinguished Professor
Department of Audiology and Speech Pathology
Memphis State University

and

Irving Hochberg, Ph.D.
Professor and Executive Officer
Doctoral Program in Speech and Hearing Sciences
Graduate School, City University of New York

University Park Press
Baltimore

UNIVERSITY PARK PRESS
International Publishers in Science, Medicine, and Education
233 East Redwood Street
Baltimore, Maryland 21202

Copyright ©1980 by University Park Press

Typeset by ComCom, Inc.
Manufactured in the United States of America by
The Maple Press Company.

Library of Congress Cataloging in Publication Data
Conference on Acoustical Factors Affecting Hearing Aid
Measurement and Performance, New York, 1978.
Acoustical factors affecting hearing aid performance.
(Perspectives in audiology series)
Bibliography: p.
Includes index.
1. Hearing aids—Congresses. 2. Hearing aids—
Fitting—Congresses. 3. Acoustical engineering—
Congresses. I. Studebaker, Gerald A. II. Hochberg,
Irving. III. Title. IV. Series.
RF300.C6   1978   617.8'9   79-25974
ISBN 0-8391-1553-9

CONTENTS

MODELING TECHNIQUES

FREQUENCY RESPONSE SELECTION TECHNIQUES

PARTICIPANTS

David A. Berkley, Ph.D.
Acoustics Research Department
Bell Telephone Laboratories
600 Mountain Avenue
Murray Hill, New Jersey 07974

Louis D. Braida, Ph.D.
Research Laboratory of Electronics
Massachusetts Institute of Technology
Cambridge, Massachusetts 02139

M. Jane Collins, Ph.D.
Ph.D. Program in Speech and Hearing Sciences
City University of New York
Graduate School and University Center
33 West 42 Street
New York, New York 10036

Robyn M. Cox, Ph.D.
Department of Audiology and Speech Pathology
Memphis State University
807 Jefferson Avenue
Memphis, Tennessee 38105

Raymond L. Dugal, S. B.
Research Laboratory of Electronics
Massachusetts Institute of Technology
Cambridge, Massachusetts 02139

A. J. Duquesnoy, M.Sc.
Faculty of Medicine
Free University
1007 MC Amsterdam, The Netherlands

Nathaniel I. Durlach, M.A.
Research Laboratory of Electronics
Massachusetts Institute of Technology
Cambridge, Massachusetts 02139

David P. Egolf, Ph.D.
Auditory and Electroacoustics Research Group
Department of Electrical Engineering
University of Wyoming
Laramie, Wyoming 82071

Norman Erber, Ph.D.
Central Institute for the Deaf
818 South Euclid Avenue
St. Louis, Missouri 63110

Craig Formby, M.S.
Central Institute for the Deaf
818 South Euclid Avenue
St. Louis, Missouri 63110

Irving Hochberg, Ph.D.
Ph.D. Program in Speech and Hearing Sciences
City University of New York
Graduate School and University Center
33 West 42 Street
New York, New York 10036

James F. Jerger, Ph.D.
Division of Audiology and Speech Pathology
Baylor College of Medicine
1200 Moursund
Houston, Texas 77025

Bertil Johansson, M.Sc.
Teknisk Audiologi
Karolinska Institutet
Teknrska Hogskolan
100 44 Stockholm, Sweden

Mead C. Killion, Ph.D.
Industrial Research Products, Inc.,
a Knowles Company
321 Bond Street
Elk Grove Village, Illinois 60007

Hugh S. Knowles
Knowles Electronics
3100 North Mannheim Road
Franklin Park, Illinois 60131

George F. Kuhn, Ph.D.
Vibrasound Research Corporation
4673 South Zenobia Street
Denver, Colorado 80236

Harry Levitt, Ph.D.
Ph.D. Program in Speech and
  Hearing Sciences
City University of New York
Graduate School and University
  Center
33 West 42 Street
New York, New York 10036

Samuel F. Lybarger, B.S.
101 Oakwood Road
McMurray, Pennsylvania 15317

James D. Miller, Ph.D.
Central Institute for the Deaf
818 South Euclid Avenue
St. Louis, Missouri 63110

Edward L. Monser, IV, B.S.
Industrial Research Products, Inc.,
  a Knowles Company
321 Bond Street
Elk Grove Village, Illinois 60007

Douglas Mook, M.S.
Massachusetts Institute of Technology
36-737
Cambridge, Massachusetts 02139

Anna K. Nábělek, Ph.D.
Department of Audiology and Speech
  Pathology
University of Tennessee
Knoxville, Tennessee 37916

Arthur F. Niemoeller, Sc.D.
Central Institute for the Deaf
818 South Euclid Avenue
St. Louis, Missouri 63110

David Pascoe, Ph.D.
Central Institute for the Deaf
818 South Euclid Avenue
St. Louis, Missouri 63110

R. Plomp, D.Ph.
Institute for Perception TNO
376g zg Soesterberg
The Netherlands

Edgar A. G. Shaw, Ph.D.
Division of Physics
National Research Council
Ottawa, Ontario, KlA OR6, Canada

Margo W. Skinner, Ph.D.
Central Institute for the Deaf
818 South Euclid Avenue
St. Louis, Missouri 63110

Gerald A. Studebaker, Ph.D.
Department of Audiology and Speech
  Pathology
Memphis State University
807 Jefferson Avenue
Memphis, Tennessee 38105

Edgar Villchur, M.S.Ed.
Foundation for Hearing Aid Research
Woodstock, New York 12498

Richard E. C. White, Ph.D.
Ph.D. Program in Speech and
  Hearing Sciences
City University of New York
Graduate School and University
  Center
33 West 42 Street
New York, New York 10036

Jozef J. Zwislocki, Sc.D.
Institute for Sensory Research
Syracuse University
Merrill Lane
Syracuse, New York 13210

PREFACE TO
PERSPECTIVES IN AUDIOLOGY

Audiology is a young, vibrant, dynamic field. Its lineage can be traced to the fields of education, medicine, physics, and psychology in the nineteenth century and the emergence of speech pathology in the first half of this century. The term "audiology," meaning the science of hearing, was coined by Raymond Carhart in 1947. Since then, its definition has expanded to include its professional nature. Audiology is the profession that provides knowledge and service in the areas of human hearing and, more broadly, human communication and its disorders. Audiology is also a major area of study in the professional preparation of speech pathologists, speech and hearing scientists, and otologists.

Perspectives in Audiology is the first series of books designed to cover the major areas of study in audiology. The interdisciplinary nature of the field is reflected by the scope of the volumes in this series. The volumes (see p. ii) include both clinically oriented and basic science texts. The series consists of topic-specific textbooks designed to meet the needs of today's advanced level student and of focal references for practicing audiologists and specialists in many related fields.

The **Perspectives in Audiology** series offers several advantages not usually found in most texts, but purposely featured in this series to increase the practical value of the books for practitioners and researchers, as well as for students and teachers:

1. Every volume includes thorough discussion of relevant clinical and/or research papers on each topic.
2. Most volumes are organized in an educational format to serve as the main text or as one of the main texts for graduate and advanced undergraduate students in courses on audiology and/or other studies concerned with human communication and its disorders.
3. Unlike ordinary texts, **Perspectives in Audiology** volumes will retain their professional reference value as focal reference sources for practitioners and researchers in career work long after completion of their studies.
4. Each volume serves as a rich source of authoritative, up-to-date information and valuable reviews for specialists in many fields, such as administration, audiology, early childhood studies, linguistics, otology, psychology, pediatrics, public health, special education, speech pathology, and/or speech and hearing science.

Electronic developments of this century have resulted in the emergence of electroacoustic auditory trainers and wearable hearing aids. During the past 50 years the use of electroacoustic amplification has emerged as a major habilitative approach to helping the hearing impaired overcome or reduce the communication handicap of a hearing loss. The 1970s were a decade of significant research and advances in the evolution of amplification.

Although several books on hearing aids have appeared during the past decade, **Acoustical Factors Affecting Hearing Aid Performance** is the first to focus on the many facets of these acoustic factors. Gerald A. Studebaker and Irving Hochberg assembled an outstanding collection of authorities from varying professional and scientific backgrounds to discuss this topic in June, 1978. The papers and discussions of that conference included in this volume provide a critical reference for all audiologists and others concerned with hearing aid measurement and selection.

Lyle L. Lloyd, Ph.D.
Chairman and Professor of
Special Education
Professor of Audiology and
Speech Sciences
Purdue University

PREFACE

This book contains the papers and discussions of the participants in the "Conference on Acoustical Factors Affecting Hearing Aid Measurement and Performance" held in New York City, June 14–16, 1978. The conference was conceived by its organizers in the fall of 1977 in response to a series of unique and significant developments in the acoustic area during the several years immediately preceding the conference—developments that appeared to presage substantial changes in how hearing aid performance will be measured and in how hearing aids will be selected for the individual user in the future. Indeed, some of these changes have already begun to occur.

In the conceptualization of this conference a debt of gratitude is owed to the "Seventh Annual Danavox Symposium on Earmolds and Associated Problems" organized by Stig Dalsgaard and held in Copenhagen in 1975 and to the conference on "Current Trends in Auditory and Hearing Prosthetics Research" organized by Vernon Larson and David Egolf and held in Laramie, Wyoming, in 1976. It was our desire to carry forward the fine work started by these conferences and to concentrate even more intensely on the acoustic factors that affect hearing aid performance.

The purpose of this book is to bring to a wider audience a progress report on the "state of the art" in the acoustic areas of hearing aid performance. It is perhaps appropriate that we offer our perspective on why we feel such a report is justified at this time.

The history of research concerning the acoustic factors affecting hearing aid performance in an uneven one at best, one that includes long periods of time when it appeared that little, if any, progress was made, and one that until recently had little, if any, impact on clinical or laboratory practices. In 1962 a landmark thesis by Wansdronk was published. Although this thesis was long ignored in the United States, it appeared to have a considerable influence on a paper by Dalsgaard, Johanssen, and Chisnall, published in 1966, and it has had a substantial influence on European research in the acoustic area, both directly and indirectly, ever since. In this country, two early papers on earmold acoustics by Lybarger (1958, 1967) stood alone until 1970.

The year 1970 probably best serves to identify the beginning of the current period in hearing aid acoustics research. While such "beginnings" must be assigned somewhat arbitrarily, it was in this year that Zwislocki released his report on the development and design of the "ear-like" coupler. It was also the year in which the number of papers published concerning earmold acoustics and related topics underwent a significant increase from its previous low rate, an increase that has continued and grown.

We believe that this increased activity and the outcomes of this research are significant because of the influence these developments have already had, and are likely to have in the future, in at least three areas: 1) how hearing aid electroacoustic performance is measured, 2) how hearing aids are selected for individual users, and 3) how the environment and/or the signal might be modified to improve speech intelligibility.

As an example, let us examine briefly the history of how hearing aids have been selected for the individual. Before 1946 hearing aid characteristics were selected principally by relating the hearing aid's electroacoustic performance to the individual's hearing test results. However, at that time the number of unknown and poorly understood factors (many of them acoustic) were too numerous to satisfy the thoughtful clinician or investigator with this approach. Therefore, seeking to circumvent these many problems, Carhart proposed what is essentially a trial-and-error procedure using speech as the test signal. Although there are a number of problems associated with this method too, principally its lack of reliability, it has persisted in wide use among audiologists until present, presumably because of the apparent face validity of working directly with speech test signals. However, it is now clear that this method also is inadequate. Thirty years of extensive research effort have failed to demonstrate that hearing aids can be chosen reliably or validly by this method in a reasonable period of time. The field of audiology has paid dearly for failing to recognize and heed the implications of published research concerning this method's weaknesses.

If, indeed, the speech test-based approach is as bankrupt as it now appears, what are the alternatives? There are several possibilities. All of them require an improved grasp of and control over the operating acoustic factors. One alternative involves fitting (according to rules yet to be defined) the hearing aid-processed signal into the user's residual auditory area. The need for accurate information concerning all of the factors that affect the signal delivered to the cochlea is obvious when using such an approach. The desirability of an electroacoustic measurement technique that takes all of these factors into account is also apparent when using this and most other methods.

A second alternative approach is that used in a series of studies done at the Central Institute for the Deaf, wherein aided and unaided sound field thresholds are measured and compared. While the need for accurate knowledge of some acoustic factors is reduced with this approach, a substantial need for this knowledge still exists unless each prospective hearing aid is tested exactly as it will be worn. The need for "insertion gain"-type measurements of the hearing aid to guide the hearing aid fitter remains as well.

A third alternative involves the use of a master hearing aid of some sort, perhaps even one synthesized by a computer. The ability to relate the results obtained by any master hearing aid to an actual wearable hearing aid on a particular individual depends heavily on a thorough understanding of the acoustic factors that affect hearing aid measurement and performance.

Other justifications could be cited for our feelings concerning the basic importance of these acoustic developments in such areas as microphone placement, environmental control, and research methodology. However, the chapters in this volume speak most eloquently on these topics. We invite the reader to study them and their implications for future practice with care.

We believe that if hearing aid selection methods can be improved, and we must believe that they can be, those improvements will be based on research like that found in these pages. For this reason this book should be of interest to researchers, clinicians, academicians, hearing aid dealers, engineers, government policy makers, directors of health insurance company programs, and anyone else interested in the implications of current research for the future of hearing aid measurement and selection methods.

The papers presented at the conference and recorded in this book were divided into the following four general areas: 1) acoustic factors that influence the hearing aid's input signal, 2) acoustic factors affecting the hearing aid's output signal, 3) modeling the hearing aid's acoustic systems, and 4) alternative hearing aid selection strategies. Each of these sessions was followed by discussion periods involving both the participants and members of the audience. A synopsis of each of these sessions is presented in Chapter 18.

A great many people contributed to the success of the conference and to the preparation of this book, so many we cannot name them all individually. Nevertheless we want to express our deepest gratitude to all of our friends and colleagues who gave so unselfishly of their time. Their help was indispensable to the success of the conference. We do, however, wish to single out and thank Linda Hoffnung for her exceptional efforts in the arduous task of transcribing and editing the discussion sessions found in Chapter 18 of this volume. Finally, we want to express our appreciation to the Bureau of Education for the Handicapped, USOE, for their partial financial support of the conference.

Gerald A. Studebaker
Irving Hochberg

REFERENCES

Dalsgaard, S. C., P. A. Johansen, and L. G. Chisnall. 1966. On the frequency response of ear-moulds. J. Audiol. Tech. 5:2–15.

Lybarger, S. F. 1958. The earmold as a part of the receiver acoustic system. Radioear Corporation. (This material was undated but was subsequently referenced by Mr. Lybarger as 1958.)

Lybarger, S. F. 1967. Earmold acoustics. Audecibel 16:9–19.

Wansdronk, G. 1962. On the mechanism of hearing. Philips Res. Rep. 62(1):1–140.

Zwislocki, J. J. 1970. An acoustic coupler for earphone calibration. Report #LSC-S-7. Laboratory of Sensory Communication, Syracuse, N.Y.

ACOUSTICAL EFFECTS
OF THE ENVIRONMENT

CHAPTER 1

NORMAL LISTENERS
IN TYPICAL ROOMS
Reverberation Perception,
Simulation, and Reduction

David A. Berkley

CONTENTS

This chapter makes three major points: 1) Normally, reverberation in small enclosures is characterized by two distinct perceptual parts that have generally been called *coloration* and *echo*. Our *preference* for rooms is dependent on both of these variables. 2) A laboratory method exists for computer simulating the physical properties of enclosures sufficiently well to evoke these percepts in a normal human listener. 3) We understand reverberation phenomena well enough to mathematically process reverberant signals and produce a result that is perceived as having considerably less reverberation than the original signals.

PHYSICAL BASIS OF REVERBERATION

Environment

The word *normally* as used in this chapter applies to the following conditions: a listener with normal one- or two-ear hearing, listening to speech in a fairly quiet enclosure, such as a one- or two-person office, a small class-room, or a home environment, and hearing that is bandwidth limited to about 100–4000 Hz. Under these conditions reverberation does not significantly reduce the intelligibility of perceived speech. However, there are other important conditions under which reverberation does significantly reduce or totally destroy speech intelligibility. Nábělek discusses some of these in Chapter 2.

It is important that reverberation under the above conditions does not significantly reduce the intelligibility of perceived speech because, first, listeners can clearly distinguish and have strong preferences for varying reverberant conditions, even in the range where little reduction is seen in intelligibility, and, second, if intelligibility is not reduced (and this has not been validated, to my knowledge, in full-scale, careful experiments), some other perceptual measure is necessary to quantify reverberant effects.

Ideal Rooms

If intelligibility is not the key to what normal listeners hear in many typical rooms, what is? In order to get to the heart of this problem let us first consider what, physically (as opposed to perceptually), reverberation is.

It is sufficient, for the purposes of this discussion, to treat a room as a rectangular, solid enclosure (Figure 1). By this idealization we ignore the effects of absorption and diffusion of sound produced by objects (and people) in the "room." However, if we allow the sides of the enclosure to absorb sound energy, experiment shows this ideal space to produce both realistic-sounding reverberant effects and physical phenomena that agree well with known results in less ideal real rooms. Speech sounds are introduced into the room by a small (or "point") sound source, which might be a person's mouth idealized to radiate sound equally in all directions. The resulting energy in the room is picked up or "heard" by a point receiver, which might be a similarly idealized "ear." If one wanted to study binaural hearing (without all the important outer-ear acoustics, which are discussed in Chapter 6), two pickup points, spaced by the human head size, would be used.

Transfer Function

How can the transmission of sound energy between source and receiver be described? Two alternative formulations of this problem have been made by acoustic physicists: 1) frequency domain approach (Morse and Ingard, 1968), in which the effect of the room on each *frequency* in the sound source is considered, or 2) time domain approach (Mintzer, 1950), in which the time course of reflections made by the sound waves hitting the room walls

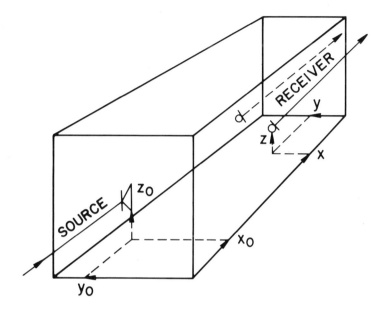

IDEALIZED ROOM

Figure 1.   Idealized rectangular solid "room."

(thought of as mirrors) is followed. Thus a pulse of sound at the source turns into many pulses at the receiver.

The frequency domain method has considerable theoretical attraction and is the method that has generally dominated theoretical consideration of room reverberation. However, an initial approach is much more attractive through the time domain (or "impulse") method, both intuitively and computationally. The frequency response of rooms is referred to only when it helps understanding of perception.

Single-Wall Echos   What happens when a pulse of sound is emitted by the source? Consider first a single wall, as shown in Figure 2. If we start with an acoustic pressure pulse, the resulting acoustic wave will propagate away from the source, with its sound pressure level *(P)* attenuated as the reciprocal of distance traveled. The wave will intersect the microphone with level $P_1 = P/r$. The wave will also reflect off the wall, as shown, being attenuated by the "pressure reflection coefficient" *(k),* and finally arrive at the pickup, after traveling the longer distance r_2, with pressure $P_2 = kP/r_2$.

If more walls are present this process may go on forever, with the wave bouncing around the room being attenuated by distance traveled and being absorbed on each reflection from a wall.

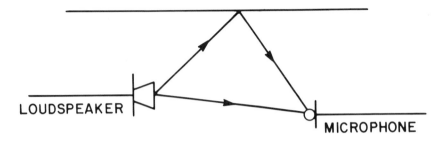

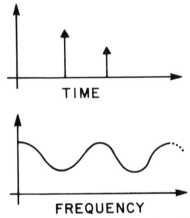

Figure 2.   Sound reflection from a single wall showing the impulse response and frequency response between source and receiver.

Impulse Response   Figure 3 shows what happens in a rectangular room. It shows the pulse response of a small office about $10' \times 12' \times 10'$ high. The first pulse to reach the pickup (a microphone in this case) is always the direct wave from the source ("direct sound"). The first few reflections are from nearby walls and are fairly well defined; however, later reflections, representing multiple reflections from various walls, are so numerous that they simply blur together.

This complicated picture has many uses. The pulse response shown in Figure 3 contains all the (linear) information available about the room derivable from this combination of source/receiver locations. In particular, using digital computer methods, we can process speech using this response so that it is the same as speech actually passed through the room (i.e., the result sounds the same within the bandwidth used). The process by which this is done is called *convolution* (Gold and Rader, 1969). Consider the one

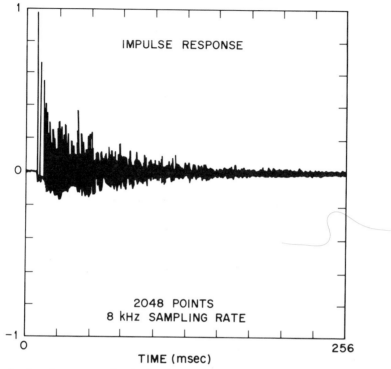

IMPULSE RESPONSE

2048 POINTS
8 kHz SAMPLING RATE

TIME (msec)

Figure 3. Impulse response of a simulated rectangular room.

wall case again (Figure 2). If a speech signal *(s(t))* were emitted by the source, the received signal would be

$$P(t) = s(t-T_1)\, P_1 + s(t-T_2)\, P_2$$

where T_1 is the time for the direct sound to travel distance r to the pickup, T_2 is the propagation time for the reflected wave traveling distance r_2, and P_1 and P_2 are the sound pressure levels of the direct and reflected signals, respectively. However, this is just what the impulse response gives—the delay and level of each reflection. Therefore, to get a resulting signal output we sum the input signal with itself, where the time delay and gain of each term are given by the corresponding values of the impulse response samples. (The resulting process is convolution, as mentioned previously. In this form it is oversimplified because the original pulse may be spread out in time by wall reflections, further smearing later portions of the impulse response. However, the convolution process in the digital domain is still basically the same and the effect of pulse spreading is generally ignored in the discussion that follows.)

PERCEPTION OF REVERBERATION

Simple Perception

Echo   The physical origin and description of reverberation sounds fairly simple (although efficient use of such impulse responses is somewhat intricate). However, how does one *perceive* the complex resulting reverberant signal? Let us return to the simple case of a single reflection. Suppose the wall is distant from the source, perhaps 25 feet, producing approximately a 50-msec delay in the reflected pulse (sound travels about 1 foot/msec), as seen in Figure 4A. (Such delay could also arise from multiple reflections.) Listening to the resulting "reverberation," a clear echo is heard.

Coloration   However, when one listens to a signal with a short echo delay (i.e., the wall is close—less than a foot away), giving approximately a 2-msec delay, the perception is one of a change in timbre of the speech, usually called coloration. When the pulses are so close together the ear cannot distinguish the time difference between the pulses, one hears the frequency response of the pulse pair, which looks like that shown in Figure 4B. Thus, the room acts like a filter, distorting (or coloring) the frequency content of the original speech signal.

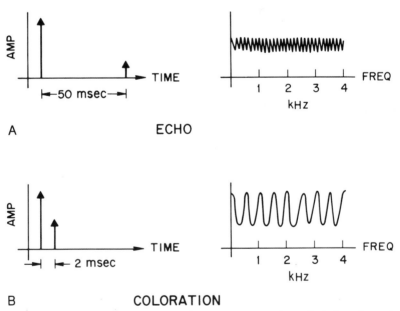

Figure 4.   A, The impulse response and frequency response for a single long-time echo ("Echo"); B, the impulse response and frequency response for a single short-time echo ("Coloration").

Ear Model Figure 5 illustrates how perception breaks into echo and coloration. It shows a very simple, but surprisingly powerful, model of the ear. An incoming signal is broken into frequency bands corresponding to the ear's critical bands—ranging from less than 50 to a few hundred Hz in bandwidth.

The effect on the two example signals clearly shows why we expect two very different types of perception. A long-delay pulse produces no variation in amplitude across filter (i.e., the rapid variations in the frequency domain are "averaged out"), but the time or echo nature of the signal is well preserved (Figure 5, top). The short-delay case, generally less than 25 msec, is just the opposite. Frequency content is well-preserved in the colored filter output, but all time information is lost (Figure 5, bottom).

Perception in Real Rooms

Dissection (Intuitive) Now let us examine these two perceptual effects in a real room. Figure 6 shows a simplified real-room impulse re-

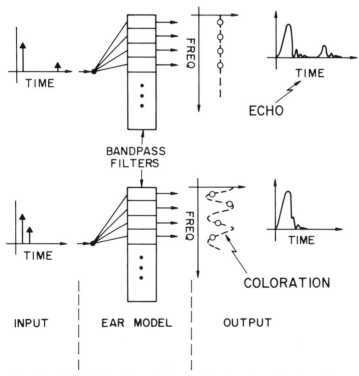

Figure 5. The effect of a long-time echo (top) and a short-time echo (bottom) on a simple hearing model.

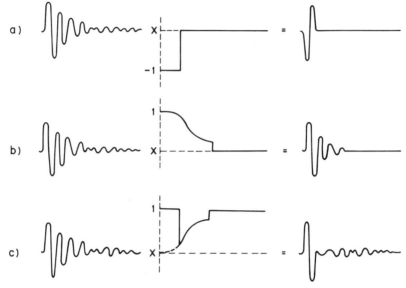

Figure 6. Simplified impulse response windowed to yield: a) direct sound, b) early echoes (less than 50 msec), and c) late echoes (greater than 50 msec).

sponse similar to the one we examined previously and a separation or dissection of the impulse into the direct (a), early (b), and late (c) portions of the response. Using the computer to perform these convolutions on an actual impulse response we are able to hear the result of this separation. The results are similar to those for a single echo, as previously discussed. Windowing at 50 msec primarily produces a change in timbre or coloration of the speech, whereas listening to direct sound plus the late response of the room yields "booming" or echoing speech (Berkley, Curtis, and Allen, 1973).

Quantification of Perception   The general picture describing perception of coloration and echo has been understood and accepted for some time. However, the stumbling block has been how to formalize these results under "realistic" conditions. Two areas of difficulty have existed:

1. The perceptual phenomena are multidimensional (i.e., *no direct psychophysical measurement method can analyze the complex underlying perceptual basis* by which individuals *discriminate* between differing reverberant conditions).

2. The underlying physical variables are also complex and, more important, many of them are difficult if not impossible to control fully or even measure accurately. Also, in real rooms, reverberant effects are often corrupted by noise or imperfect recording instruments.

A first step toward the resolution of these two problems has been taken by two workers at Bell Laboratories, Jont Allen and Barbara McDermott. The material in the next section of this chapter is drawn from a preliminary report on their elegant series of experiments entitled "The Perceptual Variables of Small Room Reverberation" and from some later experiments performed by myself and Sheryll Berggren. These are currently being prepared for publication (Allen, McDermott, and Berkley, in press).

Experimental Design   These experiments dealt with the two major problems by designing the experiment and analyzing the results within the framework of procedures collectively known as multidimensional scaling (Kruskal and Wish, 1978), and using a realistic, but completely controllable, computer simulation of the reverberant test rooms (Allen and Berkley, 1979).

Physical Variables   The simulation has significant practical implications because, to my knowledge, it is the first simulation designed for simplicity and efficiency that still provides a good approximation of actual room physics (and hence fully realistic perception). This makes it applicable to a wide range of problems (including many outside of psychophysics). Within the confines of a simple rectangular enclosure, the model can easily vary in room dimension, absorption of each room surface, and position of source and receiver.

In these experiments the room size is a constant: $12.5' \times 15' \times 16.25'$. However, by changing the surface absorptions the room reverberation time was varied from 75 msec to 480 msec. (The *reverberation time* is the time for sound level in the room to decay by 60 dB after being turned off.) The talker-microphone (source-listener) distance also was varied from 0.63 to 10.0 feet. For each condition, 512 msec of the room impulse response were calculated. (This response, on the computer, is sampled at 125-μsec intervals, which is a sampling rate of 8 kHz, allowing a 4-kHz bandwidth for the simulation.) Then 10 different sentences, each spoken by two male and two female talkers, were convolved with the calculated impulse responses, producing digital reverberant samples. These were then converted to analog tape recordings with a bandwidth of 100–4000 Hz.

Listening Tests—Difference Judgments   Sample tapes of all possible pairs of room conditions were played to 25 untrained normal listeners (balanced over talkers, order, and sentences) and they were asked to rate how different the two samples were on a scale of 0 (for no difference) to 9 (for maximum difference).

Listening Tests—Preference Judgments   In a separate experiment the same 25 subjects listened to two samples of each room condition (with different talkers and sentences) and were asked to rate the speech samples on a 9-point scale with descriptive adjectives (unsatisfactory, poor, fair, good, and excellent) labeling alternate scale points.

Finally, two "experienced" listeners rated all the conditions separately according to their judgment of the amount of echo and coloration.

Experimental Results

Multidimensional Analysis of Difference Judgments   Analyses of the difference judgments are shown in Figure 7. In order to discuss the results, it is necessary to suggest how the scaling procedures used present the data.

Each point on the plot in Figure 7A is one of the experimental room conditions. The positions in the space have been chosen so that the *distances* between pairs of points best represent the conglomerate judgments of the subjects. This set of distances is well represented by the two-dimensional plot shown in Figure 7A. This is, already, a nontrivial result. One dimension might have been sufficient or, possibly, more than two might have been required to produce a good representation.

To use a classical analogy to the subject distance judgments, consider the airline distances between cities in the United States. The multidimensional scaling programs would produce an actual two-dimensional map of the U.S. from this list of distances. However, if only East Coast cities were used, a one-dimensional map would be a fairly good representation. (A similar example is shown in Figure 8, where three points are projected onto a two-dimensional space.)

This is the beginning, rather than the end, of the required analysis. So far we do not know north/south from east/west or what may be the underlying perceptual and physical variables. However, given any other set of data relating these points, we can find the optimum vector in the space relating the experimental points and the new data, as well as get a measure of how well the data fit into the basic space. It is this method that we use to further understand the results.

Where do we stand? As expected, the underlying perceptual space is, in fact, two-dimensional, but what are the two dimensions? It is tempting to simply assume that they are coloration and echo. This is not, in fact, a bad assumption, as is seen if we place the "expert" judgments in the perception space, as Figure 7B shows. In fact, these two perceptual variables appear to be reasonably independent of each other.

What about the underlying physical variables? After considering a number of possible candidates, there appear to be two simple physical variables that represent well both axes of the perceptual space. One is *reverberation time,* and the other, an important but not commonly used room variable, is *spectral deviation.* Spectral deviation is a measure of the roughness of the frequency response of the room and is defined as the square root of the sum of the squares of the difference between actual room log

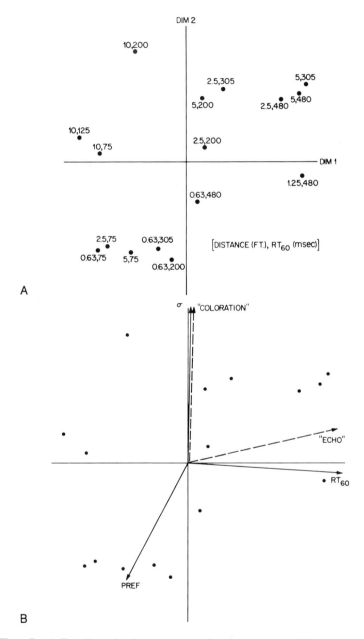

Figure 7.   A, Two-dimensional representation of subject judgments of distance between simulated reverberation samples; B, distance space with superimposed vectors representing expert
judgments (coloration and echo), physical variables (reverberation time and spectral deviation), and preference judgments.

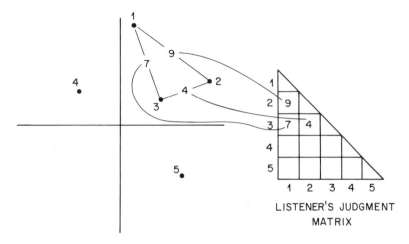

SOLUTION SPACE

Figure 8.   Example of projection of three points onto a two-dimensional solution space.

frequency response and a perfectly flat response. Reverberation time and spectral deviation define orthogonal axes of the difference space and, as seen in Figure 7B, agree well with echo and coloration.

Results—Preference   The preference results contain a useful surprise: unlike the discrimination results, preference is a one-dimensional phenomenon, which essentially means that all subjects agree on their preference for reverberant listening conditions. Projected onto the perception space, preference lies between the two axes (Figure 7B). Hence, preference is the combined result of the perceptual factors of coloration and echo. More important, preference may be represented as a linear combination of reverberation time and spectral deviation, as shown in Figure 9.

Projected this way, with the preference expressed in the form of the excellent-poor scale (as a mean opinion score), a parsimonious picture emerges of what a normal listener (with one nondirectional ear) feels a room should sound like for comfortable listening.

This same combination of reverberation time and spectral deviation may be re-plotted in the form of the *isopreference* curves in Figure 10. The concept of *critical distance* allows further interpretation of this figure. Critical distance is distance from source to receiver where the direct sound energy is equal to the total reverberant sound energy. However, more important, the actual source-receiver distance, normalized by the critical distance, is theoretically related to spectral deviation (Jetzt, 1979). This is discussed later, but the theoretical relation is shown on the upper scale of Figure 10.

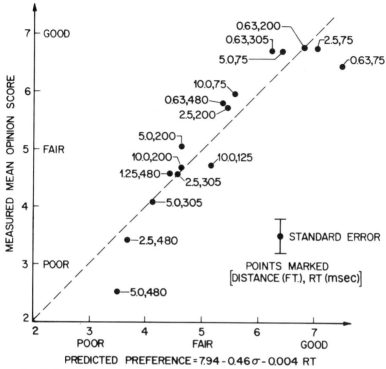

Figure 9.   Preference (mean opinion score) vs. predicted preference.

Consider an example. For the test room with a reverberation time of 125 msec, critical distance is 10 feet. (This is equivalent to an office with less-than-average reverberation.) This implies that, from consideration of Figure 10 alone, a one-ear listener with a 4-kHz hearing bandwidth will want to be less than 2 feet away from a talker for "good" listening! The effect of natural, one-ear directionality or the influence of a directional pickup (or source) will increase the "good" distance proportional to directionality.

We also know from experience that good two-ear listening allows greater distances for the same subjective quality. However, a full treatment, comparable to the one-ear case, awaits further work.

Summary

The physical bases for our perception of reverberation have been defined. Perception is a "two-dimensional" phenomenon consisting of "coloration" and "echo" dimensions caused by a combination of spectral deviation and reverberation time. The desirability of a given reverberant condition is

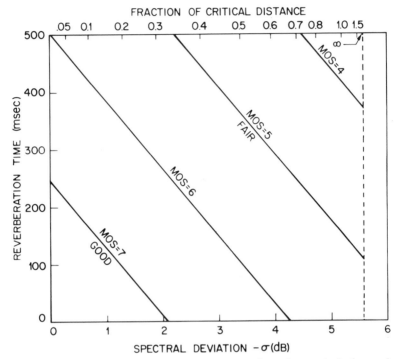

Figure 10.   Prediction equation of Figure 9 expressed as an isopreference plot in the reverber-ation-time vs. spectral-deviation (or fraction of critical distance) plane.

mediated by a simple physical relation to spectral deviation (or normalized distance) and reverberation time.

ROOM SIMULATION

The room simulation techniques used in the perceptual experiments rest on the simple principle of sound images.

Basic Physical Principles

Figure 11A shows the single reflection of a sound wave from a wall, shown originally in Figure 2, in a different form. If the wall is reflecting perfectly, the wall may be replaced by a second "image" source placed symmetrically with respect to the original source relative to the wall in analogy with an optical mirror and image. If the original source emits a sound pulse, the image emits an identical pulse at the same time. If the wall absorbs some portion of the sound wave, this may be approxi-mately accounted for by decreasing the image strength by the reflection coefficient.

With two opposite perfect walls, as in Figure 11B, there is an infinite number of images, as with two opposing optical mirrors.

A two-dimensional enclosure, shown in Figure 11C, looks still more complex, but the origin is the same. A three-dimensional structure produces the same result, but it is not easily depicted. However, mathematical expressions for the receiver output with a pulse source input can be written directly for all these cases. The resulting expressions look complicated but are only direct extensions of the equation given previously for a reflection from a single wall (Allen and Berkley, 1979).

It is remarkable that this result is exact (in the sense that it is an exact solution to the full wave equation formulation of acoustics) when the walls of the room are "hard" or reflecting perfectly. Thus, an alternative way of thinking about an impulse response is that each point of the response represents the arrival time and amplitude of a pulse emitted by one of the images. The later in the response, the farther away is the contributing image.

Even when the walls are not perfect reflectors, images should still result in a good approximation of the physical results in the room.

Computer Implementation

The image representation is powerful because after a relatively short time (about one-half of the room reverberation time) the remaining incoming pulses no longer contribute sufficient energy to be perceptible or to affect other room measurements, the energy having been lost in reflections from the walls. Thus, the complete *significant* response is built up by summing those images within a sphere, the radius of which is determined by the distance sound travels in the significant time. Although this may still be tens of thousands of images, the computation is well suited to efficient evaluation on a modern digital computer, and a reasonably high-speed machine can compute about 1000 images/sec.

Table 1 shows the number of images and the computation time required to generate the impulse response for a reasonable-size room represented by various impulse response lengths. Also shown (for the same Data

Table 1.   Computation parameters

Impulse response				Convolution rate (sec/sec)
Length (msec)	Number of points	Image count	Computation time (sec)	
64	512	585	1	12.5
128	1024	4690	8	13.8
256	2048	37,500	60	15.0

Room size is 10' × 15' × 12.5'. Sampling rate is 8 kHz.

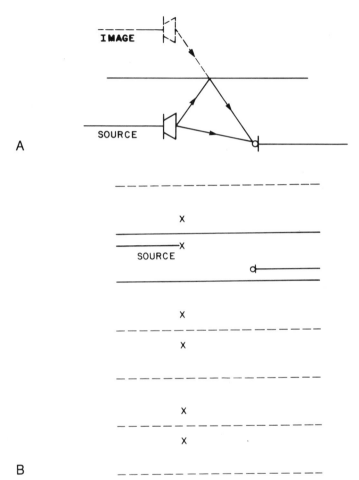

Figure 11.   A, Image produced by single wall reflection; B, images produced by two opposite walls. x = original and image sources, o = receiver. Solid lines represent physical walls, broken lines represent image walls.

General Eclipse S/200 minicomputer) are the times taken to compute actual speech samples using an efficient FFT/convolution method. These are of interest because they give the time required to prepare samples for a given psychophysical experiment.

Applications

Reverberation perception, the major application originally considered for room simulation, has been discussed. A number of other applications exist, for instance the development of a highly controllable, specifiable set of standard reverberation test conditions or test tapes for monaural or

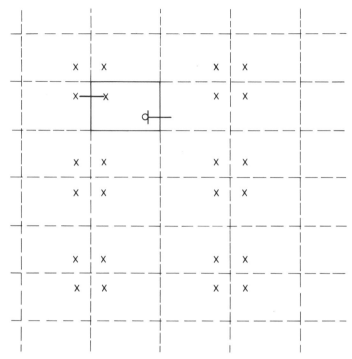

C

Figure 11. C, images produced by a two-dimensional "room." x = original and image sources, o = receiver. Solid lines represent physical walls, broken lines represent image walls.

possibly binaural hearing tests and the development of a new measurement method for room acoustics. This spectral deviation room measurement method worked out by John Jetzt at Bell Laboratories was initially tested using the simulation technique (Jetzt, 1979).

First Jetzt derived the theoretical relation between spectral deviation and source-receiver distance (normalized by critical distance). Then he verified the relationship using the room simulation, as shown in Figure 12. This is, of course, impossible in a real room where, in general, direct and reverberant energy cannot be measured separately. However, once verified, spectral deviation provides a very sensitive tool for determining critical distance, even in rooms where means for determining reverberation time are unreliable and inaccurate.

REVERBERATION REDUCTION

Much is known about reverberation: how it is produced physically, how it is perceived, what leads to individual judgments of good and bad reverberant conditions, and how to simulate and measure reverberation. How one

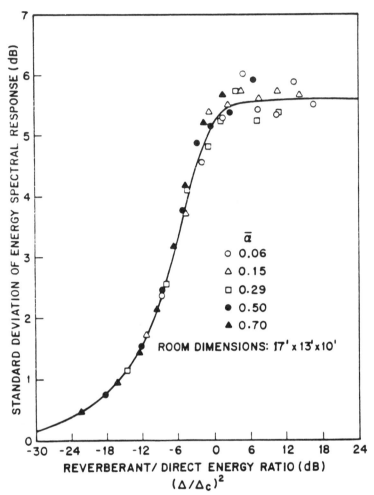

Figure 12.   Spectral deviation of computer-simulated room (17′ × 13′ × 10′) with theoretical calculation. (Reprinted with permission from Jetzt, 1979.)

goes about reducing perceived reverberation is the last topic of discussion.

The primary methods discussed in the literature for reverberation and noise control (outside of adjusting the external physical environment) are such approaches as use of directional microphones and frequency shaping. However, other interesting approaches are currently being considered, based on computer processing, that are still too complex to leave the laboratory, but point toward things to come.

The method discussed herein was developed by Allen, Berkley, and Blauert (1977). The method was suggested by the observation that two-ear listening is far superior to one ear in reverberant fields. Experiments and theory indicate that the reverberant field at the two ears is generally uncorrelated (e.g., like two independent noise generators) and the processor takes advantage of this. Consider the black box in Figure 13. Given two input signals it behaves in the following manner:

For identical inputs the output perfectly replicates the inputs.
For uncorrelated inputs the output is zero.
For identical inputs, with a small time delay (e.g., less than 1 msec, roughly the spacing between ears) the output is identical to one of the inputs, and for longer delays the output falls to zero.

The actual processor implements these operations for each of many overlapping frequency bands, allowing correlated (or direct speech) to pass in one frequency region while rejecting uncorrelated (reverberant) signals in another. Figure 14 shows a more detailed block diagram of the processor and Allen et al. (1977) and Allen and Rabiner (1977) discuss details of the process.

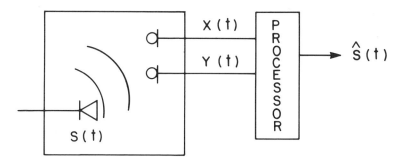

INPUTS		OUTPUT	
X	Y	$\hat{S}$	
$S(t)$	$S(t)$	$S(t)$	IDENTICAL
$N_1(t)$	$N_2(t)$	0	UNCORRELATED
$S(t)$	$S(t-\tau)$	$S(t)$	$\tau \leq 1$ msec
		0	$\tau \gg 1$ msec

Figure 13.   Black-box deverberation processor.

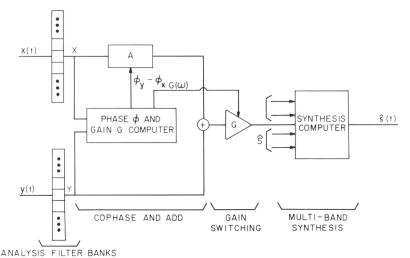

Figure 14.   Two-microphone dereverberator. (Reprinted with permission from Allen et al., 1977.)

Listening to processed speech is the best demonstration of the effectiveness of this processing. For informal listening tests, a number of people have indicated a preference for the processed/dereverberated samples over the original binaural recordings. Figure 15 shows spectrograms of reverberant, dereverberated, and nonreverberant speech samples, which were made using binaural analog tape recordings in a real room as input to the computer speech processing program. Much of the reverberant "smearing" seen in the unprocessed sample has disappeared after processing.

REFERENCES

Allen, J.B., and D.A. Berkley. 1979. Image method for efficiently simulating small room acoustics. J. Acoust. Soc. Am. 65:943–950.

Allen, J.B., D.A. Berkley, and J. Blauert. 1977. Multimicrophone signal-processing technique to remove room reverberation from speech signals. J. Acoust. Soc. Am. 62:912–915.

Allen, J.B., B.J. McDermott, and D.A. Berkley. A method for measuring subjective perception and preference of small room reverberation. J. Acoust. Soc. Am. In press.

Allen, J.B., and L.R. Rabiner. 1977. A unified approach to short-time Fourier analysis and synthesis. Proc. IEEE 65:1558–1564.

Berkley, D.A., T.H. Curtis, and J.B. Allen. 1973. Effects on speech perception of modifying the impulse response of a small room. J. Acoust. Soc. Am. 53:30A.

Gold, B., and C.M. Rader. 1969. Digital Processing of Signals, Chapters 2, 7. McGraw-Hill Book Co., New York.

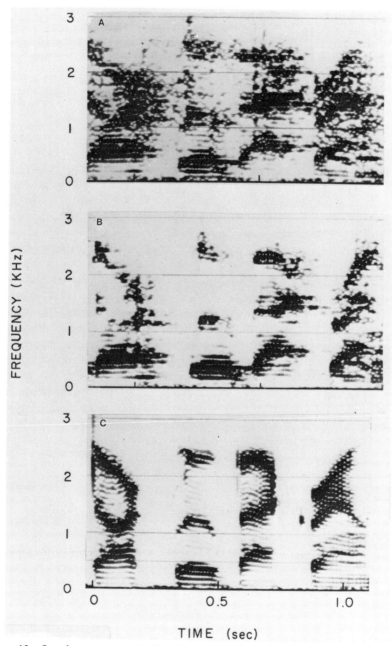

Figure 15.   Speech spectrograms. A, Reverberant original speech; B, dereverberated speech; C, nonreverberant speech (different sample).

Jetzt, J.J. 1979. Critical distance measurement of rooms from the sound energy spectral response. J. Acoust. Soc. Am. 65:1204–1211.

Kruskal, J.B., and M. Wish. 1978. Multidimensional scaling. In E.M. Uslaner (ed.), Quantitative Applications in the Social Sciences, Vol. I. Sage University Press, Beverly Hills, Cal.

Mintzer, D. 1950. Transient sounds in rooms. J. Acoust. Soc. Am. 22:341–352.

Morse, P.M., and K.U. Ingard. 1968. Theoretical Acoustics, Chapter 9. McGraw-Hill Book Co., New York.

617.89   C76a

c·1

CHAPTER 2

EFFECTS OF ROOM ACOUSTICS ON SPEECH PERCEPTION THROUGH HEARING AIDS BY NORMAL-HEARING AND HEARING-IMPAIRED LISTENERS

Anna K. Nábělek

CONTENTS

DIRECT AND REFLECTED SOUNDS

The sounds we perceive in enclosures are characterized by some amount of reverberation. The total energy present within any listening environment is a mixture of three components: the original, or direct, sound; the early reflections, occurring shortly after the direct sound; and the later, more diffuse reflections, or reverberant tails.

The two groups of reflections produce different perceptual effects. Early reflections introduce coloration into sound, whereas later reflections are responsible for the prolongation of sounds normally called "reverberation."

The preparation of this paper was supported by a grant from the National Institute of Neurological and Communicative Disorders and Stroke, U.S. Public Health Service, # RO1 NS 12035.

25

The reverberation time *(T)* is the time that would be required for the sound pressure level, originally in a steady state, to decrease 60 dB after the source stops. It is a crude measure because it does not take into account various effects on the distribution of the early reflections or the relation between the early and later reflections.

It has been suggested that the distribution of reflections in large enclosures with long reverberation time affects speech intelligibility (e.g., Santon, 1976) and quality of music (e.g., Schroeder, Gottlob, and Siebrasse, 1974). Recent studies by Patterson et al. (1975), Botros (1976), and McDermott and Allen (1976) on subjective judgments of preference in telephone listening by normal-hearing subjects indicate that the distribution of reflections in small rooms with short reverberation may also influence speech perception.

Little is known about hearing-impaired listeners' preference to reverberation. John (1957) recorded speech in rooms where $T = 0.3, 0.5$, and 0.7 sec, and obtained subjective preferences and objective word identification scores from a group of hearing-impaired adults. They preferred 0.5 sec over 0.7 sec, and expressed little preference between 0.3 and 0.5 sec. The word identification scores were 18% higher at 0.5 sec than at 0.7 sec, but no data were reported at 0.3 sec. This example indicates that some hearing-impaired listeners might express preference on T or the distribution of the early reflections, but there are no more data to support such a statement. Presently, the acoustic conditions producing the best speech intelligibility are recognized as the most desirable for hearing-impaired listeners.

EARLY REFLECTIONS

It would seem that the best intelligibility could be achieved with direct sound only. However, when the intensity of the direct sound is low enough to produce poor speech perception, the increase in loudness caused by early reflections with no reverberation may improve word identification scores. This was demonstrated for normal-hearing subjects by Lochner and Burger (1964), Santon (1976), and Walker, Delsasso, and Knudsen (1970).

The influence of early reflections on speech perception for hearing-impaired subjects was investigated by Nábělek and Robinette (1978a,b). The test consisted of 18 lists of the modified rhyme test (MRT) (Bell, Kreul, and Nixon, 1972; Kreul et al., 1968) recorded by a male speaker. These recorded lists were processed by a PDP-12 computer system programmed to serve as single (Nábělek and Robinette, 1978a) or multiple (Nábělek and Robinette, 1978b) delay lines. These delayed versions were recorded on a second track on magnetic tapes.

The test was reproduced through two loudspeakers in a small, sound-insulated room, with $T = 0.25$ sec. The listening arrangement is shown in Figure 1. The direct sound was reproduced through one loudspeaker and the delayed version, or multiple delays, was reproduced through the other loudspeaker. To reduce word identification scores, a babble of eight voices was added as a masker in both channels.

The results for the single reflection are shown in Figures 2 and 3 for binaural and monaural listening, respectively. The effect of the delay is similar for both groups of subjects. The reflection with a delay up to 20 msec did not impair word identification, which is consistent with previous data for normal-hearing subjects. However, individual results indicated that some hearing-impaired subjects were adversely affected even by the shortest delays.

An analysis of percent correct responses for some selected consonants for binaural condition (Nábělek and Robinette, 1978a) was performed. The results are shown in Figures 4 and 5 for normal-hearing and hearing-impaired subjects, respectively. Although the word identification scores were similar for both groups of subjects, the perception of individual consonants differed. The normal-hearing subjects tended to recognize consonants equally well over a range of delays up to about 80 msec, whereas the

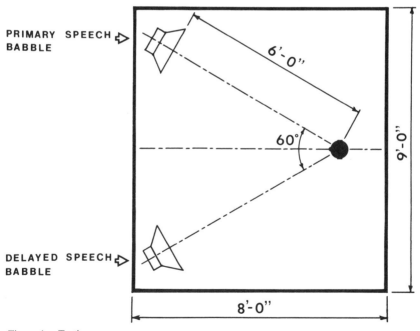

Figure 1.   Testing room.

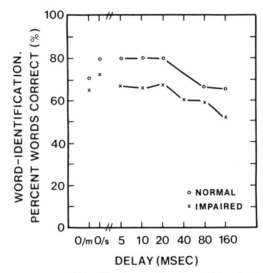

Figure 2.   Percent correct word identification as a function of delay time between direct and delayed sound in the presence of a babble of eight voices for binaural condition. (Reprinted with permission from Nábělek and Robinette, 1978a.)

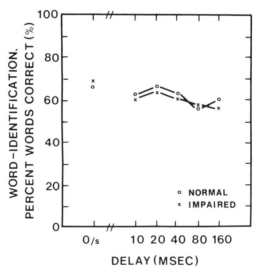

Figure 3.   Percent correct word identification as a function of delay time between direct and delayed sound in the presence of a babble of eight voices for monaural condition. (Reprinted with permission from Nábělek and Robinette, 1978a.)

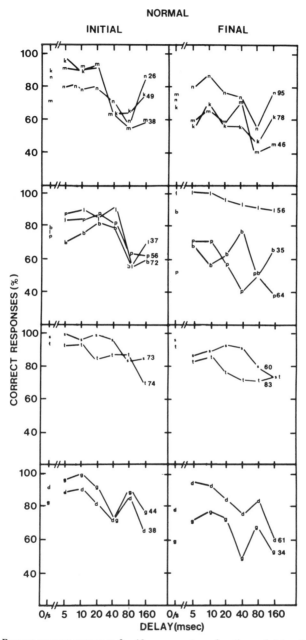

Figure 4. Percent correct responses for 10 consonants as functions of delay time between direct and delayed sound for binaurally listening normal-hearing subjects, in initial and final position. (Reprinted with permission from Nábělek and Robinette, 1978a.)

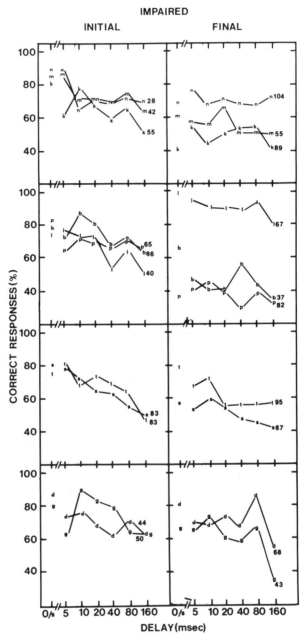

Figure 5. Percent correct responses for 10 consonants as functions of delay time between direct and delayed sound for binaural-listening hearing-impaired subjects, in initial and final position. (Reprinted with permission from Nábĕlek and Robinette, 1978a.)

hearing-impaired subjects' recognition was sometimes affected even for the shortest delays (e.g., consonants /s/ and /t/ in the initial position). This finding seems to indicate that the influence of a reflection on word identification by hearing-impaired subjects is not quite the same as for normal-hearing subjects.

In another study, Nábělek and Robinette (1978b) added five delays to the original sound in three time sequences, as shown in Figure 6. Word

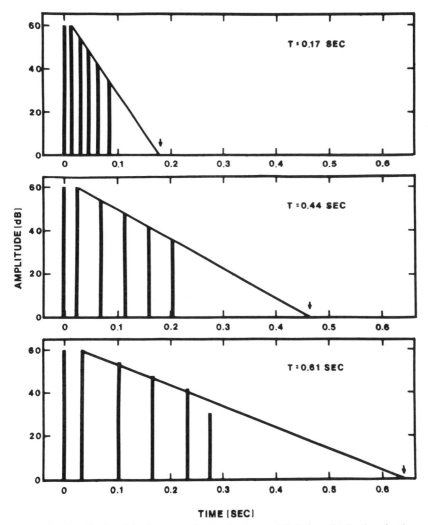

Figure 6.   Amplitudes of the direct sound and reflections and their time distributions for three simulated reverberations. (Reprinted with permission from A.K. Nábělek and L.N. Robinette, 1978b, Reverberation as a parameter in clinical testing, *Audiology* 17:239–259.)

identification scores measured with five normal-hearing and five hearing-impaired subjects remained constant for both groups for all three complexes, indicating that discrete reflections do not necessarily impair speech perception. The results were consistent with Lochner and Burger's (1964) results for complexes of delays for normal-hearing subjects.

REVERBERATION

In real environments, early reflections are followed by later reflections (reverberation), which cause time-smearing of the speech patterns and reduce speech intelligibility. In many studies word identification was measured as a function of T regardless of the distribution of the reflections.

Knudsen (1929) measured word identification as a function of T in large auditoriums. He found a maximum on each curve for T of about 1 sec for smaller auditoriums and at Ts of 1.5 sec for larger auditoriums. The reverberation time producing maximum score was called "optimum reverberation time." Although Knudsen pointed out that the maximum was caused by intensity requirements (the larger the room, the more reverberant energy that was needed to maintain an intensity level adequate for good speech perception), the optimum reverberation time concept influenced the way of thinking about room acoustics. First, since word identification for Ts up to 1 or 1.5 sec remained constant or increased in Knudsen's studies, it was assumed that short Ts do not influence speech intelligibility. Second, it was believed that each room has its best T for speech intelligibility. Later, Bolt and MacDonald (1949) calculated masking effects of reverberation and concluded that speech intelligibility is affected by Ts larger than 0.35 sec.

More recently, the influence of T in smaller rooms on speech perception has been investigated. Some results from older studies for normal-hearing and hearing-impaired subjects are shown in Figure 7. The results for normal-hearing subjects are shown in solid lines. The decrease in word identification scores in quiet is moderate and perception remains good (above 80%) up to $T = 1.2$ sec. When masking is added, perception is poorer and scores decrease more rapidly with prolongation of reverberation time.

Similarly, Plomp (1976) reported an increase of the masked threshold for speech with prolongation of reverberation for normal-hearing subjects.

Some of the earlier data for hearing-impaired listeners are shown in Figure 7 in broken lines. All the data collected in quiet indicate that sensorineural hearing-impaired listeners are also sensitive to increased reverberation. With masking noise added, the slope of the speech perception curves is about the same as in quiet.

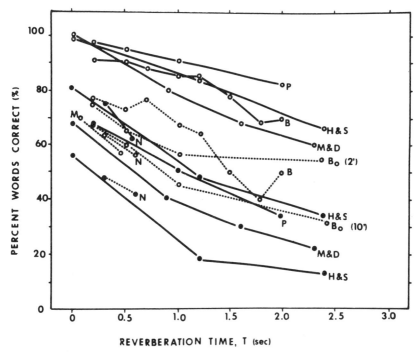

Figure 7.   Mean percentage of words correct as a function of reverberation time in quiet (open dots) and in noise (solid dots) for normal (solid lines) and impaired (broken lines) subjects. The data are from the following studies: B, Bullock (1967); B_0, Børrild (1970); H&S, Houtgast and Steeneken (1973); M, Millin (1968); M&D, Moncur and Dirks (1967); N, Nábělek and Pickett (1974b); and P, Peutz (1971). (Reprinted with permission from Nábělek, 1976.)

The differences in slopes of speech perception curves in individual studies may be attributable to differences in speech material, room sizes, and measuring procedures used. Let us examine some of these factors.

ROOM SIZE

The influence of room size on the reverberation effect, that is, a decrease in word identification with prolongation of reverberation, can be examined by comparing the results of studies by Crum and Tillman (1973) and Finitzo-Hieber and Tillman (1978) with the results of studies by Nábělek and Pickett (1974a) and Nábělek and Robinette (1978b).

Crum and Tillman and Finitzo-Hieber and Tillman used an anechoic room for $T = 0$ sec condition and another room with a variable amount of absorption for $T = 0.4, 0.8,$ and 1.2 sec conditions. The volumes of both rooms were similar and equal to 162 m^3. The speech material (CNC lists of 50 monosyllabic words) and noise (babble of eight voices) were the same. Crum and Tillman tested 12 normal-hearing adults listening binaurally and Finitzo-Hieber and Tillman tested 12 normal-hearing children listening monaurally (one ear was covered with a muff). The results of the two studies are shown in Figure 8.

Note that in quiet (upper two curves), speech perception deteriorates for $T > 0.4$ sec for both groups of subjects. However, it tends to decrease faster for the children than for the adults. There are two possible explanations for the difference: 1) the children are more susceptible to reverberation than adults, or 2) the monaural perception is more difficult than binaural in reverberation.

When masking noise was added (two lower curves, filled triangles, and circles) the reverberation effect was the same for both groups and the slopes were identical. The deterioration in speech perception was present throughout a whole range of Ts, even for the short reverberation, where there was no deterioration in quiet.

Nábělek and Pickett (1974a) and Nábělek and Robinette (1978b) used the same speech material (MRT), masking noise (babble of eight voices) and other procedures (arrangement of sources, level settings). In both studies five normal-hearing adults were tested. The main difference was in the size of the rooms used. In the first study the volume was 119 m³ and in the

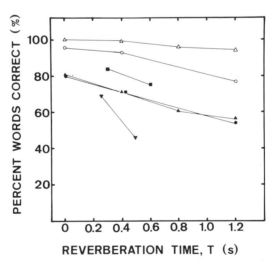

Figure 8.   Word identification scores as functions of reverberation time for normal-hearing subjects in rooms of various sizes. Open symbols are for scores measured without masking noise; filled symbols are for scores measured with masking noise. △, ▲, Crum and Tillman (1973) (adults, V = 162 m³); ○, ●, Finitzo-Hieber and Tillman (1978) (children, V = 162 m³); ▼, Nábělek and Robinette (1978b) (V = 13.5 m³); ■, Nábělek and Pickett (1974a) (V = 119 m³).

second study the volume was 13.5 m³. The results are shown in Figure 8. It appears that the slope is greater in a small room than in the large rooms. It is also noteworthy that the word identification scores were lower in a small room.

When comparing Crum and Tillman (1973) and Finitzo-Heiber and Tillman (1978) data with the data from the Nábělek and Pickett (1974a) and Nábělek and Robinette (1978b) studies with masking noise added, the slope of the curves measured in the 119-m³ room is comparable to the slope measured in the 162-m³ room. On the other hand, the greater slope measured in the small room (13.5 m³) is similar to slopes measured by MacKeith and Coles (1971) and Millin (1968) in comparatively small rooms and at similar Ts.

The data with hearing-impaired subjects are sparse and more difficult to compare. In four studies (Finitzo-Heiber and Tillman, 1978; Mason and Asp, 1976; Millin, 1968; Nábělek and Robinette, 1978b) the room volumes ranged from 13.5 to 162 m³. The average hearing level (HL) at 500, 1000, and 2000 Hz (ANSI, 1969) of the subjects was about 50 dB. The slope of the word identification curves, as functions of reverberation time, was similar for these studies without any noticeable trend for the volume.

It appears, therefore, that for the normal-hearing subjects more reverberant effect can be expected in small rooms than in large ones. The difference can be explained by differences in the distribution of the early reflections. For example, in the large room utilized by Nábělek and Pickett (1974a,b), the first reflections arrived up to 30 msec after the direct sound. In the small room used by Nábělek and Robinette (1978a,b), the smallness of the room caused the early reflections to be grouped within the first 10 msec after the direct sound. The later reflections were closely spaced in both rooms. It seems that this difference in the distribution of the early reflections was important for word identification.

Some comparative studies are needed to see if room size has an effect on the speech perception of hearing-impaired subjects.

DISTANCE FROM THE SOURCE

In some studies (Børrild, 1970; Crum and Tillman, 1973; Nábělek and Pickett, 1974a; Peutz, 1971; Plomp, 1976), the distance between the speech source and the listener is mentioned as a factor influencing the dependence between speech perception and reverberation. A so-called critical distance must be considered. The critical distance (D) is a distance at which the intensity of the direct sound is equal to the intensity of the reflected sound. Near the source there is an area in which the intelligibility decreases with increased distance, but outside that area the intelligibility remains constant, irrespective of where the listener is situated. The intelligibility in the constant area depends only on the reverberation time of the room, as was measured by Peutz (1971) in rooms of different sizes and reverberation times.

To calculate the critical distance, Peutz used the formula:

$$D = 0.2\sqrt{V/T} \tag{1}$$

where D is in m, V = volume in m³, and T = reverberation time in sec.

Klein (1971) introduced the directivity factor, Q, for the source into the formula for the critical distance:

$$D = 0.2\sqrt{VQ/nT} \tag{2}$$

where n is a number of sources (e.g., loudspeakers). It is clear from equation 2 that when the source is highly directional the critical distance is larger, more direct energy reaches the listener, and the reverberation effect is reduced. The Q factor for human voice is 2.5, but it can be larger for a loudspeaker. A use of directional sources is one of the ways to reduce the influence of room reverberation, providing that the listener sits within the critical distance.

Plomp (1976) used another formula to calculate the critical distance and obtained smaller values than calculated from Peutz's (1971) formula.

Since Crum and Tillman (1973) did not report on the directivity of their source, the critical distance cannot be calculated accurately. The examination of their data indicates that the measurements 1.8 m from the source were made within the critical distance and measurements 3.6 and 7.2 m from the source were made outside the critical distance, since there was more reverberant effect farther from the source, but no significant differences between results for two longer distances.

Therefore, the reverberant effect in any enclosure depends not only on the acoustics of the room but also on the distance between the source and the listener and directionality of the source. In order to measure the greatest possible reverberant effect in any given room, the distance from the source has to be greater than the critical one. If we wish to overcome room reverberation, we can introduce a highly directional source and locate listeners in the field of the direct sound.

MASKING NOISE

The influence of type of masking noise on the reverberant effect was investigated by Nábělek and Pickett (1974a). Two types of noise were used: 1) a recorded, quasi-steady noise made up of a babble of eight voices, and 2) and impulsive noise as produced by the loudspeaker fed by clicks at a rate of 16 clicks per sec. The temporal pattern of the babble was not greatly changed by the prolongation of reverberation, but the temporal pattern of the impulsive noise changed substantially. The periods between the pulses contained stronger and more prolonged reflections under the longer reverberation; thus the impulsive noise became more continuous. The reverberant effect measured with five normal-hearing subjects was greater for the impulsive noise than for the babble.

BINAURAL VS. MONAURAL LISTENING

The effect of reverberation on speech perception is especially interesting for comparisons between monaural and binaural listening. Many patients with unilateral deafness and many monaural hearing aid users perceive sounds with only one ear. Is that perception different than through two ears or two hearing aids? Is any advantage found using binaural hearing aids in reverberant environments?

Koenig, Allen, and Berkley (1977) showed that binaural processing provides about a 3-dB masking level difference for normal-hearing subjects in a reverberant field with background noise, and thereby serves to squelch some of the masking effects of the reverberant environment. It remains to be seen, however, whether binaural listening yields better word identification scores in a reverberant field without masking noise added. Three of five hearing-impaired subjects investigated by Nábělek and Pickett (1974b) showed better binaural than monaural scores in quiet in both tested reverberant conditions (0.3 and 0.6 sec). In our new study (in progress), 13 of 14 hearing-impaired subjects tested in quiet at $T = 0.25$ sec obtained better binaural than monaural scores without hearing aids. The average binaural advantage was 4.6%. With hearing aids, 9 of the subjects obtained better binaural than monaural scores, 2 obtained the same scores for binaural and monaural listening, and 3 obtained better monaural than binaural scores. The average binaural advantage was 2.1%.

With masking noise added, the advantage of binaural listening is easier to demonstrate, both with normal-hearing and hearing-impaired listeners. However, it is not clear if the size of the binaural advantage (expressed as a difference between the scores for the binaural and monaural listening) depends on the amount of reverberation.

The data for the normal-hearing listeners show a decrease in advantage (Hirsh, 1950; MacKeith and Coles, 1971; Nábělek and Pickett, 1974a; Plomp, 1976), an increase (Gelfand and Hochberg, 1976; Moncur and

Dirks, 1967; Nábělek and Robinette, 1978b), or constant value for those listening through hearing aids (Nábělek and Pickett, 1974a).

The data for the hearing-impaired listeners show an increase in binaural advantage (Nábělek and Robinette, 1978b) for unaided listening, but with large individual differences, and a constant value of the binaural advantage (Nábělek and Pickett, 1974b) for listening through hearing aids.

Thus, there seems to be little support for the statement of MacKeith and Coles (1971) that binaural fitting for reverberant environments is not worth the cost because the binaural advantage diminishes with reverberation.

HEARING LOSS

Nábělek and Pickett (1974b), Finitzo-Hieber and Tillman (1978), and Nábělek and Robinette (1978b) compared reverberation effects for normal-hearing and hearing-impaired subjects. When masking noise was added, the amount of reverberant effect for the impaired subjects was equal to or smaller than the effect for normal-hearing subjects. However, a noticeable reverberant effect was measured for hearing-impaired subjects in quiet or at high speech-to-noise ratios (S/Ns), conditions that produced none or very small reverberant effects for normal-hearing listeners.

The slope of reverberant deterioration for the hearing-impaired listeners is the same with and without masking noise. This finding could be taken to indicate that the effects of reverberation on speech perception in hearing-impaired listeners outweigh those of the addition of noise, since there is a decrease in the slope for listeners with normal hearing.

In most of the studies, the average hearing level (HL) of the hearing-impaired subjects was between 40 and 50 dB. Nábělek and Pickett (1974b) reported data for subjects with an average HL of 73 dB. Their data seem to indicate a trend toward a decreasing reverberant effect with increases in HL, probably because the best performance of subjects with greater hearing losses is lower than that of subjects with better hearing, and there is less room for further deterioration by additional factors.

INDIVIDUAL SUSCEPTIBILITY TO REVERBERATION

Nábělek and Pickett (1974b) reported large individual differences in reverberant effect for their hearing-impaired subjects, which were not noticed for the normal-hearing subjects (Nábělek and Pickett, 1974a). The range of the reverberation effect for five subjects was from 1% to 14%. Plomp (1976) reported a substantial difference among his 10 normal-hearing subjects in their susceptibility to reverberation. Also, Nábělek and Robinette (1978b) reported individual differences for five normal-hearing and seven hearing-impaired subjects. The range of the reverberant effect was from 16% to 31% and from 0% to 23% for normal-hearing and hearing-impaired subjects, respectively. The cause of the individual differences is not yet understood.

HEARING AIDS IN REVERBERATION

Speech perception in reverberation for unaided and aided condition was compared by Nábělek and Pickett (1974a) with five normal-hearing adults and by Finitzo-Hieber and Tillman (1978) with 12 hearing-impaired children.

Nábělek and Pickett (1974a) investigated the following factors: 1) two different conditions of reverberation (0.3 and 0.6 sec), 2) steady and impulsive masking noises, 3) monaural vs. binaural modes of listening, and 4) unaided vs. aided listening. In the aided condition, ear-level hearing aids (Zenith Coronation model) were used. The frequency band of the aids was from 0.6 to 4.5 kHz (ANSI, 1976). Each subject had custom-made standard, full earmolds. The test was performed binaurally (both ears open or two hearing aids inserted) or monaurally. In the monaural condition, the contralateral ear was plugged, muffed, and masked by white noise. In all monaural tests, the open ear was on the side of the speech source (two loudspeakers were positioned in front of a subject at $\pm$ 30° azimuth).

The speech test (MRT) was delivered through one of the loudspeakers while the masking noise was delivered through the other loudspeaker at three different levels to decrease word identification scores by between 20% and 85%. The results are shown in Figures 9 and 10 for unaided and aided conditions, respectively. Each figure shows percent words correct as a function of S/N.

The functions of speech perception vs. S/N are reasonably straight, parallel lines. For the analysis a pair of curves was selected that corresponded to two conditions that were to be compared. The change in perception between the two conditions was taken as a difference between the two S/Ns in decibels that would yield 60% scores. The hearing aids' influence is summarized in Table 1.

The hearing aids caused a deterioration in speech perception; it was larger for the impulsive noise than for the babble. For the babble, the deterioration was smaller for monaural listening than for binaural listening. For both maskers, the size of deterioration was smaller for longer reverberation (except monaural listening with babble noise). In quiet, the introduction of the hearing aids did not cause any deterioration in speech perception,

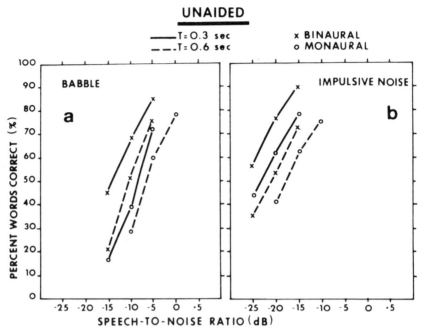

Figure 9. Percent consonants correct as functions of S/N ratio for the unaided condition (open-ear listening). (Reprinted with permission from Nábělek and Pickett, 1974a.)

as indicated by a brief set of additional tests with two normal-hearing subjects. The scores were nearly perfect for aided, unaided, monaural, and binaural listening under both conditions of reverberation.

It is interesting that the use of the hearing aids caused more deterioration of perception for the impulsive noise than for the babble. This might be because of a combination of the difference between the two noise spectra and a holding-over of transient overload distortion in the intervals between the pulses of the impulsive noise as discussed by Ingelstam et al. (1972) and Nábělek (1972).

Also, an introduction of the hearing aids had an influence on the size of the binaural gain. The binaural gain was taken as the difference between the two S/Ns that would yield 60% scores for monaural and binaural listening. In the aided condition, the binaural gain was equal to 3 dB for both conditions of reverberation and for both maskers, a value significantly lower than for unaided listening at $T = 0.3$ sec. The lower binaural gain with the aids may be because of a combination of the frequency response limits of the aids and distortion of binaural temporal cues. The aids progressively reduced the sound perceived in the range below about 800 Hz. Both the head-shadow differential and the low frequency dependence of binaural

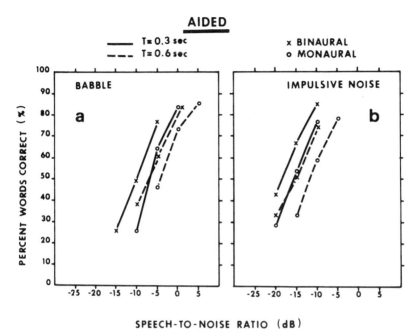

SPEECH-TO-NOISE RATIO (dB)

Figure 10.   Percent words correct as functions of S/N ratio for the aided condition (listening through hearing aids). (Reprinted with permission from Nábĕlek and Pickett, 1974a.)

masking level differences imply that the binaural gain depends more heavily on the lower frequencies than on the middle and high frequencies. Thus, the aids may be suppressing some of the more important contributions to binaural gain. Temporal distortions by the aids would be expected to affect rapid transient cues that may normally be an important factor in binaural listening.

Finitzo-Hieber and Tillman (1978) tested 12 hearing-impaired children at three reverberation times (0, 0.4, and 1.2 sec) with and without hearing aids. All tests were monaural and the nontest ear was muffed. In all the

Table 1.   Decibel changes for equivalent speech perception due to hearing aids vs. unaided listening

Listening condition	Babble		Impulsive noise	
	$T = 0.3$ sec	$T = 0.6$ sec	$T = 0.3$ sec	$T = 0.6$ sec
Binaural	4.0	3.0	7.0	5.0
Monaural	1.5	2.5	7.0	5.0

Reprinted with permission from Nábĕlek and Pickett (1974a).

Decibel changes expressed as the difference in S/Ns between the worse and better condition yielding 60% correct for two types of noise and binaural and monaural listening.

tests, speech was presented at a level 25 dB above the subject's speech reception level established in the sound field with spondees. In the aided condition, the subjects were equipped with ear-level hearing aids (Siemens, Auriculina, Model 28 E-MP-EP). The gain was adjusted to provide 30 dB of acoustic gain at 1000 Hz.

The mean results indicate that the word identification scores were somewhat lower with the aid. The differences at T of 0 sec and 0.4 sec were up to 9% lower and dependent on S/N. The largest differences were at S/N $= +12$ dB. At $T = 1.2$ sec, introduction of the hearing aid caused up to 17% reduction in the mean score. The authors concluded that reverberation and noise interacted with the tested hearing aid to reduce speech perception, specifically at the longer reverberation time of 1.2 sec.

It remains to be seen what influence hearing aids will have on speech perception in reverberation for subjects who benefit from properly fitted aids in anechoic rooms.

PHONEMIC ERRORS IN REVERBERATION

The effect of reverberation on speech perception can be measured by word identification scores or by terms of phoneme errors. Knudsen (1929) reported consonant errors made when listening in large auditoriums with long reverberation times. He noted very few vowel errors.

Nábělek and Pickett (1974a,b) and Gelfand (1977) analyzed consonant errors made in small rooms with moderate reverberation when listening to the MRT. The results of these studies give only limited information on the errors because the MRT allows choices from only six responses. A study is needed in which subjects have a much larger or unlimited choice of responses.

Nábělek and Pickett (1974a,b), analyzing consonant confusions, derived the percent correct perception of the following phonetic features: place of articulation, manner of articulation, and voicing. For the normal-hearing subjects (Nábělek and Pickett, 1974a), the reverberation effect on feature perception depended on consonant position. The place feature suffered morè by reverberation for final position than for initial position, and the manner and voicing features suffered more by reverberation for the initial position of the consonants than for final position, but these differences were small. In final position, reverberation produced much greater deterioration of the place feature than of the manner and voicing features.

All phonetic features (Nábělek and Pickett, 1974a) were perceived more poorly through hearing aids than without the aids. The perception of the place feature deteriorated more than for voicing. The deterioration with hearing aids was approximately the same for the initial and final position of consonants for all features, except for the place feature. For the place feature, the deterioration was larger for the initial position of consonants than for the final position. All of these differences were small.

For the hearing-impaired subjects, Nábělek and Pickett (1974b) reported that the mean reverberation effects on feature perception were relatively the same as for the normal-hearing subjects. No data are available on hearing aid effect in reverberation since only the aided condition was tested.

Gelfand (1977) reported unaided data on consonant feature recognition under room reverberation for 10 normal-hearing subjects. His results are in general agreement with the Nábělek and Pickett (1974a) data.

CONCLUSIONS

The general effect of reverberation is a reduction in speech intelligibility. The amount of deterioration depends on many factors, such as: 1) room size, 2) distance from the source, 3) type and amount of masking noise, 4) mode of listening (binaural or monaural), 5) amount of hearing loss, 6) individual susceptibility, and 7) hearing aids.

The clinical methods currently used do not include reverberation time as a parameter in the battery of hearing tests. Before such testing is included in the audiometric battery, further study is needed to establish the type and size of changes caused by the variables. If room reverberation is considered a parameter for clinical testing, its introduction might be more complicated than the use of simulated reverberation. An adequate type of simulation is not yet available, as discussed by Nábělek and Robinette (1978b). Some work on simulated reverberation has been reported recently by Allen and Berkley (1979) and by Houtgast and Steeneken (1978).

Finally, to facilitate listening in spaces designed for hearing-impaired listeners (such as schools for the deaf), the rooms should have some absorption added in the form of drapes, carpets, and absorptive ceilings to reduce reverberation to about $T = 0.5$ sec in medium-size enclosures. In rooms with longer reverberation, hearing-impaired listeners should be seated as closely to the source of speech as possible. In addition, the rooms should have low ambient noise levels, because noise in combination with reverberation might be very detrimental to speech understanding. Erber (1971) and Gengel (1971) recommended S/Ns of +15 to +20 dB for hearing-impaired children, about 10 to 15 dB greater S/Ns than normal-hearing children need for maximum speech intelligibility.

It seems that listeners who benefit from binaural aids in anechoic environments will benefit from them in moderately reverberant rooms as well. Only very long reverberations, of 2 sec or more, such as those in large auditoriums or churches, can eliminate the benefits of a second hearing aid.

REFERENCES

Allen, J.B., and D.A. Berkley. 1979. Image method for efficiently simulating small room acoustics. J. Acoust. Soc. Am. 65:943–950.

American National Standards Institute. 1969. American National Standard Specifications for Audiometers, ANSI-S3.6–1969. American National Standards Institute, New York.

American National Standards Institute. 1976. Specification of Hearing Aid Characteristics, ANSI-S3.22–1976. American National Standards Institute, New York.

Bell, D.W., E.J. Kreul, and J.C. Nixon. 1972. Reliability of the modified rhyme test for hearing. J. Speech Hear. Res. 15:287–295.

Bolt, R.H., and A.D. MacDonald. 1949. Theory of speech masking by reverberation. J. Acoust. Soc. Am. 21:577–580.

Børrild, K. 1970. The Acoustic Environments in Schools for the Deaf. State Boarding School for Hard of Hearing and Deaf, Fredericia, Denmark.

Botros, R. 1976. Acoustic environment for audio teleconferencing. J. Acoust. Soc. Am. 60:S9.

Bullock, M.D. 1967. The effects of different reverberation times upon the intelligibility of PB words as perceived by subjects with normal hearing and subjects with sensorineural impairments and concomitant discrimination losses. Master's thesis, Ohio State University, Columbus.

Crum, M.A., and T.W. Tillman. 1973. Effects of speaker-to-listener distance upon speech intelligibility in reverberation and noise. ASHA 15:473(A).

Erber, N.P. 1971. Auditory and audiovisual reception of words in low-frequency noise by children with normal hearing and by children with impaired hearing. J. Speech Hear. Res. 14:496–512.

Finitzo-Hieber, T., and T.W. Tillman. 1978. Room acoustics effects on monosyllabic word discrimination ability for normal and hearing impaired children. J. Speech Hear. Res. 21:440–458.

Gelfand, S.A. 1977. Recognition of some consonant features under room reverberation. ASHA 19:648(A).

Gelfand, S.A., and I. Hochberg. 1976. Binaural and monaural speech discrimination under reverberation. Audiology 15:72–84.

Gengel, R.W. 1971. Acceptable speech-to-noise ratios for aided speech discrimination by the hearing-impaired. J. Aud. Res. 11:219–222.

Hirsh, I.J. 1950. The relation between localization and intelligibility. J. Acoust. Soc. Am. 22:196–200.

Houtgast, T., and H.J.M. Steeneken. 1973. The modulation transfer function in room acoustics as a predictor of speech intelligibility. Acustica 28:66–73.

Houtgast, T., and H.J.M. Steeneken. 1978. Application of the modulation transfer function (MTF) in room acoustics. Report, Institute for Perception TNO, Soesterberg.

Ingelstam, R., B. Johansson, A. Pettersson, and H. Sjogren. 1972. The effects of nonlinear amplitude distortion, and investigation by variation of the quadratic and the cubic components. In G. Fant (ed.), Proceedings of Symposium on Speech Communication Ability and Profound Deafness, pp. 309–321. Alexander Graham Bell Association, Washington, D.C.

John, J.E.J. 1957. Acoustics and efficiency in the use of hearing aids. In A.W.G. Ewing (ed.), Educational Guidance and the Deaf Child, pp. 63/1–63/4. Manchester University Press, Manchester.

Klein, W. 1971. Articulation loss of consonants as a basis for the design and judgement of sound reinforcement systems. J. Audio Eng. Soc. 19:920–922.

Knudsen, V.O. 1929. The hearing of speech in auditoriums. J. Acoust. Soc. Am. 1:56–82.

Koenig, A.H., J.B. Allen, D.A. Berkley, and T.H. Curtis. 1977. Determination of masking-level differences in a reverberant environment. J. Acoust. Soc. Am. 61:1374–1376.

Kreul, E.J., J.C. Nixon, K.D. Kryter, D.W. Bell, J.S. Lang, and E.D. Schubert. 1968. A proposed clinical test of speech discrimination. J. Speech Hear. Res. 11:536–552.

Lochner, J.P.A., and J.F. Burger. 1964. The influence of reflections on auditorium acoustics. J. Sound Vibr. 1:426–454.

McDermott, B., and J. Allen. 1976. Perceptual factors of small room reverberation. J. Acoust. Soc. Am. 60:S9.

MacKeith, N.W., and R.R.A. Coles. 1971. Binaural advantages in hearing of speech. J. Laryngol. Otolaryngol. 85:213–232.

Mason, D., and C.W. Asp. 1976. The relationship between speech discrimination ability and self-assessed hearing-handicap of adults with sensorineural hearing losses as a function of reverberation and noise. J. Acoust. Soc. Am. 60:S124.

Millin, J.P. 1968. The effect of small room reverberation on discrimination tests. Doctoral thesis, Case Western Reserve University, Cleveland, Oh.

Moncur, J.P., and D. Dirks. 1967. Binaural and monaural speech intelligibility in reverberation. J. Speech Hear. Res. 10:186–195.

Nábělek, A.K. 1976. Reverberation effects for normal and hearing-impaired listeners. In I.J. Hirsh and S.R. Silverman (eds.), Hearing and Davis: Essays Honoring Hallowell Davis, pp. 333–341. Washington University Press, St. Louis.

Nábělek, A.K., and J.M. Pickett. 1974a. Reception of consonants in a classroom as affected by monaural and binaural listening, noise, reverberation and hearing aids. J. Acoust. Soc. Am. 56:629–639.

Nábělek, A.K., and J.M. Pickett. 1974b. Monaural and binaural speech perception through hearing aids under noise and reverberation with normal and hearing-impaired listeners. J. Speech Hear. Res. 17:724–734.

Nábělek, A.K., and L.N. Robinette. 1978a. Influence of the precedence effect on word identification by normally hearing and hearing-impaired subjects. J. Acoust. Soc. Am. 63:187–194.

Nábělek, A.K., and L.N. Robinette. 1978b. Reverberation as a parameter in clinical testing. Audiology 17:239–259.

Nábělek, I.V. 1972. On some parameters of hearing aids with compression. J. Acoust. Soc. Am. 52:183(A).

Patterson, B.R., I. Esteves, G.M. Sessler, and J.E. West. 1975. Effect of early echoes and reverberation on the subjective evaluation of speech. J. Acoust. Soc. Am. 58:S129.

Peutz, U.M.A. 1971. Articulation loss of consonants as a criterion for speech transmission in a room. J. Audio Eng. Soc. 19:915–919.

Plomp, R. 1976. Binaural and monaural speech intelligibility of connected discourse in reverberation as a function of azimuth of a single competing sound source (speech and noise). Acustica 34:200–211.

Santon, F. 1976. Numerical prediction of echograms and of the intelligibility of speech in rooms. J. Acoust. Soc. Am. 59:1399–1405.

Schroeder, M.R., D. Gottlob, and K.F. Siebrasse. 1974. Comparative study of European concert halls: Correlation of subjective preference with geometric and acoustic parameters. J. Acoust. Soc. Am. 56:1195–1201.

Walker, B.E., L.P. Delsasso, and V.O. Knudsen. 1970. Reflective surfaces for hearing of speech and music in an anechoic chamber. J. Acoust. Soc. Am. 47:99.

CHAPTER 3

THE SPEECH RECEPTION THRESHOLD OF HEARING-IMPAIRED SUBJECTS IN NOISE AS A FUNCTION OF REVERBERATION TIME

R. Plomp and A.J. Duquesnoy

CONTENTS

In recent years a method has been developed (Houtgast and Steeneken, in preparation) for calculating the combined effects of noise, single echoes, and reverberation on speech intelligibility in an enclosure. This method is based on the modulation transfer function (MTF) for sounds transmitted from the location of the speaker to the location of the listener. Leaving the other parameters (volume, distances to the sound sources) constant, contour lines of equal intelligibility can be drawn in a diagram representing reverberation time *(T)* along one axis and signal-to-noise ratio (S/N) along the other.

Some preliminary data are available in relation to the question of whether these contour lines, based on normal hearing, are also valid for impaired hearing. For an adequate evaluation of this extrapolation, some insight into how the isointelligibility contour lines are obtained seems desirable. Therefore, these lines are explained before data from hearing-impaired subjects are presented.

DERIVING ISOINTELLIGIBILITY CONTOUR LINES IN THE S/N VS. *T* PLANE

Suppose we have, in an enclosure with reverberation time *T,* a speaker and a noise source with equal sound spectra. Both are located far enough from the listener for the direct sound to be negligible compared with the indirect sound, which reaches the listener after having been reflected a variable number of times. This condition is attractive because it simplifies the equations and is the most sensitive way of testing the effect of reverberation on speech intelligibility. In living rooms this condition is usually fulfilled at distances larger than 2 m, in classrooms at distances larger than 3 m.

We replace the speaker with a noise signal with intensity $I_i(t)$, varying sinusoidally between 0 and $2\bar{I}_i$:

$$I_i(t) = \bar{I}_i(1 + cos2\pi Ft) \tag{1}$$

where $\bar{I}_i = $ *long-term average of $I_i(t)$*, measured at a distance of 1 m from the source, and $F = $ modulation frequency. The interfering noise source has a constant intensity I_n at a distance of 1 m. Equation 1 shows that the input signal is 100% modulated. Because of the noise and reverberation, the modulation index near the listener will be less than 1 and will depend on the modulation frequency. This frequency-dependent modulation index *(m(F))* represents the MTF.

The MTF can be calculated easily for the specific condition considered assuming that there is an exponentially decaying sound field. Then, the intensity $I_o(t')$ at the position of the listener at time t' is the sum of the contributions from all past moments weighted with an exponential function:

$$I_o(t') = C \int_{-\infty}^{t'} [I_i(t) + I_n] \cdot e^{-a(t' - t)}dt \tag{2}$$

in which C depends upon the properties of the enclosure, whereas a is determined exclusively by the reverberation time: $e^{-aT} = 10^{-6}$, or $a = 13.8/T$. Substituting $I_i(t)$ from equation 1 in equation 2, we get successively:

$$I_o(t') = C\bar{I} \int_{-\infty}^{t'} e^{-a(t' - t)}dt + C\bar{I}_i \int_{-\infty}^{t'} cos2\pi Ft \cdot e^{-a(t' - t)}dt$$

$$+ CI_n \int_{-\infty}^{t'} e^{-a(t' - t)}dt$$

$$= \frac{C\bar{I}_i}{a} + \frac{C\bar{I}_i}{a\sqrt{1 + (2\pi F/a)^2}}$$

$$\cdot cos(2\pi Ft' - \theta) + \frac{CI_n}{a}$$

$$= \frac{C\bar{I}_i}{a}(1 + I_n/\bar{I}_i)$$

$$[1 + \frac{1}{1 + I_n/\bar{I}_i} \cdot$$

$$\frac{1}{\sqrt{1 + (2\pi F/a)^2}} \cdot \cos(2\pi Ft' - \theta)] \tag{3}$$

in which θ is a phase lag relative to the input signal. This equation, written in a form similar to equation 1, shows that the MTF is equal to:

$$m(F) = \frac{1}{1 + I_n/\bar{I_i}} \cdot \frac{1}{\sqrt{1 + 0.207 F^2 T^2}} \tag{4}$$

We see that $m(F)$ is the product of two independent factors, one determined by S/N, the other by the reverberation time.

In Figure 1, $m(F)$ is plotted for $I_n = 0$, with T as the parameter. By multiplying the envelope spectrum of speech with $m(F)$ we get the envelope spectrum at the location of the listener. This is illustrated in Figure 2. The solid curve represents the envelope spectrum of connected discourse obtained by performing a ⅓-octave frequency analysis on the envelope for the audiofrequency octave band centered at 2000 Hz (for other octave bands similar curves are obtained). The envelope spectrum has a maximum for $F \sim 3.4$ Hz, situated between the frequency of words (2.5 words/sec) and the frequency of syllables (5 syllables/sec) of the piece of discourse on which this analysis was based. The broken curves show that for $T < 0.25$ sec the speech modulations are transferred faithfully, but they are dramatically reduced for $T > 4$ sec. This diagram may help in demonstrating that the

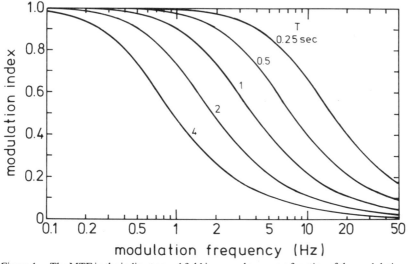

Figure 1. The MTF in the indirect sound field in an enclosure as a function of the modulation frequency, with reverberation time as the parameter and no interfering noise.

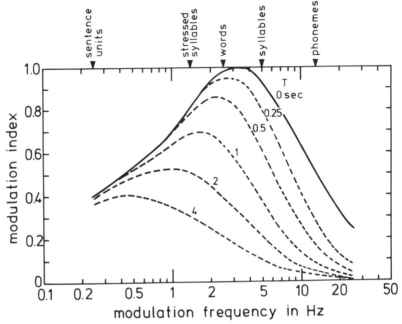

Figure 2.   The solid curve is the envelope spectrum (rms of the fluctuations within ⅓-octave bands) for the 2000-Hz octave band of a 60-sec sample of connected discourse of a single speaker. The broken curves are obtained by multiplying the solid curve with the $m(F)$ curves reproduced in Figure 1, with T as the parameter.

MTF represents a promising measure for investigating the quality of speech transmission.

Houtgast and Steeneken (in preparation) developed a weighting function, the modulation transfer index (MTI), to determine the contributions for the various modulation frequencies in order to get the best possible agreement with intelligibility scores measured using speakers and listeners. This weighting function was optimized by considering a large number of combinations of noise, single echoes, and reverberation times. Essentially, the calculation of the MTI is very simple. It consists of the following steps:

1.   For each of the 18 F values from 0.4 to 20 Hz, measured at intervals of ⅓ octave, the equivalent signal-to-noise ratio $(S/N(F))$ is determined:

$$S/N(F) \text{ in dB} = 10 \log \frac{m(F)}{1 - m(F)} \qquad (5)$$

2.   The resulting 18 S/N values are averaged after replacing all values exceeding $\pm$ 15 dB with this limiting value.

3. This average, $\overline{S/N}$, is normalized:

$$\text{MTI} = \frac{\overline{S/N} + 15}{30} \tag{6}$$

so that $0 \leqslant \text{MTI} \leqslant 1$. In the case considered here, T and S/N are independent of audio frequency, and MTI = speech transmission index (STI). (In other cases this procedure has to be repeated for 7 audiofrequency octaves to obtain STI as a weighted average of the MTIs for those bands.) Figure 3 shows how STI is related to the loss in intelligibility for consonants measured with speakers and listeners for a large number of combinations of S/N, T, and the presence of strong echoes.

By using this procedure in the computation of STI for various values of T and S/N ($= 10 \log \overline{I}_i / I_n$) in equation 4, we obtain the contour lines of equal STI, or equal speech intelligibility, plotted in Figure 4. In binaural listening an STI value below 0.4 is considered to represent unsatisfactory conditions, STI values between 0.4 and 0.6 acceptable conditions, values between 0.6 and 0.8 good conditions, and values beyond 0.8 excellent conditions.

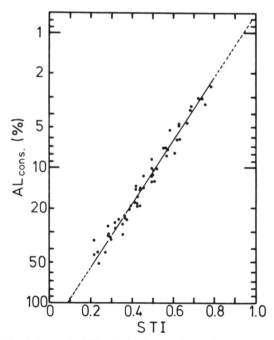

Figure 3.   Relation between the STI, calculated according to the procedure described in the text, and the loss in intelligibility for consonants in nonsense words (consonant-vowel-consonant) measured with speakers and listeners for a wide variety of auditorium-like conditions (Houtgast and Steeneken, in preparation).

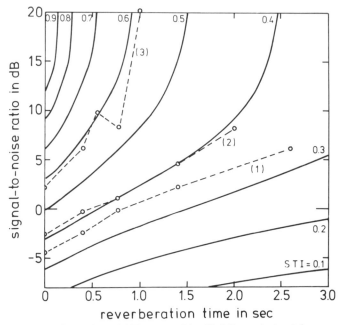

Figure 4.   Contour lines of equal STI, or equal intelligibility, calculated from equation 4. The data points represent median SRT values, relative to the level of the interfering noise, for three groups of listeners: 1) 10 young subjects, 2) 27 elderly subjects who could not understand the sentences for $T = 2.6$ sec, and 3) 13 elderly subjects who could not understand the sentences for $T = 1.4$ sec.

COMPARISON WITH THE SRT OF HEARING-IMPAIRED SUBJECTS

The monaural speech reception thresholds (SRTs) of subjects with pres-bycusis were investigated using lists of sentences that recently have been developed. As has been reported elsewhere (Plomp and Mimpen, 1979), with these lists very reliable SRT values can be obtained using an adaptive procedure for the presentation levels of the 13 sentences in each list. The lists, as well as the interfering noise, were reproduced by the same loud-speaker in an anechoic room and in a reverberation room having adjustable reverberation times, with the microphone positioned in the indirect sound field. The resulting recordings were used by the second author for measuring SRT as a function of reverberation time. The noise, which had the same spectrum as the long-term average spectrum of the sentences, was presented at a constant level of 50 dBA.

The data points in Figure 4 represent median SRT values, relative to the level of the interfering noise, for three groups of listeners. We may conclude from these preliminary data that they are described well by the isointelligibility contour lines. Apparently, the SRT of a hearing-impaired subject in a reverberant sound field can be expressed as a single number, the required STI value. This means that a theoretical model on the SRT in noise, developed for $T=0$ (Plomp, 1978), can be easily extended to include the effect of reverberation.

A more detailed description of the current experiments is in preparation (Duquesnoy and Plomp, in preparation).

REFERENCES

Duquesnoy, A. J., and R. Plomp. Effect of reverberation and noise on the intelligibility of sentences in case of presbyacusis. In preparation.

Houtgast, T., and H.J.M. Steeneken. 1978. Application of the Modulation Transfer Function (MTF) in Room Acoustics. Report, Institute for Perception TNO, Soesterberg.

Houtgast, T., and H.J.M. Steeneken. Predicting speech intelligibility in rooms from the Modulation Transfer Function. I. General acoustics. In preparation.

Plomp, R. 1978. Auditory handicap of hearing impairment and the limited benefit of hearing aids. J. Acoust. Soc. Am. 63:533–549.

Plomp, R., and A.M. Mimpen. 1977. Improving the reliability of testing the speech reception threshold for sentences. Audiology 18:43–52.

CHAPTER 4

SOME EFFECTS OF MICROPHONE LOCATION, SIGNAL BANDWIDTH, AND INCIDENT WAVE FIELD ON THE HEARING AID INPUT SIGNAL

George F. Kuhn

CONTENTS

The acoustic input signal to a hearing aid is a function of numerous variables. Experiments were conducted to systematically investigate the effects of these variables on the hearing aid input signal and on the signal at the coupler microphone in an anthropomorphic manikin. From these measurements, comparisons can be made between the received sound at the eardrum(s) of normal-hearing subjects and the input signal at the head-worn and body-worn hearing aids. These measured results are also compared to theoretical predictions of the pressure transformation from a progressive plane wave field and from a diffuse wave field to the surface of a rigid sphere and a rigid cylinder. These seemingly simplistic theoretical models are useful for predicting the interaural time differences, the time differences for head- and body-worn aids, the horizontal directivity for head-worn aids, and the low frequency horizontal

The experimental work described in this chapter was conducted while the author was with the National Bureau of Standards, Sound Section, Washington, D.C.

directivity for body-worn aids when measured on a manikin. These theoretical models, together with the experiments, are very useful for establishing order of magnitude effects on the input signal(s) to hearing aids resulting from the head size, the head shape, the fine features of the face, the pinna, the torso, and clothing absorption.

In order to keep this chapter brief, only abbreviated summaries of the individual topics, with some substantiating experimental and theoretical results, are presented. The major topics discussed deal with:

1. The pressure transformation in the azimuthal plane to the coupler microphone and to the side of the head where head-worn aids are located
2. The pressure transformation in the azimuthal plane to the torso
3. The pressure transformation from a diffuse sound field to the coupler microphone, to the head surface, and to the torso surface

THE EXPERIMENTS

The pressure transformations from an acoustic free field to the ear, head surface, and torso surface were made with an anthropomorphic manikin (Burkhard and Sachs, 1975) in an anechoic room. The azimuthal directivities were recorded automatically by rotating the manikin around the vertical axis while keeping the acoustic source stationary. The pressure transformations from a diffuse sound field to the coupler microphone, the head surface, and the torso surface were measured in a reverberation room. The measured acoustic pressures were normalized to the incident pressure(s) when measured with the manikin or the person removed. These normalized pressures were converted to levels and are called *pressure level transformations* in this chapter.

Pressure Transformation from a Free Field to the Coupler Microphone and to Head-worn Hearing Aids for Frontal Sound Incidence

Figure 1 shows the pressure level transformation for a frontally incident sound wave to the blocked ear canal entrance when the pinna is replaced by a conformal rigid plate. These measurements were made by inserting a "¼-inch" microphone into the ear canal and sealing the perimeter with a rubber "O-ring." The microphone diaphragm is flush with the head surface in this instance. Figure 1 also shows the "average" pressure level transformation to the eardrum for normal-hearing individuals (Shaw, 1974). The results shown by the hatched area represent the range of theoretical predictions for the pressure level transformation to the side of the sphere (between 90° and 105° relative to the sphere's frontal pole). Figure 1 therefore dramatizes the *difference* in

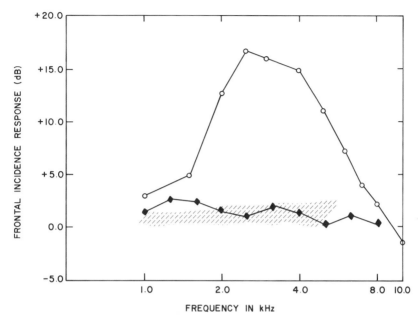

Figure 1.   Pressure level transformation for frontal sound incidence to the external ear and the head surface. ◯, Shaw's composite at the eardrum of human subjects; ◆, measurements at the blocked ear canal entrance with the pinna removed; hatched results are the range of theoretical predictions for a point receiver on a rigid sphere between 90° ≤ θ ≤ 105°. (Data in part from Shaw, 1974.)

pressure levels between the eardrum and the side of the head. The pressure level at the side of the head is as much as 16–17 dB less than the eardrum pressure level at the fundamental resonance (∼ 2.7 kHz) of the external ear. Figure 1 also shows that the pressure at the head surface can be calculated approximately on the basis of diffraction by a rigid sphere. Thus, the hearing aid input signal can be predicted approximately by a simple analytical model.

The sound pressure level at the coupler microphone and at the head-worn hearing aid is affected by the torso because of the constructive and destructive interference between the wave scattered by the torso and the direct wave to the ear. Figure 2 shows the coupler microphone pressure level averaged over six pinnas relative to the incident pressure at the center of the head measured when the manikin is removed. The pressure level at the coupler microphone between 0.25 kHz and 0.8 kHz is increased by as much as 1½ dB because of the constructive interference between the wave(s) scattered by the torso and the directly incident wave to the ear. Destructive interference of the wave scattered by the torso with the direct wave to the ear lowers the pressure level by as much as 3½ dB between 0.9 and 2.0 kHz. The effect of the torso on the pressure level at 8.0 kHz is not

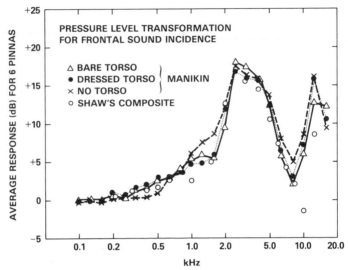

Figure 2.   Pressure level transformation for frontal sound incidence to the coupler micro-phone of a manikin and the eardrum of human subjects. (The pressure measurements at the coupler microphone were made with six different-size pinnas and the results were averaged.)

significant since this difference is caused by only one particular pinna of the six pinnas used.

Five of the six pinnas were molded from the ears of live subjects. The sixth pinna was purchased commercially. These pinnas varied approximately $\pm$ 2 standard deviations from the average length of the human pinna. The pinnas varied approximately $\pm$ 1 standard deviation from the average width of the human pinna.

The effect of the torso on the pressure levels beside the head, where head-worn aids are located, was not measured directly. However, because the acoustic wavelength for frequencies $\leqslant$ 2.0 kHz is large compared with the effective head radius and both the path length difference between the torso and hearing aid locations, neither the head diffraction nor the scattering of waves by the torso is expected to cause large changes in the pressure level beside the head. However, at high frequencies, based on theoretical predictions for scattering by a sphere (e.g., Schwarz, 1943), large changes in pressure level may be expected, particularly in the geometric shadow of the head and pinna.

Measurements of the pressure level distribution beside the head of the manikin (with its torso attached) were made along six contours, shown in Figure 3. Contours 1 to 3 were spaced 5.0 mm apart vertically, at 2.0, 5.0, and 7.0 mm, respectively, away from the head surface. Contour 4 conforms with pinna "X" (the commercial pinna), and contours 5 and 6 are spaced

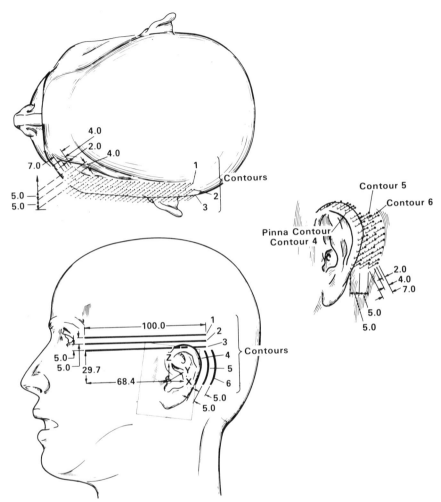

Figure 3.   Measurement contours on the manikin's head (all dimensions in mm). (Reprinted with permission from Kuhn and Burnett, 1977.)

5.0 mm apart from each other on a line normal to contour 4. (Kuhn and Burnett (1977) supply further experimental details and results.)

The pressure level distributions relative to the incident pressure at the center of the manikin's head with the manikin removed are shown in Figures 4–6. The 0-cm position in Figures 4 and 5 is vertically above the ear canal axis; the −6.8-cm position is the most forward, and the +3.2-cm position is the most rearward measurement location. Each dot represents a measurement. The height between any two measured curves in Figure 4

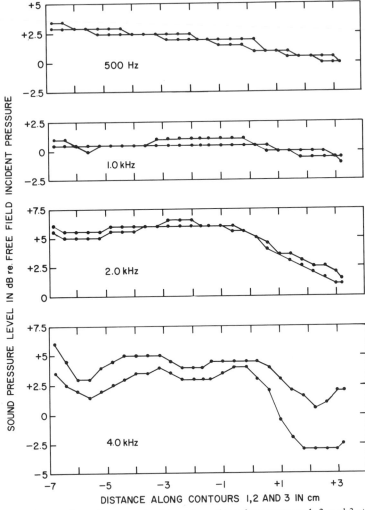

Figure 4.   Range of sound pressure level transformations along contours 1, 2, and 3 at 2, 4, and 7 mm, respectively, from the head surface for discrete frequencies. Source-to-ear canal distance is 1.0 m.

represents the maximum difference in pressure level at a particular x coordinate over a vertical range of $\pm$ 5.0 mm, centered at contour 2, and over a horizontal range from 2.0 to 7.0 mm away from the head surface. It is clear from the results shown in Figures 4 and 5 that the input signal to a head-worn aid (for frequencies less than 2.5 kHz) is not strongly dependent on its x location when it is forward of the ear canal axis. As the frequency

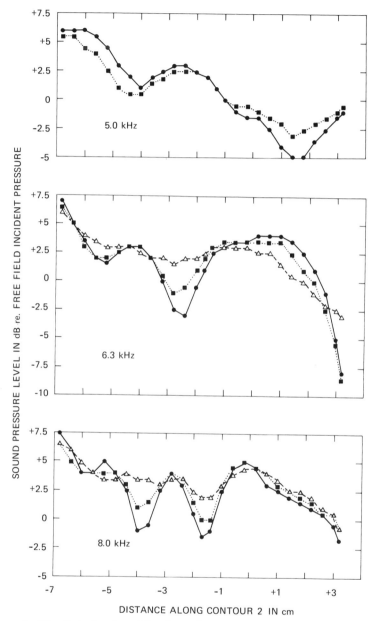

Figure 5.   The effect of random pink noise on the sound pressure level transformation along contour 2 at 4 mm from the head surface. ●, Tone; ■, 6% bandwidth pink noise; △, 29% bandwidth pink noise. Source-to-ear canal distance is 1.0 m.

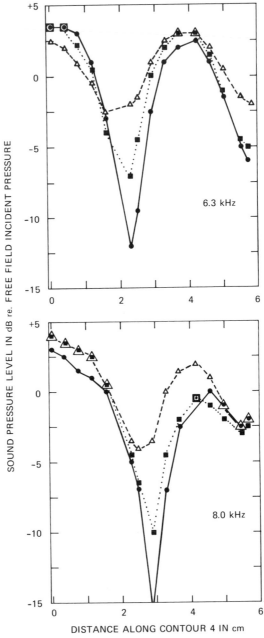

Figure 6.   The effect of random noise on the sound pressure level transformation along contour 4 at 4 mm from the head surface. ●, Tone; ■, 6% bandwidth pink noise; △, 29% bandwidth pink noise. Source-to-ear canal distance is 1.0 m.

increases, the pressure changes more severely with distance x, since diffraction and interference effects become greater with increasing frequency. For frequencies greater than 4.0 kHz, large pressure level gradients are produced by head diffraction, pinna shadows, and transverse acoustic modes excited in the pinna. Such large pressure level gradients can create difficulties for the wearer because the effective gain of the aid will depend on head orientation as well as frequency. Large pressure level gradients create difficulties also for the test or calibration laboratory since extremely accurate location will be required. These pressure level gradients and pressure level minima can be reduced through the use of 6% or 29% bandwidth (of the center frequency) pink noise, as shown in Figures 5 and 6. The particularly severe pressure level gradients that exist behind the pinna, shown in Figure 6, result from transverse modes, which are strongly excited at 6.3 kHz and 8.0 kHz. Phase measurements (not shown) show the expected 180° phase jump across the null (nodal line) of the standing waves at 6.3 and 8.0 kHz. This particular mode is analogous to the (1,0) cylindrical duct mode. (Further description of transverse modes are given by Shaw (1975) and Kuhn (1978).)

The results in Figures 4, 5, and 6 show that: a) the hearing aid input signal level is as much as 7½ dB above the incident pressure level and is a maximum for microphones near the front of the head; b) the pressure level gradients increase as frequency increases, particularly above 4.0 kHz; c) pressure level gradients are most severe for hearing aids located near the pinna, because of the coupling of transverse resonances in the pinna (at frequencies above 4 kHz) to the hearing aid; and d) hearing aid tests above 5.0 kHz using discrete frequencies or noise bandwidths of less than ⅓ octave require very exact microphone placements since the pressure level changes by as much as 9 dB over a 2.0-mm distance.

Figures 5 and 6 compare the pressure level distribution as a function of frequency bandwidth. Using 6% and 29% (of the center frequency) bandwidth pink noise "smooths" the pressure minima by as much as 5.0 dB for contours 1–3. The sound pressure levels along contours 4–6 vary smoothly with position for frequencies equal to or less than 5.0 kHz. However, at 6.3 and 8.0 kHz sharp pressure minima occur behind the pinna. These minima are raised by more than 5 and 10 dB if pink noise, with 6% and 29% bandwidth, respectively, is used. The pressure minima in Figure 6 are produced by the higher order acoustic modes excited in the pinna and are closely coupled to the area behind the helix.

Azimuthal Directivity at the
Coupler Microphone and at the Head Surface

It was seen in Figure 1 that the input signal level to the head-worn hearing aid for frontal incidence differs considerably from that to the eardrum of a normal subject. Spatial perception and localization in the azimuthal plane depend on cues of the interaural time and level differences. Thus, a comparison of the azimuthal directivity for head-worn hearing aids and the directivity at the coupler microphone is a measure of the relative strength of localization cues available to the aided and unaided listener. Azimuthal directivity measurements were made from 0.5 kHz to 16.0 kHz. However, for the sake of brevity, directivities are shown only at a few frequencies where the results are particularly significant.

Azimuthal directivities are (arbitrarily) normalized to 25.0 dB at frontal or 0° incidence, as shown in Figures 7–9. Angles of incidence of 90° and 270° represent sound waves incident to the ear from the irradiated side and from the shadowed side of the head, respectively.

Measurements at 1.6 kHz (Figure 7) and measurements at lower frequencies (not shown) indicate that there is good agreement between all the results, except in the shadow region defined by a sector between approximately 240° and 310°. Therefore, the azimuthal directivity for angles of incidence less than 180° is relatively insensitive to the head shape and to the effects of the external ear. Furthermore, the azimuthal directivity of a head-worn aid can be predicted within approximately 1½ dB up to 1.6 kHz using diffraction theory for a rigid sphere.

However, Figure 8 shows that as the frequency increases to 3.15 kHz, the pinna, together with the head, forms a directional pattern with a major lobe at an angle of incidence of 45°; a directional null, although not very sharp, is formed between 90° and 135°. The directivity in the geometric shadow zone, between 240° and 300°, is very sensitive to the head and pinna geometry as well as to the (hearing aid) microphone location.

The azimuthal directivity at 5.0 kHz is shown in Figure 9. Again, the pinna, together with the head, forms a major lobe, with a gain of 6 to 7 dB for angles of incidence between 45° and 60°, and forms a directional null in the sector between 135° and 150°. The azimuthal directivity for an aided listener, approximated by the directivity for the "flat plate," differs considerably from the directivity pattern for an unaided listener at frequencies greater than 3.15 kHz. Nevertheless, for the sector between 330° and 30°, the azimuthal directivity for an aided listener is approximately the same as that for the unaided listener. Thus, if the acoustic source lies within a ± 30° sector relative to frontal incidence, the aided listener has nearly the same azimuthal directivity, the same interaural level difference, and therefore approximately the same high frequency localization cue as the unaided listener.

Azimuthal directivities with an eyeglass and an over-the-ear (OTE) type of hearing aid were measured and compared with the azimuthal directivity measured at the coupler microphone between 1.0 kHz and 12.5 kHz.

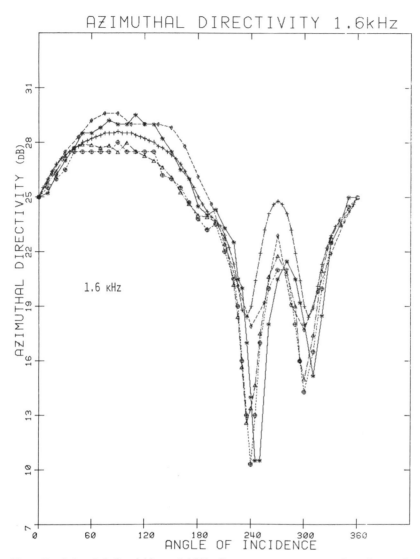

Figure 7.   Azimuthal directivities at 1.6 kHz. Pressure measurements made at the coupler microphone of a manikin: *, pinna replaced by a flat plate; ⊕, pinna "X"; △, average for five pinnas not including "X"; ◆, Shaw's composite for human subjects; +, theoretical predictions for a point receiver on a rigid sphere (radius = 0.477 wavelength).

AZIMUTHAL DIRECTIVITY 3.15 kHz

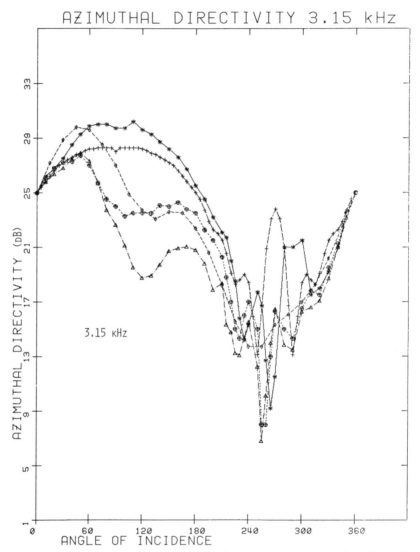

Figure 8. Azimuthal directivities at 3.15 kHz. Pressure measurements made at the coupler microphone of a manikin: *, pinna replaced by a flat plate; ⊕, pinna "X"; △, average of five pinnas not including "X"; ◆, Shaw's composite for human subjects; +, theoretical predictions for a point receiver on a rigid sphere (radius = 0.955 wavelength).

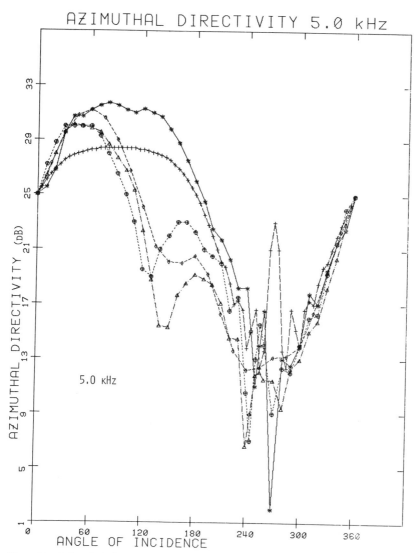

AZIMUTHAL DIRECTIVITY 5.0 kHz

5.0 kHz

Figure 9.   Azimuthal directivities at 5.0 kHz. Pressure measurements made at the coupler microphone of a manikin: *, pinna replaced by a flat plate; ⊕, pinna "X"; △, average of five pinnas not including "X"; ◆, Shaw's composite for human subjects; +, theoretical predictions for a point receiver on a rigid sphere (radius = 1.59 wavelength).

Measurements show that the azimuthal directivity of both the eyeglass aid and the OTE aid agree with the directivity at the coupler microphone within 1½ dB up to 2.0 kHz, except the shadow region between approximately 230° and 300°. Therefore, only results for higher frequencies within the operating range of most hearing aids are shown.

Figures 10–12 show the measured azimuthal directivities for the hearing aids, the coupler microphone, and the theoretical directivity of a point receiver on a rigid sphere with a perimeter equal to that of the manikin's head. (Only the commercial pinna was used in this instance for these measurements.) The solid dot at one side of the head indicates the location of the microphone for the hearing aid or coupler. Up to 4.0 kHz the major features of the azimuthal directivity for both hearing aids are predicted reasonably well by the theoretical predictions for the sphere. The directivity patterns *appear* to be different in the shadow region of the head. However, since the OTE aid and the eyeglass aid are displaced forward by 2.0 cm and 4.5 cm, respectively, relative to the ear canal axis, their directivity patterns must be rotated by corresponding angular displacements of 13.5° and 30°. After this angular rotation is made, the directional nulls and peaks of the

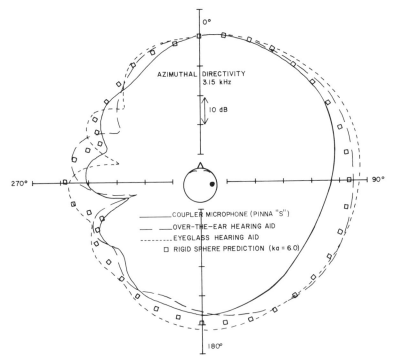

Figure 10.   Azimuthal directivities at 3.15 kHz. Radius of rigid sphere = 0.955 wavelength.

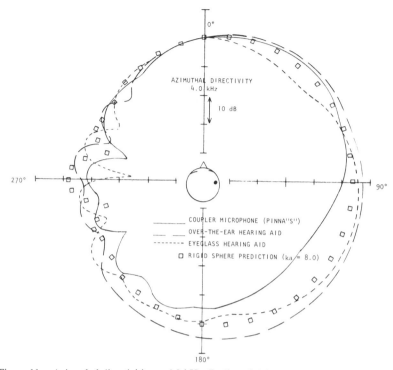

Figure 11.   Azimuthal directivities at 4.0 kHz. Radius of rigid sphere = 1.273 wavelength.

directivities agree quite well. At frequencies greater than 4.0 kHz, good agreement between directivities is found only in the frontal sector between approximately 315° and 45°. Larger differences exist elsewhere.

The azimuthal directivity for head-worn aids is reasonably well predicted analytically up to 4.0 kHz. Up to 1.6 kHz the azimuthal directivity of the head-worn aids agrees well with the directivity at the coupler microphone. For angles of incidence between 330° and 30° and for frequencies between 1.6 kHz and 4.0 kHz the azimuthal directivity of the hearing aids agrees well with that at the coupler microphone. In this forward sector, the interaural pressure level difference that is the primary localization cue above 1.6 kHz is well preserved up to 4.0 kHz. This interaural pressure level difference is larger at the head-worn aid(s) than at the coupler microphone outside of this sector. Thus, a strong localization cue is available to the aided listener, although different in magnitude from that for the unaided listener. Therefore it seems reasonable to expect ipsilaterally aided individuals to localize well in the azimuthal plane, except when the aid(s) is placed too far forward or when the aid or the ear is shadowed by the head.

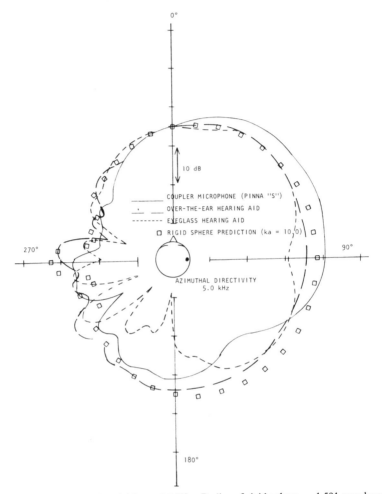

Figure 12.   Azimuthal directivities at 5.0 Khz. Radius of rigid sphere = 1.591 wavelength.

It is useful to compare the interaural time difference (ITD), the primary low frequency localization cue, available to the aided subject with that available to the unaided subject. Measurements of the ITD at the coupler microphone and at a position at the front of the tragus are shown in Figure 13. These results show that the ITDs are nearly the same for head-worn hearing aids located immediately in front of the ear canal axis as for the unaided listener. However, it is to be expected that a forward or backward movement of the hearing aid will change the ITD and therefore present a different localization cue. In this instance, poor localization accuracy and possible confusion will arise.

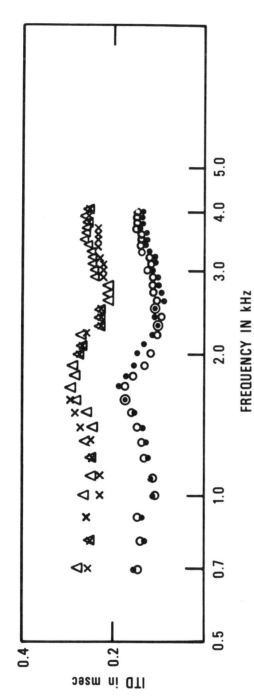

Figure 13. Measured values of ITD as a function of frequency and angle of incidence, θ_{INC}. ●●● $\theta_{INC} = 15°$, Measurements at the coupler microphone; ○○○ $\theta_{INC} = 15°$, measurements at head surface; xxx $\theta_{INC} = 30°$, measurements at the coupler microphone; △△△ $\theta_{INC} = 30°$, measurements at head surface.

71

Pressure Transformation from a Free Field to the Body Surface

Figure 14 shows a comparison of the pressure level transformations for frontally incident sound to the side of the head, where head-worn aids are located, and to the center of the chest of a manikin, where body-worn aids are sometimes worn. Theoretical predictions for the pressure level transformations to the side of a rigid sphere of equivalent perimeter to the head and to the pole of a rigid cylinder of approximately equivalent perimeter to the manikin's chest are also shown. Thus, the theory for a rigid cylinder or for a rigid sphere predicts the major features of the pressure level transformation to the torso and head of the manikin, respectively. Differences between the theory and the experiment result primarily from differences between the idealized geometries for the theoretical models and the actual geometry of the torso and the head.

However, the existing data in the literature and data from S. Lybarger (personal communication) show that the pressure level transformation to the torso of live subjects has a "notch" in the frequency range between 1.0 kHz and 2.0 kHz. Measurements of the pressure transformation for torso-worn aids on subjects are shown in Figure 15, as well as the effect of clothing on the pressure level transformation (Kuhn and Guernsey, in preparation). The acoustic absorption of the clothing reduces the pressure level on the torso surface. The sharp notch in the response, shown by Lybarger's results,

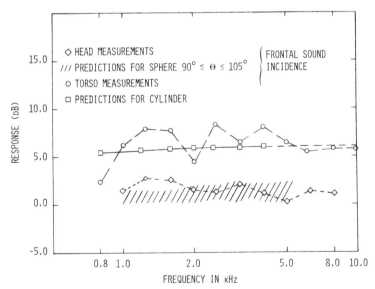

Figure 14. Comparison between head and torso measurements and predictions for a sphere and cylinder.

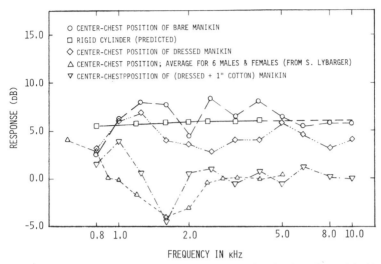

Figure 15.   Pressure level transformation to the torso surface for frontal sound incidence. (Reprinted with permission from Kuhn and Guernsey, in preparation.)

is not evident until a 1-inch thick, loosely woven cotton pad, approximately 6 inches square, was placed under the microphone. It is not clear from these experiments what physical mechanism generates this notch in the frequency response. Unfortunately, because of time limitations, it was not possible to investigate the reponse further.

These results point to the need for: a) thorough investigation of the acoustic properties to be assigned to a manikin's torso designed for hearing aid tests and b) analytical models to be applied to the human torso. Furthermore, hearing aid tests with clothing will be unreliable until knowledge of the acoustic properties of the clothing and the torso is gained. Although the major features, that is, the frequency response for frontal incidence and azimuthal directivity of the pressure level transformation to a manikin's torso, show good agreement between measurements and theory, there is poor agreement between measurements on manikins and measurements on human subjects.

Figures 16 and 17 show comparisons between the theoretically predicted azimuthal directivities for a rigid cylinder (of equivalent perimeter to the torso) and the directivity measured at the center chest position of the bare manikin. The agreement between theory and experiment is good for frequencies up to 2.0 kHz. However, at frequencies of 4.0 kHz and greater, the predictions based on rigid cylinder diffraction theory differ considerably from the measured results. Thus, a predictive model for the pressure level in the azimuthal plane based on diffraction by a rigid cylinder is only useful

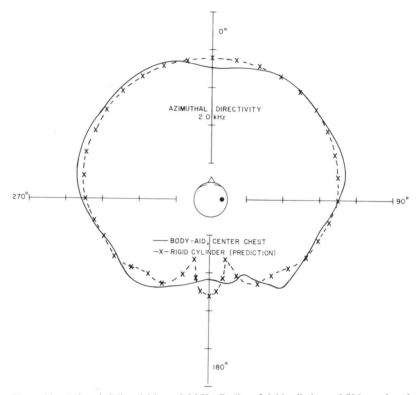

Figure 16.   Azimuthal directivities at 2.0 kHz. Radius of rigid cylinder = 0.796 wavelength.

up to approximately 3.0 kHz. Absorptive properties of the human torso and their effect on the azimuthal directivities were not investigated. The results, however, show clearly once again that the torso impedance must be included in an analytical model intended to predict the response of a body-worn hearing aid.

Pressure Transformation from a Diffuse Field to the Head, to the Ear, and to the Torso

The environment in which hearing aids are used is generally semi-diffuse. In a semi-diffuse field the acoustic waves are partially progressive and partially diffuse. The pressure level transformation from a diffuse sound field to the head surface (Kuhn, 1979) is shown in Figure 18. These results show that the pressure level transformation from the diffuse field to the head of a manikin or human subject can be predicted quite well by the theory of diffraction for a rigid sphere of equivalent perimeter to the head. For an ideal sphere, the pressure transformation from the diffuse field is indepen-

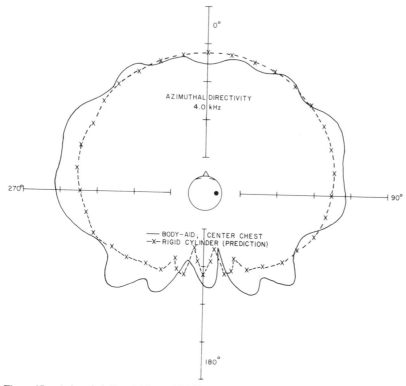

Figure 17.   Azimuthal directivities at 4.0 kHz. Radius of rigid cylinder = 1.591 wavelength.

dent of the location of the measurement point, so that the results of Figure 18 are generally valid, within ± 1 dB for head-worn hearing aids, regardless of their specific location. The maximum gain at the head-worn hearing aid lies between 3 and 4 dB at the upper frequencies.

The pressure level transformation from a diffuse sound field to the coupler microphone is shown in Figure 19. These results show a maximum gain of approximately 15–16 dB at the fundamental resonance of the external ear at 2.7 kHz. Thus, there is as much as 13-dB difference between the hearing aid input signal level and the signal level at the coupler microphone thereby simulating the signal level at the eardrum of a person with normal hearing.

The pressure level transformations from a diffuse field to the coupler microphone were measured with six different pinnas. Measurements of the radiation impedance and efficiency, together with the measured and predicted pressure transformation (Shaw, 1976), show that the pressure level transformation and radiation efficiency are proportional to the pinna size

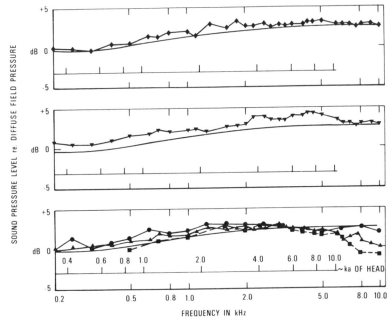

Figure 18.   Sound pressure level transformation from the diffuse field to the head surface. A, For the manikin's head without torso; B, for the manikin's head with its bare torso; C (▲), experimental results for the manikin's head with a dressed torso; ●, ■, experimental results for the heads of subjects 1 and 2, respectively. (Reprinted with permission from Kuhn, 1979.)

below approximately 6 kHz. Between 6 and 10 kHz the pressure level transformation and radiation efficiency are inversely proportional to the pinna size. Measurements (not shown here) of the pressure distribution in the pinna suggest that the first higher order transverse mode dominates in the 6–10 kHz region and governs both the radiation efficiency and the pressure level transformation to the coupler microphone.

Finally, Figure 20 shows the pressure level transformation from the diffuse field to the torso surface of an acoustically hard manikin, a dressed manikin, and the torsos of two human subjects. The pressure level transformation to the hard torso of a manikin is predicted well by diffraction theory for a rigid sphere (Waterhouse, 1963). However, the absorption by the clothing and the torso reduces the pressure buildup for both the manikin and the human subjects. A thorough study is required to determine an appropriate mathematical and physical model of the human torso. Again, it is clear that the pressure level transformation to a torso-worn hearing aid will be measured most reliably if the manikin is acoustically hard and unclothed. Large effects on the pressure level transformation from a *free* field to the torso of a dressed manikin have already been discussed in an earlier part of this paper.

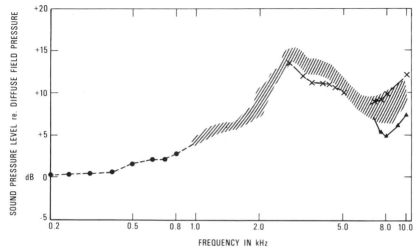

Figure 19.   Sound pressure level transformation from the diffuse field to the eardrum. ●, Low frequency experimental results; hatched area represents the range of measurements for all pinnas except pinna "X" and pinna "R," shown by X and ▲, respectively.

The pressure level buildup on the torso between 0.315 and 0.63 kHz is larger than predicted. Similar measurements by R. Guernsey (personal communication) with a "1-inch" microphone at the torso of human subjects show a substantial increase in pressure level also. It is not clear at this point what acoustic mechanism causes this pressure buildup at these low frequencies.

At high frequencies the effect of absorption by the clothing is particularly pronounced. Figure 20 shows that the human torso's absorption lowers the pressure level by as much as 8 dB at 8 kHz relative to the level on a rigid torso. Since the acoustic impedance, torso size, and microphone location on the torso surface can vary, a shaded region is shown in Figure 20. It is expected that the pressure level buildup on a torso in a diffuse sound field will fall within this shaded region. More accurate predictions are not possible at this time because of the unknown acoustic properties of the human torso and clothing.

CONCLUSIONS

Since the scope of this work is rather broad, only a few specific conclusions are drawn:

1.   For frontally incident sound, the input sound pressure level to a head-worn hearing aid is as much as 16–17 dB less than that at the eardrum of a normal-hearing subject at 2.7 kHz, the fundamental resonance of the external ear.

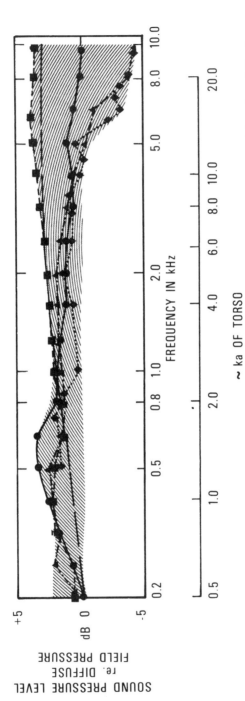

Figure 20. Sound pressure level transformation from the diffuse field to the torso. Solid line represents theoretical predictions for a rigid sphere; ■, ●, experimental results for the manikin's bare and dressed torso, respectively; ▲, ◆, experimental results for the torsos of subjects 1 and 2, respectively. The hatched area represents the range of pressure levels to be expected in practice for various torso conditions and microphone locations.

2. The torso increases the pressure level at the coupler microphone, via constructive interference of the scattered and direct wave, by as much as 1½ dB between 0.25 and 0.8 kHz. Between 0.9 and 2.0 kHz the destructive interference between the waves scattered by the torso and' the direct wave reduces the sound pressure level at the coupler microphone by as much as 3½ dB.

3. The pressure level transformation from a frontally incident sound wave to an aid worn at the side of the head is reasonably well predicted by scattering from a rigid sphere.

4. For frontally incident sound, the hearing aid input signal level is greater for aids worn forward of the ear canal axis than for aids with microphones to the rear of the ear canal axis.

5. For frontally incident sound, the signal input to hearing aid microphones located immediately behind the pinna (helix) is strongly coupled to the higher order acoustic modes of the pinna. Therefore, these aids are subject to large pressure level variations as a function of frequency and location.

6. Up to approximately 4.0 kHz, the azimuthal directivities of head-worn aids can be predicted reasonably well by scattering theory for a rigid sphere. For frequencies up to 1.6 kHz, the azimuthal directivity of a head-worn aid agrees within 1½ dB with the directivity measured at the coupler microphone. It is clear that the pinna plays only a minor role below 1.6 kHz.

7. The low frequency localization cue, the ITD, is nearly the same at the hearing aid microphone as at the eardrums if the aid is worn very near to the ear canal entrance. If the hearing aid microphone is located considerably forward or to the rear of the ear canal axis, localization errors or "confusion" may occur.

8. For angles of incidence between 330° and 30° and frequencies greater than 1.6 kHz, the interaural pressure level difference for the aided listener is approximately the same as for the unaided listener. For frequencies greater than 4.0 kHz there is considerable difference between the aided and unaided interaural pressure level. Thus, considerable adaptation would be necessary for an aided listener to localize as well as the unaided listener (in the horizontal azimuthal plane).

9. Because of the unknown acoustic impedance of the torso and clothing, pressure level transformations from the free field to the torso surface are not generally analytically predictable. Diffraction theory for a rigid infinite cylinder can be used to predict approximate directivities, but not to predict absolute pressures.

10. Pressure level transformations from a *diffuse* field to the torso surface are also strongly dependent on the acoustic impedance of the torso

and clothing. Therefore, analytical predictions or measurements of the pressure level transformation to an aid worn on a human torso are unreliable. However, when the aid is worn on the rigid torso of a manikin, the measured and predicted pressure level transformations are in good agreement.

11. The pressure level transformation from a diffuse field to a head-worn aid reaches a maximum of 3–4 dB at the high frequencies. This pressure level transformation is predicted well by diffraction theory for a rigid sphere. The pressure level transformation to the eardrum of a normal-hearing subject is 15–16 dB at the fundamental resonance of the external ear. Thus, in a diffuse field the hearing aid input signal level is as much as 13 dB less than that at the eardrum of a subject with normal hearing.

REFERENCES

Burkhard, M.D., and R.M. Sachs. 1975. Anthropomorphic manikin for acoustic research. J. Acoust. Soc. Am. 58:214–222.

Kuhn, G.F. 1977. Model for the interaural time differences in the azimuthal plane. J. Acoust. Soc. Am. 62:157–167.

Kuhn, G.F. 1978. Some acoustic properties of the human pinna. J. Acoust. Soc. Am. 63:S75(A).

Kuhn, G.F. 1979. The pressure transformation from a diffuse sound field to the external ear and to the body- and head-surface. J. Acoust. Soc. Am. 65:991–1000.

Kuhn, G.F., and E.D. Burnett. 1977. Acoustic pressure field alongside a manikin's head with a view towards in situ hearing-aid tests. J. Acoust. Soc. Am. 62: 416–423.

Kuhn, G.F., and R.M. Guernsey. Draft manuscript on the pressure distribution about the human head and torso. Vibrasound Research Corporation, Denver. In preparation.

Schwarz, L. 1943. Zur Theorie der Beugung einer ebenen Schallwelle an der Kugel. [The theory of diffraction of a plane sound wave by a sphere.] Akust. Zeits. 8:91–117.

Shaw, E.A.G. 1974. Transformation of sound pressure level from the free field to the eardrum in the horizontal plane. J. Acoust. Soc. Am. 56:1848–1861.

Shaw, E.A.G. 1975. The external ear: New knowledge. In S.C. Dalsgaard (ed.), Earmoulds and Associated Problems, pp. 24–28. Seventh Danavox Symposium. The Almqvist and Wiksells Periodical Co., Stockholm.

Shaw, E.A.G. 1976. Diffuse field sensitivity of external ear based on reciprocity principle. J. Acoust. Soc. Am. 60:S102(A).

Waterhouse, R.V. 1963. Diffraction effects in a random sound field. J. Acoust. Soc. Am. 35:1610–1620.

CHAPTER 5

THE EFFECT OF ENVIRONMENT ON THE DIRECTIONAL PERFORMANCE OF HEAD-WORN HEARING AIDS

Gerald A. Studebaker,
Robyn M. Cox, and Craig Formby

CONTENTS

The directional characteristics of hearing aids normally are measured in an anechoic chamber, mainly because an anechoic chamber is a well-defined condition that will reveal the directional characteristics of the hearing aid being tested independent of the characteristics of the room itself. However, people normally live and wear hearing aids in reverberant spaces, not in anechoic spaces. Therefore, it seems necessary that we should understand the effects of commonly encountered environments on the electroacoustic performance of hearing aids.

The primary purpose of the investigations described here was to study the effect of environmental acoustics, as revealed by reverberation

The work reported in this chapter was supported in part by Public Health Services Grants #NS 13514 and NS 12588 from the National Institute of Neurological and Communicative Diseases and Stroke.

81

time, on the directional characteristics of head-worn hearing aids with "directional" and "nondirectional" microphones. Before conducting these investigations, we wished to learn whether the manikin KEMAR could serve as an adequate substitute for human subjects across environments and azimuths.

MEASUREMENTS OF HEAD DIFFRACTION

General Method

The procedure used to collect all of the head diffraction data reported in this chapter used a broadband thermal noise as the test signal. The noise was presented to head-mounted hearing aids by a loudspeaker located in the room under study. After processing by the hearing aid under investigation, a real time spectrum analyzer was used to derive a time-averaged intensity-by-frequency display of the processed noise signal. The spectrum was averaged over a period of 30 to 60 sec, thus providing a smooth, repeatable result. This averaged spectrum was then transferred to a special purpose digital computer, where it could be compared with, subtracted from, or averaged together with spectra generated under other acoustic conditions. The results were printed out using an X-Y recorder. In some cases the signals were recorded on tape for analysis later in the laboratory. We have referred to this method as the noise subtraction technique (Studebaker, 1976).

Placement of the Hearing Aids on KEMAR

In these studies of environmental effect the hearing aids were placed in the normal manner over the right pinna of KEMAR. In order to ensure that KEMAR adequately represented the head diffraction effects of a real head over the conditions to be studied, a preliminary investigation was carried out.

Procedure Two nondirectional ear-level hearing aids were used, one with a forward-facing microphone and one with a downward-facing microphone. The hearing aids were placed on KEMAR and on each of four to six subjects in the normal over-the-ear (OTE) position. Measurements were made in four different environments at four azimuths. The hearing aid's output was, in the case of this study, directed via a length of tubing to a 2-cm^3 cavity. The cavity was held near the neck of the subject or KEMAR. Each subject and KEMAR were rotated successively into the four cardinal azimuths relative to the signal source for data collection. Comparisons were thus made between KEMAR and the human subjects in each of the 32 conditions.

Results   Figure 1 shows an example of the results. These particular results were obtained from the hearing aid with the downward-facing microphone and with the signal source at 0° azimuth. The results for KEMAR are shown as the dotted line and the mean results from the four subjects are shown as the broken line. The solid line at the bottom shows the difference between the two sets of results at the top. The curves differ in one place by as much as 5 dB, but they are generally within 2.5 to 3.0 dB of each other.

Figure 2 shows a second example, in this case from the aid with the forward-facing microphone at 0° azimuth. The agreement was even better in this case, as the difference line at the bottom of the figure reveals. The

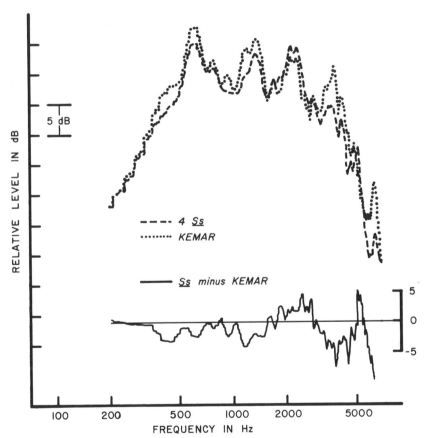

Figure 1.   Mean output spectrum obtained from a down-facing microphone hearing aid placed on four subjects and the output spectrum obtained from the same aid on KEMAR at a 0° azimuth. The solid line shows the difference between the two results.

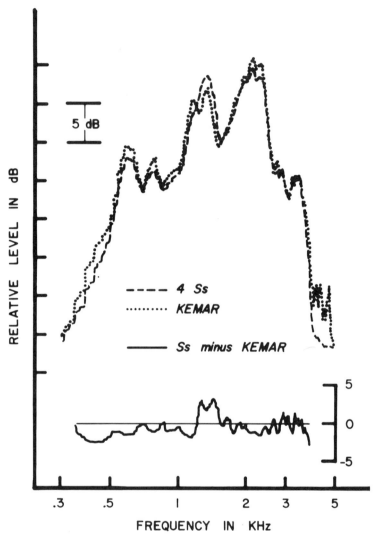

Figure 2.   Mean output spectrum obtained from a forward-facing microphone hearing aid placed on four subjects and the output spectrum obtained from the same aid on KEMAR at 0° azimuth. The solid line shows the difference between the two results.

results in these two figures were typical with respect to the extent of the agreement seen and in that when differences between microphone locations appeared, the downward-facing microphone produced somewhat poorer agreement than seen with the forward-facing microphone. In the 32 comparisons made between KEMAR and the mean of four subjects, differences larger than those in Figure 1 were seen on only four occasions and the agreement was as good as shown in Figure 2 in about 50% of the 32 occasions. On this basis it was concluded that KEMAR was a good substitute for the real thing under all of the conditions of environment and azimuth studied. Recently, Cox carried out a study in which better control was maintained over head position and more subjects were used. The results revealed essentially the same discrepancies between forward-facing and downward-facing microphones as shown in these figures. Some of these data are reported in Chapter 9.

MEASUREMENTS OF REVERBERATION TIME

In the investigations of head-worn hearing aid directivity, measurements were made at 1 m from the loudspeaker signal source in five environments: an anechoic room, an audiometric test room, a living room, a school classroom, and a church classroom. The acoustic characteristics of all the rooms except the anechoic room were evaluated by making reverberation time measurements.

Procedure

Abruptly terminated octave-band thermal noise signals centered at octave intervals from 250 to 4000 Hz were tape recorded in each environment. After slowing the tape recording down to $\frac{1}{16}$ or $\frac{1}{32}$ of the original speed through the use of an FM tape recorder, the decaying signal was read out onto strip chart paper using a graphic-level recorder. Although slowing of tape-recorded signals is a commonly used technique, the use of the FM tape recorder in this case made possible measurements of reverberation times down to the 30- to 50-msec region, with control runs down to less than 10 msec.

Results

Table 1 is a summary of the results of the reverberation time measurements for three of the rooms. It shows mean values measured at a speaker-microphone distance of 1 m and also at a greater measurement distance,

Table 1. Mean reverberation times in three environments

Environment	Room volume (m^3)	Microphone-speaker distance (m)	Mean reverberation time (msec)
Audiometric test room	11.6	1.0	112
Living room	60.7	1.0	361
		2.74	495
Church classroom	181.0	1.0	618
		3.96	1025

which was dependent on room size. The values for 1 m are those that represent the reverberant conditions under which the head-worn hearing aid's directional performance was measured. The reverberation time of the anechoic room was not measured. The averages of the reverberation times for the other rooms were audiometric test room, 112 msec; living room, 361 msec; and church classroom, 618 msec. The reverberation time for the school classroom, not shown in the table, was 379 msec.

As can be seen, at greater distances the measured reverberation times were substantially in excess of those observed at 1 m. These larger values represent those one would use to characterize the room as a whole, and the smaller values represent the conditions at the distance at which the data presented later in this paper were collected.

EFFECT OF ENVIRONMENT: HEARING AIDS WITH NONDIRECTIONAL MICROPHONES

The effect of the azimuth of the signal source relative to the head on the performance of nondirectional head-worn hearing aids has been studied quite extensively in anechoic space by Kasten and Lotterman (1967), Wansdronk (1959), and others. However, few data were available on this effect in reverberant environments except for the audiometric test room (Olsen and Carhart, 1975).

Anechoic vs. Reverberant Conditions

Procedure   A nondirectional hearing aid was placed on KEMAR in each of two environments: an anechoic room and a school classroom with a reverberation time of 379 msec. In each environment, KEMAR was rotated successively through the four cardinal azimuths relative to the loudspeaker, and the hearing aid's output was measured at each azimuth. The loudspeaker was located 1 m from the center of KEMAR's head.

Results   Figure 3 shows the output spectra obtained in the anechoic room at four azimuths. The results are typical of those from a head-worn, nondirectional, forward-facing microphone hearing aid in that the 90° condition produced the highest level; the 0° and 180° conditions produced similar results, which may differ by a few decibels depending upon exact microphone location and frequency; and the 270° condition produced results lower in level than the other azimuths and results that decreased relatively with increased frequency to a difference of some 25 to 35 dB in the high frequency region.

As a contrast, Figure 4 shows results from this same hearing aid in the school classroom. Every condition was the same except for the change in environment. The overall relationships remained the same, but the range of

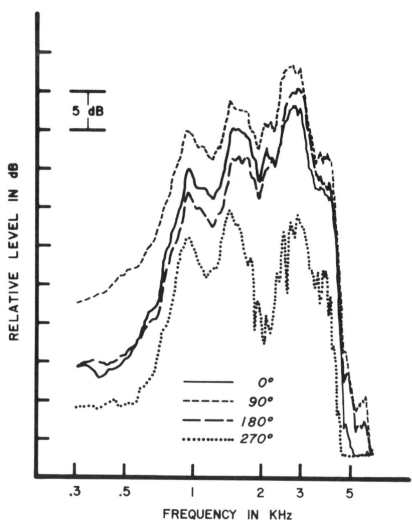

Figure 3.   The output spectra of a nondirectional hearing aid on KEMAR at four azimuths in an anechoic room.

difference was reduced substantially. In some ways, the differences between Figures 3 and 4 are more apparent than real. First, note that the higher level produced by a signal at the 90° azimuth was again prominent in the classroom, just as it was in the anechoic room. Although the head baffle effect was slightly smaller in the more reverberant environment, this effect was a very persistent finding across all environments studied, as will be seen. In relation to the field response of the hearing aid, the head baffle effect typically produces a 4- to 6-dB increase in level in the 90° azimuth result

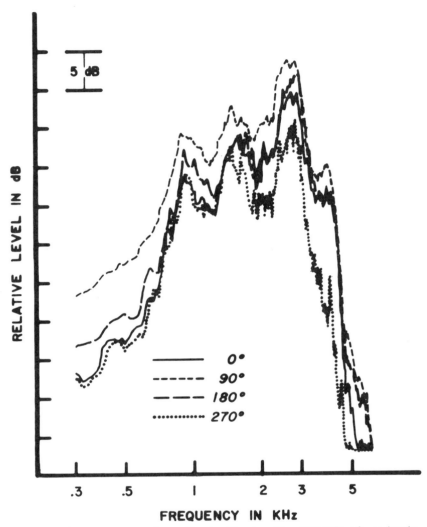

Figure 4.   The output spectra of a nondirectional hearing aid on KEMAR at four azimuths in a school classroom.

in an anechoic space and about a 3- to 5-dB increase in more reverberant space. It is generally relatively constant in magnitude across frequency, although this feature may vary somewhat with environment, presumably depending upon the room's absorption characteristics in various frequency regions.

It was evident that in the classroom environment the 0° condition and 180° conditions were somewhat more similar than in the anechoic space. However, the big change was seen in the 270° results. The head shadow was substantially reduced in the relatively reverberant classroom. The change in the 270° result was the single factor that accounted for most of the apparent difference between the results in Figure 4 and those in Figure 3.

Effect of Increased Reverberation Time

Procedure Two nondirectional hearing aids, one with a forward-facing microphone and one with a downward-facing microphone, were placed on KEMAR in each of the four reverberant environments described earlier. An anechoic environment was not used in this study. The KEMAR was rotated through the four cardinal azimuths relative to the loudspeaker, which was located at 1 m from the center of KEMAR's head in all cases. The output of each hearing aid was measured at each azimuth. In addition, the field response of each hearing aid was measured with the hearing aid placed at the location of the center of KEMAR's head with the manikin absent.

Results In Figures 5–10, the results from the head-mounted aid are plotted relative to the field response of the same aid. Comparison of the data

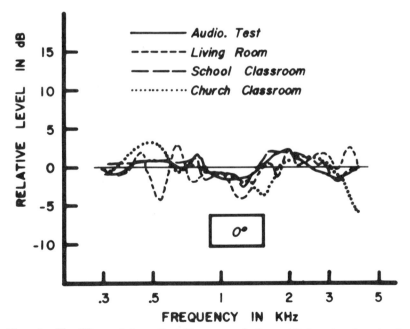

Figure 5.   The difference between the field response of a forward-facing microphone hearing aid and its response at 0° on KEMAR in four environments.

obtained from these nondirectional hearing aids across the four nonanechoic environments, with reverberation times ranging from 112 msec for the audiometric test room to 618 msec in the church classroom, revealed much smaller changes than between anechoic and nonanechoic conditions. The data obtained in all of these environments showed relationships across azimuths like those shown in Figures 3 and 4, but there was little or no systematic change in the directional characteristics as reverberation time varied in this range. A single exception was that in moving from the audiometric test room to the more reverberant rooms a slight additional reduction in the head shadow was seen with the downward-facing microphone hearing aid (see Figure 10).

Figures 5–8 illustrate the small changes observed in head diffraction effects at a given azimuth as reverberation time varied. For illustration, only the results for the front-facing microphone hearing aid are shown. Figure 5 shows the head diffraction effects measured in each of the four environments for a 0° azimuth signal. Although the individual curves are far from identical, there is a clear tendency for the data to vary above and below the reference condition in a similar fashion in all environments. Figure 6 shows the results obtained in each environment for a 90° azimuth signal. There was

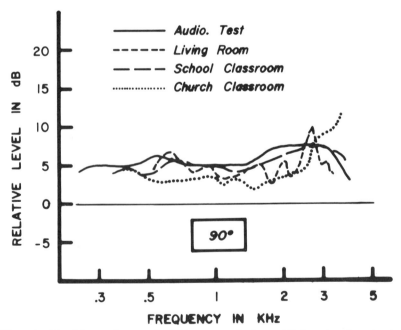

Figure 6.   The difference between the field response of a forward-facing microphone hearing aid and its response at 90° on KEMAR in four environments.

a small but perhaps systematic decrease in the head baffle effect seen at this azimuth as reverberation time increased. The results obtained with the downward-facing microphone hearing aid (not shown) were similar to these results with respect to effects of environment. However, there were some differences between the two microphone locations, which are discussed later.

Figure 7 shows the effect of environment on the head diffraction effects measured for a 180° azimuth signal with the front-facing microphone hearing aid. Note that the 180° azimuth result in all four environments was consistently elevated relative to the field response of the hearing aid by 1 to 3 dB and that this put these results somewhat higher than those obtained at the 0° azimuth, especially in the midfrequency region, even with the front-facing microphone. There was no noticeable effect attributable to reverbation time.

Results from the downward-facing microphone aid at 180° were somewhat less consistent across environments, but three of four environments showed an approximately equal increase in level of 2 to 4 dB relative to the field response at this azimuth. The increase became gradually greater, ranging from 2 dB at 300 Hz to 6 to 7 dB above 3000 HZ. Wansdronk's (1959)

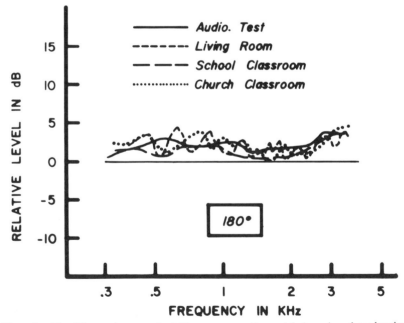

Figure 7. The difference between the field response on a forward-facing microphone hearing aid and its response at 180° on KEMAR in four environments.

results for his backward-facing microphone were similar to these except that the extent of the effects were somewhat greater in his study. Kasten and Lotterman's (1967) 180° results agree with ours at the higher frequencies in placing the levels from the backward-facing microphone 2 to 3 dB above those from the forward-facing microphone at this azimuth. Olsen and Carhart's (1975) results, which were obtained in the audiometric test room, showed a similar relationship between front- and backward-facing microphone results above 400 Hz. In absolute terms, the head diffraction effects were slightly smaller at this azimuth in the Olsen and Carhart study than in this study with both microphone types.

It was concluded, on the basis of these data and those reported in the literature, that forward-facing microphones produce outputs 0 to 3 dB above the hearing aid's field response for signals from 180°, with perhaps a slight relative increase at the higher frequencies. Backward-facing microphones also produce outputs a few dB higher than the field response, which steadily increases in size across the frequency range. The high frequency roll-off seen with this microphone location at other azimuths is not seen at 180° (see Figure 10). There was perhaps a greater rate of increase in level above 2500 Hz. Finally, forward-facing microphones in general produce 1 to 3 dB less output than backward-facing microphones for signals of rearward origin from about 800 to 900 Hz and upward.

Figure 8 shows the results obtained at 270° across the four environments with the front-facing microphone hearing aid. Note that again there was little or no effect attributable to environment. However, the head shadow effect, which reaches some 6 to 7 dB at 3000 Hz in this figure, was much smaller than that found by Wansdronk (1959), by Kasten and Lotterman (1967), or by us in anechoic space (Figure 3). All of the studies carried out in anechoic rooms produced head shadow effects of at least 15 dB and sometimes up to 30 or 35 dB in narrow frequency regions for signals from 270°. Olsen and Carhart (1975), on the other hand, working in an audiometric test room similar to that used here, showed much smaller effects, never exceeding 10 dB. In fact, with a forward-facing microphone they produced results nearly identical to the values reported in Figure 8.

With a downward-facing microphone, we found, in agreement with Olsen and Carhart but in disagreement with Kasten and Lotterman, a somewhat larger notch at 270° than with the forward-facing microphone. However, even this difference between microphone orientations was seen only in the audiometric test room and not in more reverberant spaces.

Effect of Azimuth Averaged across Environment

Because the four reverberant environments had little effect on the hearing aids' directional characteristics, the results were combined across environments to show the median effect of the head shadow and baffle at each azimuth on the response of these two nondirectional hearing aids in moderately reverberant spaces. Figure 9 shows the median results derived from the four environments obtained with the forward-facing microphone hearing aid: the enhancement at 90°, the slight increase in level at 180°, the close-to-reference-line result obtained at 0°, and the head shadow effect at 270°. (All data were plotted relative to the aid's field response at the same location.)

Figure 10 shows the same result for the downward-facing microphone hearing aid. It is apparent that the overall relationships for this microphone orientation are not greatly different, except for some fine detail already discussed and the notch that appears at each azimuth except 180° in the 2 to 3 kHz region.

The results for these two microphone locations seem more similar to each other in these data than in some other studies. Whether this relative similarity occurred because of the more reverberant conditions used or

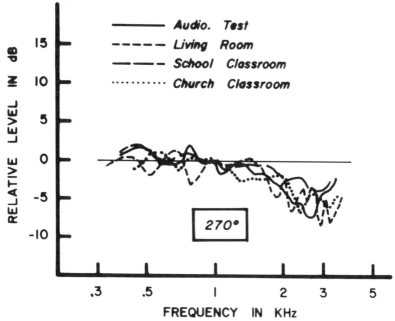

Figure 8.   The difference between the field response of a forward-facing microphone hearing aid and its response at 270° on KEMAR in four environments.

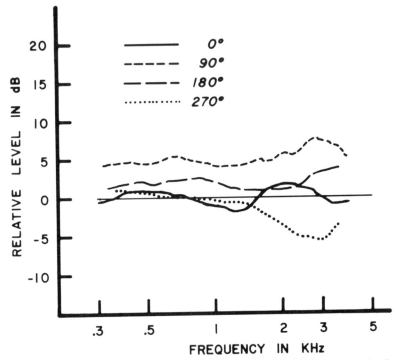

Figure 9.   The difference between the field response of a forward-facing microphone hearing aid and its response at four azimuths on KEMAR averaged across four nonanechoic environments.

because the hearing aids used here differed in some other ways from those used by other investigators is not known.

Summary

The head shadow effect measured for signals from 270° was significantly greater in an anechoic room than in a more reverberant audiometric test room. The head diffraction effects for signals from other azimuths were also affected by the move from anechoic to nonanechoic space, but in relatively minor ways, depending, to some degree, upon microphone location. The head baffle effect at the 90° azimuth decreased slightly and the 180° and 0° azimuth results became slightly more similar to each other in nonanechoic as opposed to anechoic space. Changes in the head shadow effect, revealed in the results from 270°, produced most of the apparent difference between the results obtained in anechoic and nonanechoic spaces. Measurements made in spaces progressively more reverberant than the audiometric test room showed very little additional change in head diffraction effects. The

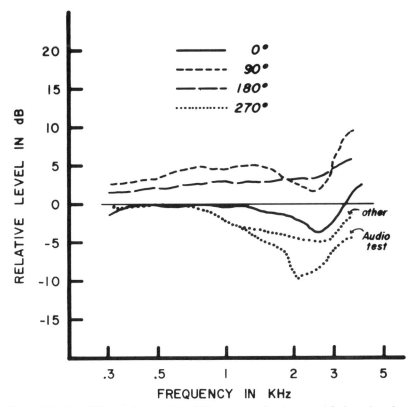

Figure 10. The difference between the field response of a downward-facing microphone hearing aid and its response at four azimuths on KEMAR averaged across four nonanechoic environments.

head shadow at 270° decreased a bit more for the downward-facing microphone, but otherwise changed only slightly.

EFFECT OF ENVIRONMENT: HEARING AIDS WITH DIRECTIONAL MICROPHONES

Procedure

In a separate study, the authors investigated the effect of environment on the directional characteristics of three head-worn directional hearing aids. In this study, the three directional hearing aids were mounted in turn on KEMAR in five environments, one of which was an anechoic room. The four reverberant rooms were the same rooms as those used in the study of nondirectional aids. Environment had a substantially greater effect on the performance of these directional hearing aids than on the nondirectional aids.

Because of the complexity of the data, qualitative visual displays of the results seem most revealing. In the interest of brevity, results are shown for only four of the eight azimuth conditions that were actually run, and results from only four of the five environments are presented. The school classroom environment is not shown because it produced results, both on reverberation measures and on the hearing aid azimuth performance measures, that were highly similar to those results obtained in the living room environment that is reported. Finally, results from only two of the three aids are presented.

Results

Figure 11 shows the results obtained at the four cardinal azimuths in an anechoic room for a directional hearing aid designated hearing aid "A." This aid had an adjustable directional characteristic. Although this hearing aid had a directional pattern characteristic of directional aids, the magnitude of the directivity in this case was relatively small, with front-to-back ratios not exceeding about 10 dB. Perhaps this was related to its being an adjustable directional hearing aid. Notable features were as follows. The 0° and 90° results were close together. The 270° and 180° results were similar, with the 270° result falling above the 180° result between about 700 and 1500 Hz. The 180° and 270° results were about equal up to around 3000 Hz, above which the 270° result fell below the 180° result, presumably as a result of head shadow. By contrast, as noted earlier, nondirectional hearing aids with front-facing microphones typically produced patterns in which the observed levels in the 90° condition exceeded those seen in the 0° condition and in which the 0° and 180° conditions produced similar results.

Figure 12 shows the results from the directional microphone hearing aid designated hearing aid "B," also tested in an anechoic room under the same set of conditions used in Figure 11. Note that in the anechoic room this hearing aid, although it too had a user-adjustable directional control, was considerably more directional than the last aid, with front-to-back ratios as large as 15 to 20 dB. However, the pattern of the results was much the same as that seen with the aid in Figure 11.

Starting with the anechoic room result for hearing aid "B" seen in Figure 12, look successively at Figures 13, 14, and 15, which report results obtained with this aid in progressively more reverberant spaces. Note the degree to which the vertical spread of the lines showing the results obtained at the various azimuths decreases in successive figures, reflecting the extent to which directivity decreased as reverberation time increased.

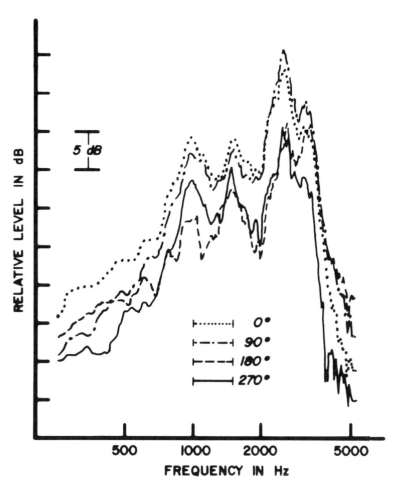

Figure 11.   The output spectra of directional hearing aid "A" on KEMAR at four azimuths in an anechoic room.

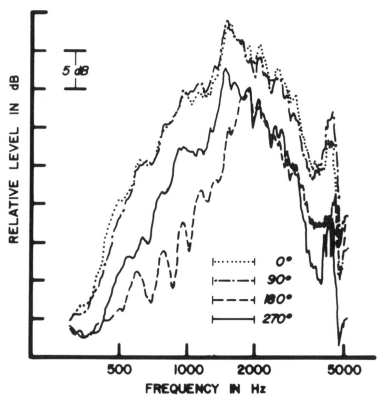

Figure 12.   The output spectra of directional hearing aid "B" at four azimuths in an anechoic room.

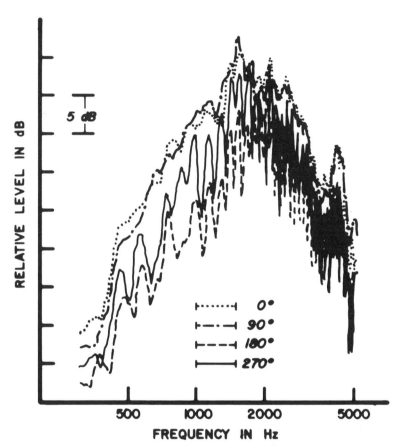

Figure 13.   The output spectra of directional hearing aid "B" at four azimuths in an audiometric test room.

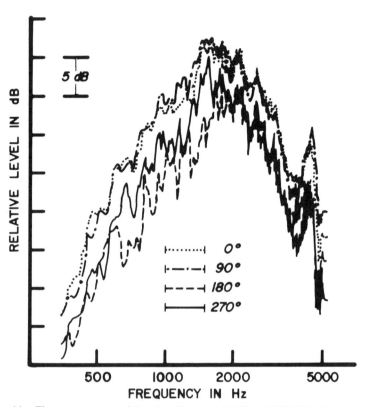

Figure 14.   The output spectra of directional hearing aid "B" on KEMAR at four azimuths in a living room.

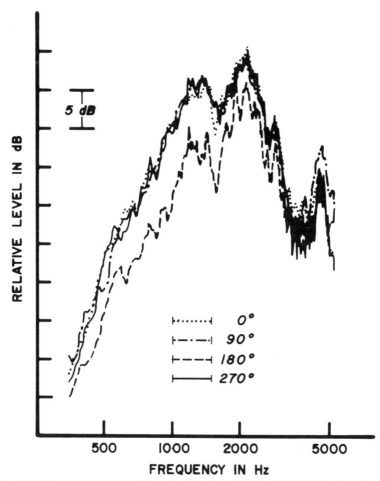

Figure 15.   The output spectra of directional hearing aid "B" on KEMAR at four azimuths in a church classroom.

Figure 13 shows the results for hearing aid "B" in the audiometric test room (size about 2.7 × 2.1 m). A very prominent feature in these results is the standing wave effect. This pattern was particularly prominent with all three directional aids in the audiometric test room. This same pattern was seen, although to a lesser degree, with the nondirectional hearing aids in this same room. These standing wave effects were not seen with such prominence in any other environment.

Figure 14 presents the results obtained from hearing aid "B" in the living room. Note that the standing waves smoothed out and there was a further reduction in directivity. Figure 15 presents the results for hearing aid "B" obtained in the church classroom, the most reverberant room used. Note here that directivity has nearly disappeared from the results obtained at 270°, with only a small amount remaining at the 180° azimuth.

Hearing aid "A" provides another example of the effect of environment on a head-worn directional hearing aid. Note again the results for hearing aid "A" in the anechoic room shown in Figure 11. Figure 16 shows the results for this hearing aid in the audiometric test room. Again, the standing waves are very prominent in this environment. It cannot be seen clearly whether the average amount of directivity has been reduced in this environment relative to that in the anechoic room

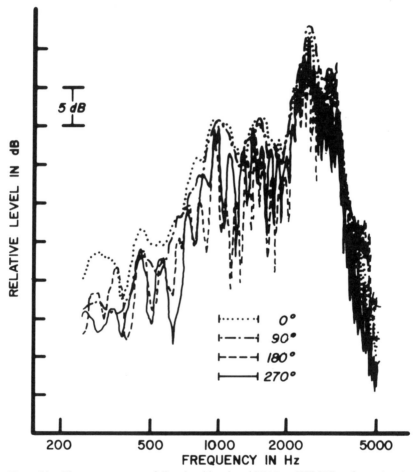

Figure 16.   The output spectra of directional hearing aid "A" on KEMAR at four azimuths in an audiometric test room.

(Figure 11). The conclusion depends on some arbitrary choices that one must make in attaching numbers to the standing wave patterns in Figure 16. Figure 17 shows the results obtained with hearing aid "A" in the living room. Directivity is clearly reduced in this environment. Finally, Figure 18 shows the results obtained with "A" in the church classroom. Directivity in this condition is nearly lost.

Summary

These results fall short of providing a basis for far-reaching general conclusions about the directional behavior of head-worn directional hearing aids as a function of environment. However, the results do demonstrate the following: 1) that the directivity of different directional hearing aids varies

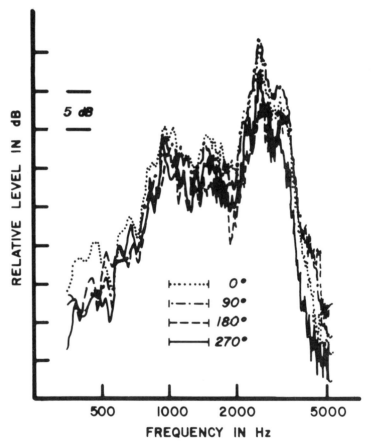

Figure 17.   The output spectra of directional hearing aid "A" on KEMAR at four azimuths in a living room.

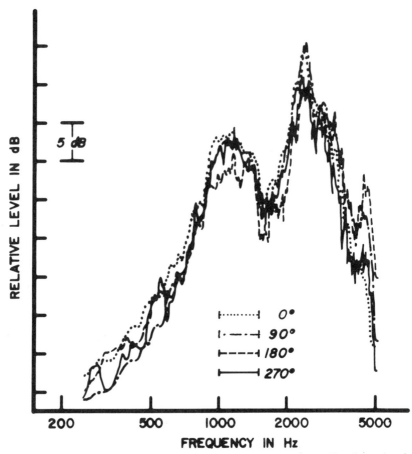

Figure 18.   The output spectra of directional hearing aid "A" on KEMAR at four azimuths in a church classroom.

substantially, 2) that the effective directivity of directional hearing aids may be substantially decreased in the nonanechoic environments in which they normally are used, some becoming functionally nondirectional in commonly encountered environments, and 3) that the audiometric test room with perforated metal walls normally used for hearing aid evaluations across the United States are unique places, unlike more normal environments, with regard to the generation of standing wave patterns. These results suggest that the unique characteristics of the audiometric test environment may be especially troublesome in the evaluation of directional hearing aids, particularly if competing signals presented from rearward or other azimuths are made a part of the evaluation procedure. Finally, the data suggest that environment has an effect on directional aids that extends progressively over a relatively broader range of environments than in the case of the nondirectional aids reported earlier. However, significant differences do exist between the results obtained in anechoic and nonanechoic space for both nondirectional and directional hearing aids.

REFERENCES

Kasten, R.N., and S.H. Lotterman. 1967. Azimuth effects with ear level hearing aids. Bull. Prosthet. Res. 10:50–61.

Olsen, W., and R. Carhart. 1975. Head diffraction effects on ear level hearing aids. Audiology 14:244–258.

Studebaker, G.A. 1976. Further applications of the noise-subtraction method. Paper presented at the annual convention of the American Speech and Hearing Association, November, Houston.

Wansdronk, C. 1959. On the influence of the diffraction of sound waves around the human head on the characteristics of hearing aids. J. Acoust. Soc. Am. 31: 1609–1612.

The External Ear,
the Earmold,
and the Earphone

CHAPTER 6

THE ACOUSTICS OF THE EXTERNAL EAR

Edgar A.G. Shaw

CONTENTS

It was only recently that a consistent description of the human acoustic antenna system emerged from the diversity of data collected over many years. We now know that the external ear is an efficient sound collector above 2 kHz, with remarkable directionality at the higher frequencies, providing us with rich perceptions of acoustic space in ways not yet fully understood. From another point of view, the external ear is an intruding filter of complex and uncertain characteristics, shedding doubt on the meaning and jeopardizing the accuracy of auditory and acoustic measurements. Since an altogether different set of characteristics comes into play when the ear is closely coupled to an earphone, it is hardly surprising that the external ear is also seen as "an acoustic factor affecting hearing aid performance."

TRANSFORMATION FROM FREE FIELD TO EARDRUM

If there is one parameter above others that characterizes the behavior of the external ear as a normal component of the auditory system, it is the transformation of sound pressure level from the free field to the eardrum. Since direct pressure measurements at the human eardrum are difficult, few such measurements have been made. However, by piecing together data from many experiments, it has been possible to delineate the *average* response of the human antenna system with considerable confidence for plane waves in the horizontal plane (Shaw, 1974c). The principal results are presented in Figure 1. The primary resonance of the external ear at 2.6 kHz and the rapid rise in response on the low frequency side of the resonance are clearly of primary importance. With frontal incidence (0° azimuth) this resonance produces a pressure gain of 17 dB. At 45° azimuth the gain is greater yet (21 dB), and even in the shadow zone the sound pressure level at the eardrum exceeds that of the incident field. As we pass the peak, a strong resonance in the concha sustains the response for almost 1 octave. The downward slope beyond 5 kHz is peculiar to the horizontal plane. Notice also the difference in gain between 45° and 135° azimuth, which amounts to 11 dB at 4.5 kHz. This lack of front-back symmetry, which provides a physical basis for localization in the lateral sector, is attributable to diffraction by the pinna extension.

FUNCTIONAL COMPONENTS

A descriptive diagram of the external ear is presented in Figure 2, which also shows some of the positions at which pressure measurements are made. We need to pay particular attention to the fossa, which is acoustically connected to the cymba, and the crus helias, which separates the cymba from the cavum. These structures are individually significant at the higher frequencies. On the other hand, other structures that surround the concha, such as the helix, the antihelix, and the lobule (the pinna extension), seem to function together as a simple flange (Shaw, 1975a).

BLOCKED MEATUS RESPONSE

Much of our knowledge of external ear function has been gained through experiments on real ears and replicas under carefully controlled conditions (Shaw, 1972, 1975a; Shaw and Teranishi, 1968) and through the construction of physical models with simple geometry (Shaw, 1974a; Teranishi and Shaw, 1968). It is fortunate that the ear canal proper is only 7–8 mm in diameter; this means that the wave motion in the canal, except in the vicinity of the entrance, is essentially plane up to approximately 18 kHz. Furthermore, careful measurements on simple models have shown that the addition of the ear canal makes little difference to the directionality of the

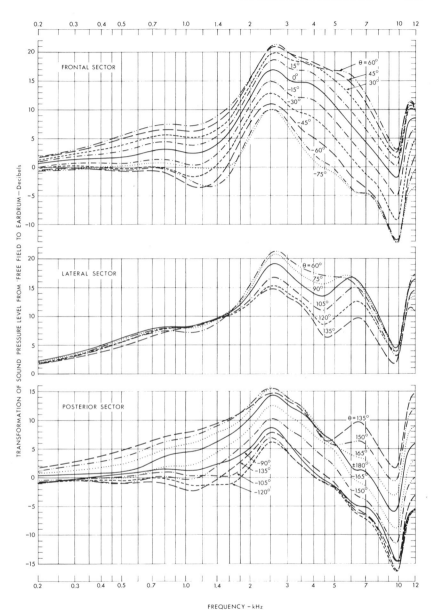

Figure 1.   Average transformation of sound pressure level from free field to eardrum as a function of frequency at 15° intervals of azimuth. Curves are fitted to data from 12 studies. (Reprinted with permission from Shaw, 1974c.)

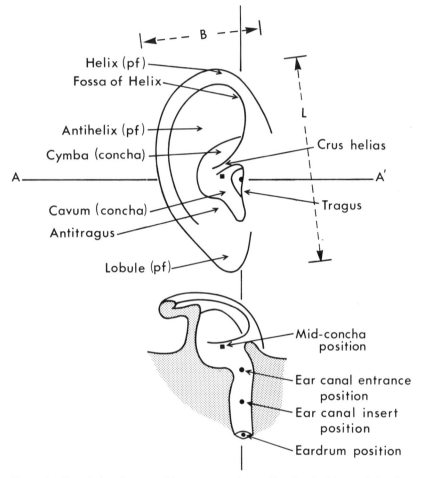

Figure 2.   Descriptive diagram of human external ear. (Reprinted with permission from Shaw, 1975a.)

ear up to high frequencies (10–15 kHz). Hence we can learn a great deal about the directionality of real ears under blocked meatus conditions. To make such measurements the ear is excited with a special source, which is designed to produce clean, progressive waves at grazing incidence (parallel to the circumaural plane) and is placed sufficiently close to the ear to avoid head diffraction effects, without disturbing the free field characteristics of the ear. The response is measured with a probe microphone anchored at the center of a carefully fitted plug, closing the canal entrance (Shaw, 1972).

The results for a typical human ear at eight angles of incidence ($\alpha =$ $-15°$ to $\alpha = 90°$) are shown in Figure 3. As can be seen, with sound waves

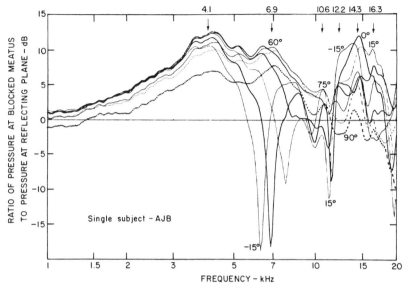

Figure 3.   Blocked meatus response of typical ear excited by progressive wave source at eight angles of incidence. Source is in front of ear at 0°, above at 90°. Arrows indicate frequencies of normal modes. (In part from Shaw, 1975a, with permission.)

approaching from the front $(-15°, 0°, 15°)$, the response falls rapidly above 5 kHz and passes through deep minima between 6 and 11 kHz. Between 11 and 16 kHz the excitation with frontal incidence is again strong. With sound waves approaching from above (angles of incidence between 60° and 90°), the situation is reversed: excitation is strong between 6 and 11 kHz but weak between 11 and 16 kHz. When the response curves for 10 subjects are averaged, as shown in Figure 4, the very deep minima are lost, but the principal features seen in the single-subject curves remain.

NORMAL MODES OF THE CONCHA

To understand Figures 3 and 4 we need to identify, separate, and measure the normal modes of the system, since these are the building blocks that shape the acoustic performance of the ear. Separating the modes is not always easy because their tuning curves are generally broad and therefore tend to overlap. Figure 5 shows the average transverse pressure distributions for five modes of the human external ear (numbers 2 to 6) under blocked meatus conditions. These distributions are based on the mode patterns observed in the 10 ears referred to in Figure 4. The circle to the right of each pattern indicates the average response of that mode on a relative scale, with optimum orientation of the progressive wave source as

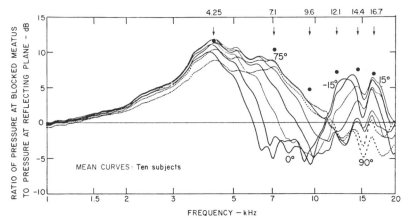

Figure 4. Average blocked meatus response of group of 10 subjects. Ears are excited by progressive wave source at eight angles of incidence. Source is in front of ear at 0°, above at 90°. Arrows indicate average frequencies of normal modes. Solid circles indicate average response for each mode at most favorable angle of excitation. (After Shaw, 1975a, with permission.)

indicated by the arrow. No pattern is shown for the first mode (4.3 kHz) since it is characterized by uniform pressure across the base of the concha. The second mode (7.1 kHz) has a single nodal surface near the crus helias and is excited most strongly at 75° elevation. The sound pressure in the cavum is, of course, out of phase with the pressure in the cymba and fossa, as indicated by the difference in sign. The third mode (9.6 kHz) also is best excited at 75°, but is poorly coupled to the sound field in the majority of ears. It has two nodal surfaces: one slightly below the crus helias and the other between the cymba and fossa. Unlike the second and third modes, the fourth (12.1 kHz) is best excited from the front. It is characterized by positive pressure zones at the front of the cavum and cymba and negative zones at the rear and in the fossa.

EXTERNAL EAR PARAMETERS AND MODELS

Modeling the Concha

Using the data presented in Figure 5, it is possible to construct model ears that, while simple in geometry, have approximately the same mode frequencies, pressure distributions, directionality, and excitation as the average human ear under blocked meatus conditions. The essential characteristics of the first mode, the depth resonance of the concha, can be matched in a variety of ways. Initially, as shown in Figure 6A, we

Average Patterns: based on 10 Subjects

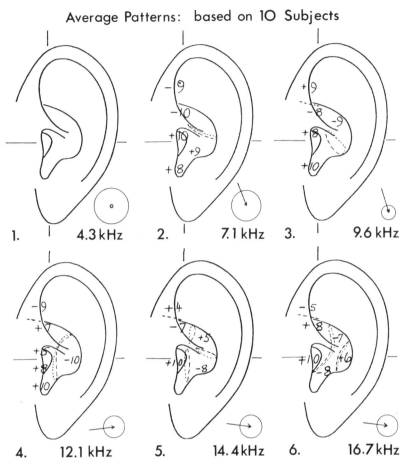

Figure 5. Average transverse pressure distribution patterns for six modes observed under blocked meatus conditions. Mode 1 has uniform pressure across base of concha. Circles indicate average response of modes, arrows most favorable direction. (After Shaw, 1974b, with permission.)

chose an inclined cylindrical cavity 22 mm in diameter and 10 mm in depth to represent the concha. If the first mode is to be properly excited, it is necessary that the cavity be embedded in a flange representing the pinna extension. The shape of the flange is not, however, critical (Teranishi and Shaw, 1968). The response curves shown in Figure 6 can be compared directly with those for human ears (Figures 3 and 4) since they were measured under identical experimental conditions. It is apparent that the representation in Figure 6A, although successful around 4 kHz, fails at higher frequencies. In particular, it provides only two reso-

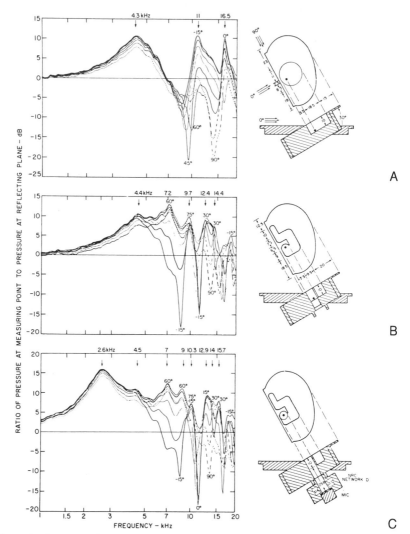

Figure 6. Response of model ear at three stages of development. Progressive wave source provides excitation at eight angles of incidence. Source is in front of ear at 0°, above at 90°. Arrows indicate frequencies of normal modes. Dimensions of model are in mm. A shows blocked meatus response of model with cylindrical concha. B shows blocked meatus response of model with rectangular concha, barrier enfolding canal entrance (crus helias), and channel (fossa). C shows response of complete model at eardrum position; canal is terminated by NRC network D and Brüel and Kjaer-type 4134 microphone.

nance frequencies between 5 and 17 kHz compared with the five found in the human ear under similar conditions.

The situation is improved when the concha is represented by a suitably proportioned rectangular cavity, but an adequate match can be achieved only by introducing two additional elements: a partial barrier in the concha to represent the crus helias and a channel above the concha to represent the fossa (Shaw 1974b, 1975a). Comparing Figure 6B with Figure 4 it is apparent that the first five modes in this model are very well matched to the human average with respect to frequency and fairly well matched with respect to directionality and excitation. Models based on other geometrical forms, such as the semicircle, have been almost equally successful.

Ear Canal Parameters

To complete our model of the external ear we must add an acoustic simulation of the ear canal and a terminating impedance that matches the impedance at the human eardrum. We have reliable figures for the average canal volume (1 cm^3) and approximate values for the physical length (22–25 mm, allowing for uncertainties in definition). Whether the average canal is appreciably tapered remains an open question. For simplicity we shall assume, in keeping with the prevailing view, that the canal can be represented adequately by a uniform cylinder, 7.5 mm in diameter. The length of the cylinder must be chosen to bring the complete ear into resonance at the correct frequency: approximately 2.6 kHz, according to Figure 1. This means that its "acoustic" length ($\lambda/4$) is approximately 32.5 mm, which is clearly much greater than the physical length of the canal proper. Most of the difference can be attributed to a large end correction associated with the junction between the canal and the cavum. There may also be a small contribution from the reactive component of eardrum impedance. It follows that the canal length in a model ear must be chosen to take account of the reactance of the eardrum simulator.

Eardrum Impedance

Clinical measurements of eardrum impedance are now routine, but are generally confined to low frequencies where middle ear pathology is most apparent. Until recently, reliable impedance data above 1 or 2 kHz have been lacking (Rabinowitz, 1977), but the gap has been partly bridged through the development of a remarkable four-branch eardrum simulator, with an impedance curve that is well matched to the measured values of impedance up to approximately 2 kHz and is consistent with other, more reliable but less direct data at higher frequencies (Zwislocki, 1971). This development (surely a tour de force) and the valuable engineering studies that have followed (Burkhard and Sachs, 1975) have provided a firm basis for further work in this field. As a consequence, the simulator—originally intended as a research tool—has come into widespread use and acquired de facto many of the attributes of a standard.

Bearing this in mind, let us look briefly at the current state of knowledge. The data in Figure 7 show the mean and median values of eardrum impedance and ear canal standing wave ratio drawn from many sources (see Shaw, 1975a, 1977). Below 1.5 kHz the fitted curves (broken lines) follow the trend of the impedance data. At higher frequencies the curves are determined primarily by the ear canal standing wave data on the assumption that the ear canal is a uniform cylinder that is 7.5 mm in diameter. The hump in the resistance curve around 2.7 kHz seems necessary if the open ear is to have sufficient pressure gain at resonance (Shaw, 1975a).

Is the fitted curve in Figure 7 also consistent with our knowledge of middle ear mechanics? In 1957 and 1962 Zwislocki presented a middle ear

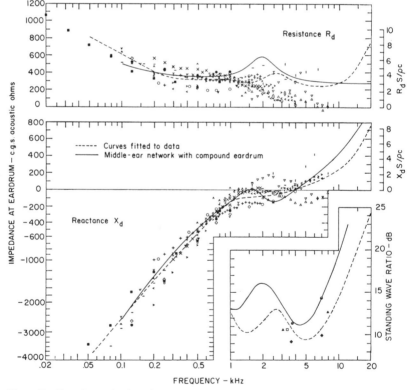

Figure 7.   Experimental points show mean and median measured values of eardrum impedance and ear canal standing wave ratio from 20 studies. Broken lines show curves fitted to data (Shaw, 1975a). Solid lines show middle ear network curves (Shaw, 1977).

acoustic network that was in excellent agreement with the available eardrum impedance data for normal and pathologic ears up to 1 or 2 kHz, but not sufficiently developed at higher frequencies. Recently this network has been modified to take into account the compound behavior of the eardrum, which, according to recent work, behaves as a simple elastic shell at low frequencies (Funnell and Laszlo, 1978), but breaks up into isolated zones at high frequencies (Tonndorf and Khanna, 1972). The modified network (Shaw, 1977), which must still be considered rudimentary, yields an impedance curve (solid line in Figure 7), that is at least plausible at most frequencies and can probably be improved with further adjustment. In any event, it seems clear that only a small fraction of the tympanic membrane is coupled effectively to the ossicular chain at high frequencies.

Modeling the Complete Ear

In Figure 7C the geometric model is embellished with an experimental two-branch eardrum simulator: NRC Network D (Shaw, 1975a) and a cylindrical canal of appropriate length. (Network D has a somewhat smaller negative reactance at 2.7 kHz than the Zwislocki four-branch simulator, and therefore requires a somewhat greater canal length to bring the ear to resonance at the correct frequency.) Comparing Figure 7B and C it can be seen that the presence of the canal has increased the number of resonances from five to eight, but, as noted earlier, the directionality of the ear has changed very little. There are as yet no measurements on complete human ears that can be directly compared with Figure 7C. However, when due allowance is made for head diffraction effects, there is satisfactory agreement between the average frontal incidence free field transformation curve shown in Figure 1 and the corresponding curve (0°) in Figure 7C.

DIFFUSE FIELD RESPONSE AND THE RECIPROCITY PRINCIPLE

We have noted that the pressure generated at the eardrum of the open external ear in a free sound field and the resistive component of eardrum impedance are strongly interdependent at the principal resonance frequency of the ear. Underlying this interdependence there is a powerful unifying principle that comes into sharp focus when we study the diffuse field response of the ear (Shaw, 1976). By invoking the acoustic reciprocity theorem, it can be shown that the diffuse field response of a complex acoustic receiver, such as the external ear, can be expressed essentially in terms of two impedances: the load impedance, in this case the eardrum impedance, and the impedance looking outward from the load, predominantly the radiation impedance. (Where there are appreciable radiation losses, the response is multiplied by the radiation efficiency, which is probably between

80% and 100% for the human ear.) It follows that the diffuse field response is an excellent indicator of the performance of the ear as a sound collector. The diffuse field response of ear replicas and models when mounted in a reflecting plane can be synthesized readily from many free field measurements in the hemispheric space above the plane. The upper panel of Figure 8 shows the hemispheric diffuse field response curves obtained in this manner for the NRC geometric model ear (IRE), which was the subject of Figure 6, and seven pinna replicas when terminated with the same eardrum simulator (NRC Network D). Two of the replicas (KEL and KER) are the original KEMAR left and right pinnas (Burkhard and Sachs, 1975) and five are other replicas (very kindly provided by Dr. G. Kuhn of the National Bureau of Standards) (see Kuhn, 1978). It can be seen that the response of the geometric model ear lies near the center of the group at most frequencies, but falls slightly below the median at 5 kHz and slightly above at 7 kHz.

The lower panel of Figure 8 shows how these data can be used to make a preliminary estimate of the diffuse field response of a median human ear.

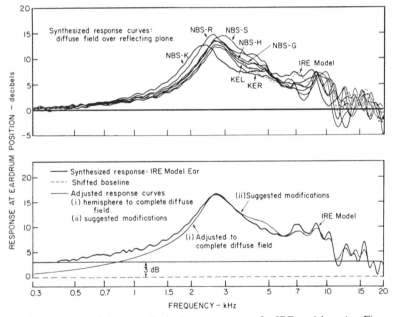

Figure 8.   Upper panel shows synthesized response curves for IRE model ear (see Figure 6 C) and seven ear replicas at eardrum position excited by diffuse sound field over reflecting plane. Lower panel shows response for IRE adjusted to complete diffuse field condition (i) with additional adjustments (ii) to correct minor differences between IRE model and median human ear.

The curve IRE from the upper panel is first plotted with a 3-dB shift in baseline, which is necessary to take account of two aspects of the measurement technique: that the response is measured in a hemispherical as opposed to a spherical space and with respect to the pressure generated at a hard reflecting plane as opposed to the free field pressure of the incident waves (Shaw, 1976). However, this shift is only valid at high frequencies where the side of the human head behaves like a plane. The adjustment (i in Figure 8) is an estimate of the correction needed to take account of head diffraction effects. The suggested modifications at high frequencies (ii in Figure 8) are intended to correct the minor deficiencies and excesses of mode excitation already noted. Strictly speaking, further adjustments are needed to allow for energy losses between the ear and the far field (e.g., losses over the body surface). However, it seems unlikely that these further adjustments will exceed 1 dB at any frequency.

INTERPRETATION OF FREQUENCY-RESPONSE CURVES

Frontal Incidence vs. Diffuse Field

In Figure 9 the adjusted diffuse field response (curve D from Figure 8) is compared with the average transformation of sound pressure level from the free field to the eardrum with frontally incident waves (curve C from Figure

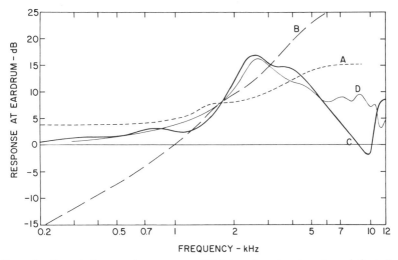

Figure 9.   Curve A shows eardrum response of high impedance insert receiver relative to 2-cc coupler response (Sachs and Burkhard, 1972). Curve B is derived from curve A by adding 6 dB/octave. Curve C shows average transformation from free field to eardrum at 0° azimuth (see Figure 1). Curve D is adjusted diffuse field response from lower panel of Figure 8.

2). The marked difference between these curves in the 7- to 10-kHz region is consistent with the vertical directionality of the ear in this frequency region, which, as we have seen in Figures 3–6, discriminates against sound waves traveling horizontally.

The diffuse field response of the ear may well determine the spectral quality of sound perceived in enclosed spaces, since in most cases the energy density in such spaces is predominantly reverberant. For hearing aid users, however, it is face-to-face speech communication that is normally of primary importance, and it is probably curve C in Figure 9 that best portrays the average response of the open ear in this situation. Hence, a hearing aid normally can be said to have a "flat" frequency response when it transforms sound pressure level from the free field to the eardrum in accordance with curve C.

Coupler Measurements

With the advent of refined instruments, such as the Zwislocki four-branch eardrum simulator, we have a practical way of measuring the output of hearing aid receivers with load and transfer impedances closely matched to those of the median occluded ear. The response curves so obtained can therefore be directly compared with transformation curves, such as curve C in Figure 9. In contrast, the conventional 2-cc coupler not only fails to load the earphone correctly, but suffers, by association, from a prevalent misunderstanding of the nature of coupler measurements in general (Lybarger, 1978). The cumulative effect of these factors is illustrated by curve A in Figure 9, which shows the response at the eardrum to be expected when a median external ear is driven by a hearing aid that has been adjusted to produce a flat response in the 2-cc coupler (Shaw, 1975b, based on Sachs and Burkhard, 1972). The difference between curves A and C expresses the resultant "loss of external ear function," which has received much attention recently (Dalsgaard and Jensen, 1974; Pascoe, 1975).

Open-mold and No-mold Operation

When hearing aids are operated in the no-mold or open-mold configuration, the radiation impedance of the open or partially open ear shunts the output. In any calibration system, this additional parameter must be matched accurately over a wide frequency range, and especially at the principal resonance frequency of the external ear. However, as Lybarger (Chapter 10, this volume) has reported, it may not be necessary to simulate the complete external ear. He proposes that a carefully proportioned, conical "concha simulator" be added to the Zwislocki eardrum simulator to provide a reliable and realistic means of measuring the response of insert earphones under open canal conditions. We must note, however, that this device, like the eardrum simulator, reproduces the characteristics of the median ear only. Eventually, we may need a family of simulators if we are to do justice to the diversity of shapes and sizes of ear canal and concha to be found in a representative human population (see, for example, Berland, 1975).

DIRECTIONALITY, PERCEPTION, AND COMMUNICATION

We have already noted that the human external ear possesses remarkable directionality at the higher frequencies. A superficial examination of Figures 1 and 4 is sufficient to indicate that this directionality has at least three major characteristics: a large reduction in response in the 4- to 5-kHz band as the source direction is changed in azimuth from 45° to 135°, a large increase in response between 7 and 10 kHz as the source elevation is increased from 0° to 75°, and a large decrease in response between 12 and 15 kHz with the same change in elevation. It follows that the spectrum of sound received from a broadband source will vary systematically with the source direction. This variation must surely be related to our ability to perceive acoustic space and hence to discriminate between sound sources that would otherwise merge. Butler (1969, 1970) has in fact shown that vertical localization is impaired where there is high frequency hearing loss. This prompts us to ask whether we should try to simulate the natural directionality of the human acoustic antenna system if, by amplification, we are able to restore the ability to hear high frequency sounds.

In answering this question we must surely take into account the radical nature of sensorineural hearing loss. When many neural receptors have been lost or no longer function normally, the ear has suffered a reduction in filter capacity that cannot be fully restored by amplification. In particular, there is impairment of the ability to process complex signals, such as speech in the presence of noise. Plomp (1978) treats this impairment as equivalent to a loss of signal-to-noise (S/N) ratio —a loss that is roughly proportional to the amount of sensorineural hearing loss. If this is so, then priority must be given to improving the S/N ratio where noise is understood to encompass all incoming sound, including reverberation, that does not convey useful information.

When we compare curves C and D in Figure 9 it is apparent that the frontal and random incidence response curves are not markedly different up to about 5 kHz, which means that nature has provided little discrimination against reverberant sound. A large increase in forward directivity, if attainable, would therefore offer the hearing aid user the additional S/N ratio that is so badly needed for effective communication in reverberant or crowded places. (Steerable directivity would, of course, be even better.) It must be admitted that a practical means of providing the 10–20 dB of directivity, which may be required, is not yet within our grasp. However, the cardioid hearing aid microphone is surely a significant step in this direction. (See Chapter 5, this volume.)

As we search for ways of improving the effective S/N ratio it may be useful to draw attention to a commonplace visual analogy: At distances of a few meters we can easily read nuances of facial expression that are imperceptible when the distance is increased by a factor of 10, as in a theater. The explanation is obvious: although the brightness is unchanged, the 10-fold increase in distance has reduced approximately 100-fold the number of neural receptors available for image formation. In the theater, a solution is found in the liberal use of grease paint to enhance the visual factors that convey information. By analogy, we can imagine a hearing aid in which relevant parameters, such as the frequency response and the directionality, are adjusted, perhaps from moment to moment in an adaptive fashion, to maximize the information content of the sound that reaches the seriously impaired inner ear.

REFERENCES

Berland, O. 1975. No-mold fitting of hearing aids. In S.C. Dalsgaard (ed.), Earmolds and Associated Problems, pp. 173–193. Proceedings of the Seventh Danavox Symposium. Scand. Audiol. (suppl. 5).

Burkhard, M.D., and R.M. Sachs. 1975. Anthropometric manikin for acoustic research. J. Acoust. Soc. Am. 58:214–222.

Butler, R.A. 1969. Monaural and binaural localization of noise bursts vertically in the median sagittal plane. J. Aud. Res. 9:230–235.

Butler, R.A. 1970. The effect of hearing impairment on locating sound in the vertical plane. Int. Audiol. 9:117–126.

Dalsgaard, S.C., and O.D. Jensen. 1974. Measurement of insertion gain of hearing aids. In R.W.B. Stephens (ed.), Proceedings of the 8th International Congress on Acoustics, Vol. 1, p. 205. Institute of Acoustics and Institute of Physics, London.

Funnell, W.R.J., and C.A. Laszlo. 1978. Modeling of the cat eardrum as a thin elastic shell using the finite element method. J. Acoust. Soc. Am. 63:1461–1467.

Kuhn, G.F. 1978. Some acoustic properties of the human pinna. J. Acoust. Soc. Am. 63:S75.

Lybarger, S.F. 1978. Selective amplification—A review and evaluation. J. Am. Audiol. Soc. 3:258–266.

Pascoe, D.P. 1975. Frequency responses of hearing aids and their effects on the speech perception of hearing-impaired subjects. Ann. Otol. Rhinol. Laryngol. 82(5) (suppl. 23).

Plomp, R. 1978. Auditory handicap of hearing impairment and the limited benefit of hearing aids. J. Acoust. Soc. Am. 63:533–549.

Rabinowitz, W.M. 1977. Acoustic reflex effects on the input admittance and transfer characteristics of the human middle-ear. Doctoral thesis, Department of Electrical Engineering and Computer Science, Massachusetts Institute of Technology, Cambridge.

Sachs, R.M., and M.D. Burkhard. 1972. Zwislocki Coupler Evaluation with Insert Earphones. Report 20022. Industrial Research Products, Inc., Elk Grove Village, Ill.

Shaw, E.A.G. 1972. Acoustic response of external ear with progressive wave source. J. Acoust. Soc. Am. 51:150.

Shaw, E.A.G. 1974a. Physical models of the external ear. In R.W.B. Stephens (ed.), Proceedings of the Eighth International Congress of Acoustics, Vol. 1, p. 206. Institute of Acoustics and Institute of Physics, London.

Shaw, E.A.G. 1974b. The external ear. In W.D. Keidel and W.D. Neff (eds.), Handbook of Sensory Physiology, Vol. V(1), pp. 455–490. Springer-Verlag, Berlin.

Shaw, E.A.G. 1974c. Transformation of sound pressure level from the free field to the eardrum in the horizontal plane. J. Acoust. Soc. Am. 56:1848–1861.

Shaw, E.A.G. 1975a. The external ear: New knowledge. In S.C. Dalsgaard (ed.), Earmolds and Associated Problems. Proceedings of the Seventh Danavox Symposium, pp. 24–50. Scand. Audiol. (suppl. 5).

Shaw, E.A.G. 1975b. The external ear: Implications for hearing aid design and calibration. In S.C. Dalsgaard (ed.), Earmolds and Associated Problems. Proceedings of the Seventh Danavox Symposium, pp. 280–297. Scand. Audiol. (suppl. 5).

Shaw, E.A.G. 1976. Diffuse field sensitivity of external ear based on reciprocity principle. J. Acoust. Soc. Am. 60:S102.

Shaw, E.A.G. 1977. Eardrum representation in middle-ear acoustical networks. J. Acoust. Soc. Am. 62:S12.

Shaw, E.A.G., and R. Teranishi. 1968. Sound pressure generated in an external-ear replica and real human ears by a nearby point-source. J. Acoust. Soc. Am. 44:240–249.

Teranishi, R., and E.A.G. Shaw. 1968. External ear acoustic models with simple geometry. J. Acoust. Soc. Am. 44:257–263.

Tonndorf, J., and S.M. Khanna. 1972. Tympanic membrane vibrations in human cadaver ears studied by time-averaged holography. J. Acoust. Soc. Am. 52: 1221–1233.

Zwislocki, J. 1957. Some impedance measurements on normal and pathological ears. J. Acoust. Soc. Am. 29:1312–1317.

Zwislocki, J. 1962. Analysis of the middle-ear function. I. Input impedance. J. Acoust. Soc. Am. 34:1514–1523.

Zwislocki, J. 1971. An Ear-like Coupler for Earphone Calibration. Special Report LSC-S-9, Syracuse University, Syracuse, N.Y.

CHAPTER 7

AN EAR SIMULATOR
FOR ACOUSTIC MEASUREMENTS
Rationale, Principles, and Limitations

Jozef J. Zwislocki

CONTENTS

Although my work never included research on any parts of hearing aids or the evaluation of hearing aids, a turn of events produced a link between some of my work and hearing aids. This work concerned a coupler for the calibration of earphones, which I called "ear-like coupler" and which now is called an "ear simulator" by experts in the field of earphone calibration. Since the coupler seems to have started a new way of approaching the calibration of earphones, in particular hearing aid earphones, its rationale and short history are presented.

HISTORY

The idea began in the Committee on Hearing, Bioacoustics, and Biomechanics (CHABA) of the National Research Council, when a working group was organized in 1966 to evaluate standard couplers that were in current use then (unfortunately, they are still in use). The group's report contained the following conclusions (Zwislocki, 1967):

1. Presently used standard couplers for earphone calibration are inadequate for a number of reasons, the most important of which are:
   a. Inadequate specification of sound pressure in the outer ear
   b. Inappropriate load for earphones, which leads to large sound pressure differences between the real ear and the couplers, especially at low and high frequencies

    c. Ambiguity of pressure calibration at high frequencies, which is particularly disturbing because it coincides with the range where the noise-induced hearing loss occurs most frequently

    d. Necessity of time-consuming and inherently inaccurate loudness balance determinations for transfer of calibration from one earphone to another

    e. Lack of fundamental acoustic information necessary for meaningful interpretation of measurements

    f. Aggravation of the problems listed above when circumaural earphone cushions are used

2. In order to solve the problem of earphone calibration adequately there is need for:

    a. Further basic research on acoustic properties of the ear, including the acoustic impedance at the eardrum and the acoustic properties of the ear canal, the pinna, and the side of the head surrounding it

    b. Development of a coupler with a well-defined reference point for sound pressure measurement and with acoustic properties that match those of the real ear

3. The task of research and development should be undertaken by at least two independent laboratories in order to provide adequate checks.

It would be an exaggeration to say that the report created a great stir. Since no laboratories seemed eager to undertake the required research and development work, I decided to take my own advice and attempt to develop a coupler according to the recommendations of the working group. The decision was prompted by the need in our laboratory to compare auditory measurements made with supraaural and insert phones and was facilitated by the then ongoing acoustic impedance measurements in the ear canals. NASA agreed to support the venture financially.

Coupler Designs—Past and Present

The substance of these events has been synthesized by Shaw (1974) (see Figure 1). The upper left corner of Figure 1 shows the probably best-known and most-used NBS type 9A coupler. It is limited to calibration of supraaural phones and produces neither the correct load nor geometry for the earphone output. The sound pressure generated in it for a given voltage at the earphone terminals cannot be expected to match that generated in real ears, and the pressure difference may be expected to be different for every earphone type. The HA-1 coupler shown in the upper right corner is limited to calibration of insert phones. It has essentially the same deficiencies as the 9A coupler. The lower left corner shows a coupler of a more recent vintage

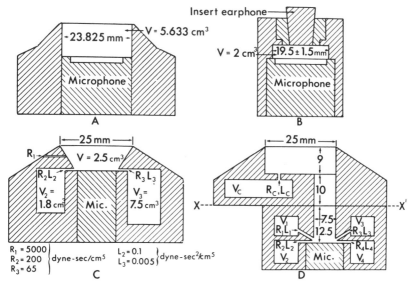

Figure 1. Couplers for earphone calibration. Upper left, NBS type 9A; upper right, HA-1; lower left, IEC artificial ear; lower right, ear-like coupler. (Reprinted with permission from E.A.G. Shaw, The external ear, in W.D. Keidel and W.D. Neff (eds.), *Handbook of Sensory Physiology*, Vol. V(1), pp. 455–490, © 1974, Springer-Verlag, Berlin.)

—the current IEC standard artificial ear. It has the advantage of presenting to supraaural phones the same acoustic load as a median real ear. Accordingly, the sound pressure developed in it by various earphones should approximate the sound pressure developed in real ears. However, this is only true in the vicinity of the earphone face, because the geometry of the coupler differs fundamentally from that of real ears. Because of this difference, two earphones developing the same sound pressure in the coupler at the microphone location may develop different sound pressures at the eardrum of an average ear. Finally, the coupler shown in the lower right corner of the figure is the "ear-like coupler," which may be regarded as the progenitor of the growing family of ear simulators. Its configuration resembles somewhat a greatly simplified configuration of the real ear. In particular, it exhibits a relatively long tube that mimicks the ear canal. The tube opens at the top asymmetrically into a cavity having the approximate dimensions of an average concha of the outer ear, and is closed at the bottom at a location corresponding to the eardrum location by the measuring microphone. Near the microphone, four small cavities are connected to the main tube by narrow tubes. These structures act as acoustic resonators, and their combined effect is to produce an acoustic input impedance approximating the median impedance measured at the eardrum. There is one additional cavity connected to the top cavity by a narrow passage; it is intended to replicate the compliance, inertia, and resistance of the pinna. The ear-like coupler has three features in common with the IEC artificial ear: a cavity mimicking the concha cavity, an acoustic resonator replicating the mechanical properties of the pinna, and a correct acoustic impedance at the face of the earphone. The latter is made approximately equal to the median impedance measured at the same location in real ears. It differs fundamentally from the IEC artificial ear in that the sound pressure is measured at the eardrum location, the acoustic impedance at this location approximates the corresponding real-ear impedance, and the coupler's geometry resembles that of a real ear. Because of the latter two features, sound pressure measured at any point within the concha or ear canal analogs of the ear simulator should approximate the sound pressure measured at a corresponding point in a median ear, when the sound pressure at the eardrum location is the same. This has substantial practical advantages. Sound pressure measured at the eardrum location may be transferred to any other location, for instance, the ear canal entrance or the face of the earphone. This means that measurements made on the ear simulator may be compared with measurements made on the IEC artificial ear. Furthermore, and more important, earphone calibration does not depend on the type of earphone. Two different earphones producing the same sound pressure at the microphone of the simulator should produce, on the average, the same sound pressure at the eardrum and, therefore, the same threshold of audibility. As a consequence, psychophysical measurements required in the past for a transfer of calibration from one earphone to another should be unnecessary. This even includes transfer from a supraaural to an insert phone. Of course, all of these advantages can be realized only if an ear simulator simulates the acoustic characteristics of real ears sufficiently well, and we should examine what has been achieved thus far.

First, let us take a more quantitative look at some performance characteristics of the three older coupler types. The characteristics of the most direct interest are those comparing the sound pressure developed in the coupler to that developed in the ear. For supraaural phones the sound pressure is measured in the ear near the earphone face, a location that does not correspond exactly to the microphone location in the coupler. This lack of correspondence is the first disadvantage of the simple 6-cc couplers. In Figure 2 the sound pressure levels developed by two different earphones in a 6-cc coupler are compared with the average sound pressure levels developed in 11 ears. The measurements were made by Wiener and Filler (1945) and were reproduced in Beranek (1949). Note that the sound pressure levels measured in the coupler differ somewhat from those measured in the ears, and that the differences are not the same for the two earphones. As a

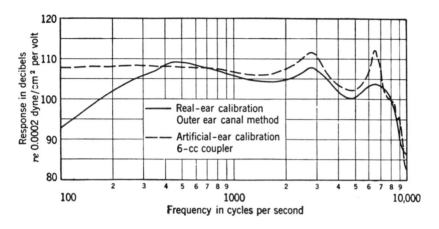

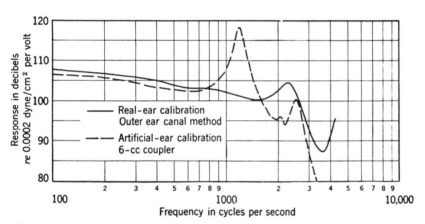

Figure 2.   Comparison of sound pressures generated by a dynamic (upper) and magnetic (lower) earphone in real ears and in the standard 6-cc coupler. (Reprinted with permission from L.L. Beranek, *Acoustic Measurements,* © 1949, John Wiley & Sons, New York.)

consequence, when the earphones produced the same sound pressure in the coupler, they, on the average, did not do so in the ears. To compare the performance of different earphone types, direct sound pressure measurements in real ears or psychophysical tests were required.

Figure 3 shows a similar situation for insert phones and a 2-cc coupler. The curves indicate the differences in sound pressure level produced in real ears and the coupler by several earphones, when the earphone input was kept constant. At some frequencies the differences are on the order of 10

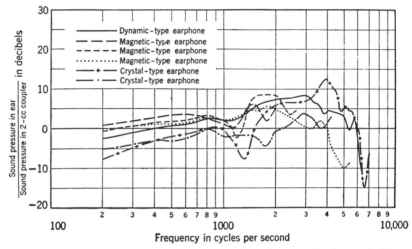

Figure 3.   Ratios between sound pressures generated in real ears and in the standard 2-cc coupler by several insert phones. (Reprinted with permission from L.L. Beranek, *Acoustic Measurements,* p. 732, © 1949, John Wiley & Sons, New York.)

dB, and the differences among the earphones are even larger. The measurements in real ears were made with a probe tube ending at the tip of a molded insert. They were performed by Nichols et al. (1945) and the results were reproduced in Beranek (1949).

As is evident from Figure 4, the more recent IEC artificial ear, with its input impedance matched to the approximately corresponding average impedance of real ears, improved the situation somewhat in the low and midfrequency ranges. The figure shows measurements performed by Delany et al. (1967) with one earphone on three couplers. They measured the input voltage to the earphone at the threshold of audibility, then the sound pressure in the couplers and ears for the same voltage. The figure shows the difference in sound pressure level between the couplers and the ears. The location of the probe-tube microphone near the earphone face matches more closely the location of the probe tube used in real-ear measurements. It is evident that, under the conditions of measurements used, the IEC artificial ear was superior to the other couplers at medium and low frequencies. At those frequencies the coupler sound pressure nearly equaled the real-ear sound pressure. However, above 4 kHz the advantage disappeared. In this frequency range the lengths of sound waves are of the same order of magnitude as the linear dimensions of the outer ear, and the geometry plays an important role.

This is confirmed by the probe-tube measurements in the artificial ear, which yielded a much closer agreement with the real-ear measurements. To

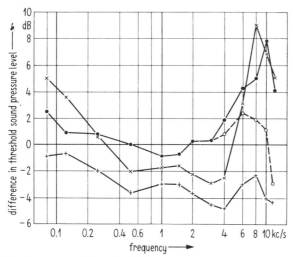

Figure 4.   Ratios between equivalent threshold sound pressures generated by one supraaural earphone in three couplers and in real ears. ●, IEC artificial ear developed by Delany; ○, IEC artificial ear measured with a probetube microphone at the earphone face; ×, NBS 9A; +, British standard artificial ear. (Reprinted with permission from Delany et al., 1967.)

test interearphone variability, Delany et al. (1967) measured the sound pressure developed in the IEC artificial ear, in the NBS 9A coupler, and in the British Standard artificial ear by several laboratory-quality earphones. The sound pressure was determined for input voltages required for the threshold of audibility. The IEC artificial ear produced a somewhat smaller interphone variance than the other couplers over the whole frequency range between 0.1 and 10 kHz, but the improvement was not impressive.

The main advantage of the simple couplers, such as the 9A and HA-1 couplers, is their simplicity. Their design does not necessitate much knowledge of the dimensions and acoustic properties of real ears. The IEC coupler requires knowledge of the input impedance of real ears near the earphone face, and the recent ear simulators require knowledge of the input impedance at the eardrum, the acoustic properties of the pinna, and relevant ear geometry. What geometric features are relevant can be the subject of much research, as Shaw (1974) and his associates have demonstrated. The simplest approach is to make an impression of an ear that is typical in appearance and replicate as exactly as possible its geometry in the simulator. Such an approach was used in the 1930s by Inglis, Gray, and Jenkins (1932; reproduced in Beranek, 1949, p. 740). Figure 5 shows their ear simulator as reproduced in Beranek (1949). In addition to replicating the geometric features of the concha and the ear canal, the simulator was also endowed with an acoustic network analog of an acoustic leak between the earphone

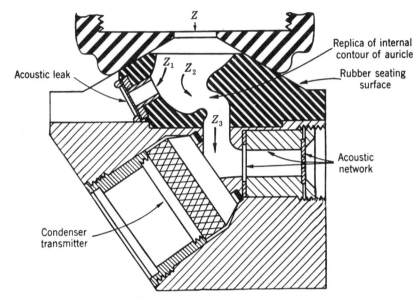

Figure 5. "Artificial ear" of Inglis, Gray, and Jenkins. (Reprinted with permission from L.L. Beranek, *Acoustic Measurements*, © 1949, John Wiley & Sons, New York.)

cushion and the pinna and with another acoustic network for the input impedance at the eardrum. Since the impedance at the eardrum was not well known, the latter network was designed to produce a correct acoustic impedance at the earphone face. Figure 6 indicates how well the simulator functioned; the sound pressure produced by one earphone in the simulator is compared with the average sound pressure produced by the same earphone for the same input current in eight ears. The agreement is within about 3 dB over most of the range tested. When the earphone transfer function exhibited stronger peaks, there was less agreement. In spite of its adequate acoustic performance, the Inglis simulator was not found suitable for a standard because of a complex shape that did not lend itself to rigorous specification. In addition, it could not be regarded as representing the geometric features of a median or average ear.

Ear-like Coupler Development

It became possible to specify the impedance at the eardrum with increasing accuracy and to determine the acoustic effects of various geometric features of the outer ear. This allowed a radical simplification of the simulator geometry without sacrificing acoustic performance. As a consequence, it became possible to design a simulator whose dimensions and acoustic properties could be rigorously specified.

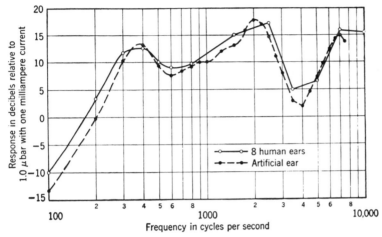

Figure 6.   Comparison of sound pressure levels measured in real ears and in the artificial ear of Inglis, Gray, and Jenkins. (Reprinted with permission from L.L. Beranek, *Acoustic Measurements,* © 1949, John Wiley & Sons, New York.)

Since the performance of an ear simulator depends substantially on accurate simulation of the acoustic impedance at the eardrum, it may be worthwhile to gain an idea of how well this impedance is known. The main body of data used in the design of the first of the current population of ear simulators, the ear-like coupler, was gathered with the help of an acoustic bridge designed specially for impedance measurements in the ear (e.g., Zwislocki, 1963). Since the bridge allows compensation of the residual volume of air in the ear canal, its dial readings refer directly to impedance components at the eardrum. The bridge could be calibrated accurately up to 2 kHz, but only approximately for higher frequencies (Zwislocki, 1970). It seems to have been the first instrument to allow reactance measurements at the eardrum up to 7kHz. Perhaps the most accurate instrument for impedance measurements at high frequencies in current use is the impedance tube devised by E. Shaw (personal communication). However, no measurements seem to have been made with this instrument thus far on real ears. The bridge is shown in its position of operation in Figure 7. Note that it is held in place by means of an adjustable arm assembly. This method of securing the bridge in the ear canal greatly reduced the variability of impedance measurements.

The median acoustic reactance and resistance measured with the bridge at the eardrums of 12 female and 10 male ears (22 subjects) are shown in Figure 8 (Zwislocki, 1970). The data for the resistance are shown only up to 2 kHz since they could not be regarded as accurate above this frequency. The one point at 3.5 kHz was determined from direct sound

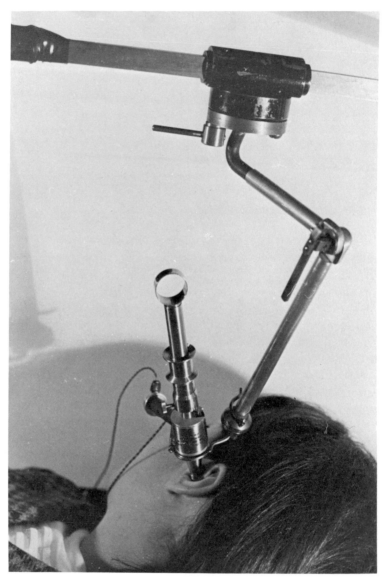

Figure 7.   Acoustic bridge held in position of operation by a three-jointed holder. (Reprinted with permission from Zwislocki and Feldman, 1970.)

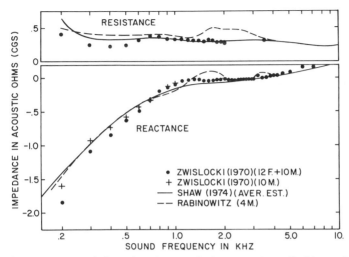

Figure 8. Average acoustic input impedance at the human eardrum. ●, Measured on 22 ears (12 female and 10 male) with the acoustic bridge (except the resistance point at 3.5 kHz; +, measured for 10 male subjects alone with the acoustic bridge; ———, a best estimate based on many studies; — — —, measured by the probe-tube technique on four male subjects.

pressure measurements in the ear canal. The crosses indicate the corresponding reactance data for the male ears alone. Note that, at low frequencies, the median absolute reactance of male ears is somewhat smaller than of the combined group. For comparison, the solid lines show what, according to Shaw (1974), are the best estimates of median acoustic reactance and resistance at the eardrum. The broken lines refer to the most recent measurements of these impedance parameters by Rabinowitz (1977) who, unfortunately, used only four ears, all male. Whether the three sets of the reactance and resistance values of the figure are found to be in agreement or not depends on how they are to be used. The discrepancies among them are small compared with differences among individual ears. One could argue that the most representative of the average impedance values are the estimates of Shaw, since they include the largest number of measurements performed with various instruments. This may not be entirely true, however. At low frequencies, for example, Shaw's reactance estimate agrees best with the data for male ears alone. This suggests that it is not representative of an average of male and female ears. Also, between about 1 and 2 kHz, it indicates a more negative reactance than do either Rabinowitz's or my data. However, these differences proved to be unimportant from the point of view of the design of ear simulators.

In addition to the impedance data measured at the eardrum location, the design of ear simulators requires average dimensions of the ear canal and cochlea. The length of the ear canal was obtained for the ear-like coupler as follows (Zwislocki, 1971). A probe-tube assembly was manufactured for sound pressure measurements in the outer ear. It consisted of a Brüel and Kjaer condenser microphone together with its probe tube and preamplifier, a holder for the microphone that was connected to a calibrated micrometer-like device, and an adjustable arm assembly for holding the device rigidly in place. The arm assembly was the same as that for the acoustic bridge. Figure 9 shows the probe tube assembly in its position of operation. The probe tube was bent so that its main length was parallel to the direction of motion of the micrometer device. After positioning the probe tube over the ear canal entrance so that it pointed toward the eardrum, the depth of the probe-tube insertion in the ear canal was adjusted by means of the micrometer device until the probe-tube tip gently touched the eardrum. Subsequently, the probe tube was withdrawn to desired insertion depths. The location of the tip of the umbo in the eardrum always served as the point of reference for all distance measurements. The length of the ear canal was determined by withdrawing the probe tube from the eardrum to a point where its tip was level with the concha floor. The measurement was performed in seven ears, four male and three female,

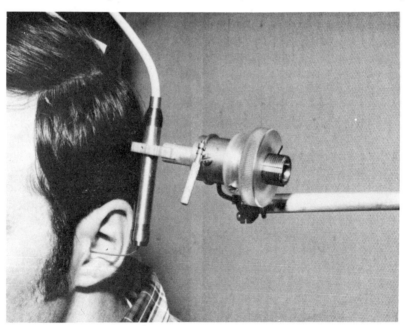

Figure 9.   Probe-tube microphone held in position of operation by a micrometer device and the three-jointed holder. (Reprinted with permission from Zwislocki, 1970.)

yielding a mean value of 2.3 cm for the entire ear canal length. This value was in good agreement with earlier determinations of Wiener and Ross (1946) and with acoustic measurements of Teranishi and Shaw (1968).

The ear canal varies somewhat in cross-sectional area from its entrance to the eardrum and, in addition, has a tortuous course. As a consequence, direct measurements of its cross-sectional dimensions are bound to be inaccurate. To obtain a representative average effective cross-sectional area, therefore, we filled the ear canal with alcohol, measured the required volume, and divided it by the length of the fluid column. The procedure yielded an average cross-sectional area of 0.44 cm^2, which is equivalent to an average diameter of 0.748 cm. Békésy and Rosenblith (1951) had estimated this diameter to be only slightly smaller.

The effective dimensions of the concha were derived from three sources of information: Delany's (1964) impedance measurements at the face of supraaural earphones and his network model based on these measurements, our own impedance and direct length measurements, and Teranishi and Shaw's (1968) experiments with simplified ear replicas. From Delany's measurements we determined an equivalent combined air volume for the concha, the ear canal, and the impedance at the eardrum. It approximated 6 cm^3. By subtracting from it a value of 1 cm^3 for the ear canal and 0.65 cm^3 for the impedance at the eardrum, as suggested by our measurements, we obtained for the concha an effective volume of 4.35 cm^3. Rough measurements of the depth of the concha in six subjects yielded a rather stable value of approximately 0.9 cm. Combining this value with the effective volume of the concha, we derived a value of 2.5 cm for the effective diameter of the concha. Having the effective depth and diameter of the concha, we were able to calculate the frequency of the first longitudinal mode of the concha and to compare it with the experimental results of Teranishi and Shaw. The two agreed within 5%, being 4.5 kHz and 4.7 kHz, respectively. It may be of some interest to point out that the effective concha diameter of 2.5 cm coincides with the diameter of the orifice of the IEC artificial ear.

To make the ear simulator applicable to supraaural phones it is necessary to include an additional set of parameters referring to the mechanical properties of the pinna. These parameters were taken from Delany's network analog underlying the structure of the IEC artificial ear.

With the data just specified it was possible to design an ear simulator that could be applied to both supraaural and insert phones. If it functioned properly, psychoacoustic measurements obtained with one kind of earphone should be directly comparable with psychoacoustic measurements obtained with the other. The basic design of the simulator is shown in Figure 10 (Zwislocki, 1970, 1971). It consists of two main sections—the lower corre-

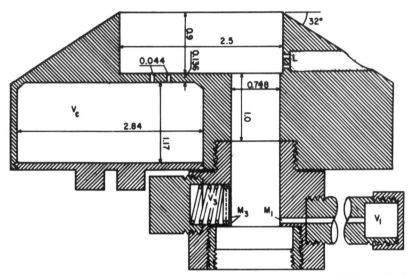

Figure 10.   Longitudinal section through the ear-like coupler. V_1 and V_3 indicate the air volumes and M_1 and M_3 the openings of two of the four resonators matching the acoustic impedance at the eardrum. V_c indicates the air volume of the resonator simulating the mechanical properties of the pinna. (Reprinted with permission from Zwislocki, 1971.)

sponding to the deepest part of the ear canal containing the eardrum, and the upper to the outer part of the ear canal, the concha and the pinna. All parts are made of metal to assure stability. The measuring microphone is screwed in from the bottom and its membrane coincides with the eardrum location. Right above the microphone are four acoustic resonators, whose combined impedance replicates approximately the impedance at the eardrum. Only two of the resonators are shown. For reasons that should be obvious it is essential to have the orifices of these resonators coincide as closely as possible with the eardrum location. For calibration of insert phones the top part of the simulator is unscrewed and, instead of it, an appropriate holder for such phones is attached. The tip of the ear insert should coincide with the top of the lower part of the simulator canal. When calibrating supraaural earphones, such earphones are attached to the top part of the simulator, as is usual in in the standard NBS-9A coupler or the IEC artificial ear. Note the set of small tubes marked "L" on the top right side of the simulator. They simulate the leak that usually arises between the earphone cushion and the pinna. The large cavity (V_c) on the left side of the simulator together with the two tubes that connect it to the concha cavity simulates the mechanical properties of the pinna. Figure 11 shows the simulator mounted on a ½-inch Brüel and Kjaer microphone assembly, and Figure 12 shows the lower part of the simulator.

Evaluation of the Simulator

For the initial tests of the simulator we made two sets of comparative measurements on the simulator and real ears, in each case using the same equipment (Zwislocki, 1971). First, the impedance at the eardrum location was measured with the acoustic bridge. In Figure 13 the solid line corresponds to the reactance of the simulator and the circles to the median reactance determined in a group of 10 males and 12 females. The two sets

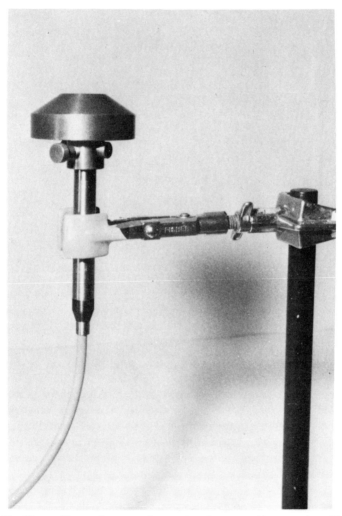

Figure 11.   The ear-like coupler mounted on a microphone assembly as seen in a side view. (Reprinted with permission from Zwislocki, 1971.)

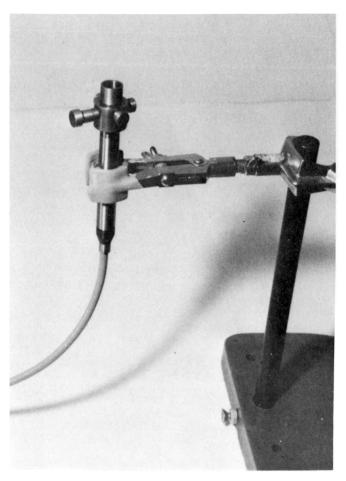

Figure 12.   A side view of the lower part of the ear-like coupler adapted for calibration of insert phones. (Reprinted with permission from Zwislocki, 1971.)

of data do not coincide completely, but their disagreement is within the uncertainty of impedance measurements on real ears. The agreement in resistance values seen in Figure 14 is slightly worse; however, a good part of the discrepancy seems to be because of a measurement artifact, since most recent measurements by E. Shaw (personal communication) suggest a closer coincidence.

More direct tests of the simulator from the viewpoint of earphone calibration consist of sound pressure measurements at its various locations and comparison of the results with corresponding median data ob-

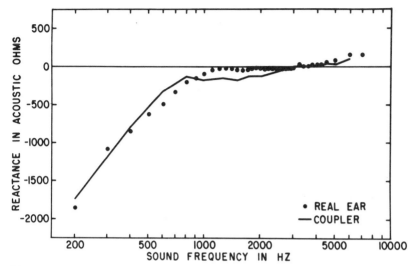

Figure 13.   Acoustic reactance of the ear-like coupler at the eardrum location (solid line) and corresponding median reactance of 22 human ears. (Reprinted with permission from Zwislocki, 1971.)

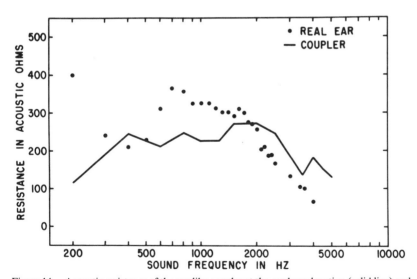

Figure 14.   Acoustic resistance of the ear-like coupler at the eardrum location (solid line) and corresponding median reactance of 22 human ears. (Reprinted with permission from Zwislocki, 1971.)

tained in real ears (Zwislocki, 1971). For insert phones the most important relationship is between the sound pressure at the eardrum and the tip of the ear insert. In Figure 15 the solid curve shows this relationship in terms of a decibel difference for the simulator, and the circles do the same for a sample of seven ears (four male and three female). Up to 6 kHz the differences between the simulator and the average for the real ears are random and on the order of 1 dB. This is within the error of measurements. Significant differences can be seen only at 7 and 10 kHz, and they are on the order of 3 dB. Sound pressure ratios in decibels between the entrance of the ear canal and the eardrum location are shown in Figure 16. Again, the solid line belongs to the simulator and the circles to the sample of seven ears. Because of the variability among individual ears, none of the differences can be considered significant—the greatest, around 4.5 kHz, is below 2 dB.

The most direct test of the simulator performance consists of the comparison of sound pressure that an earphone develops in the simulator and in real ears. Such a test was performed on an only slightly modified ear-like coupler by Sachs and Burkhard (1972), using an insert phone. The ear measurements were performed with the help of a probe-tube microphone on 11 subjects, 6 male and 5 female. All the data were referred to the eardrum location. The results are shown in Figure 17. Sachs and Burkhard commented on the outcome of their experiments: "At low frequencies, pressure in the Zwislocki coupler is essentially identical to pressure in real

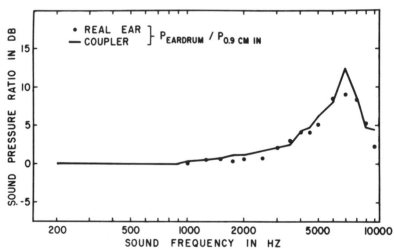

Figure 15.   Median sound pressure ratios between the eardrum and a point in the ear canal about 1 cm away from the entrance of the canal, and corresponding sound pressure ratios in the coupler. (Reprinted with permission from Zwislocki, 1971.)

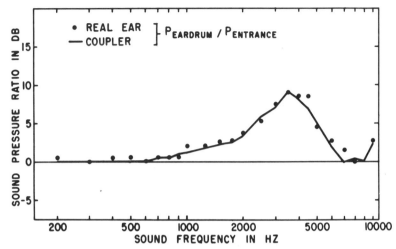

Figure 16.   Median sound pressure ratios between the eardrum and the ear canal entrance, and corresponding sound pressure ratios in the coupler. (Reprinted with permission from Zwislocki, 1971.)

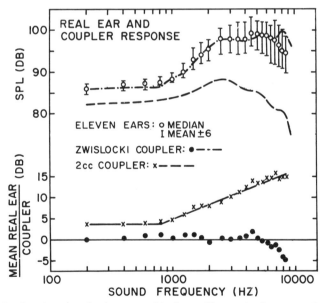

Figure 17.   Sound pressure levels generated in the ear-like coupler (— · — · —) and in 11 ears (◯) at the eardrum location by an insert phone for a constant electrical input. The lower graph shows the decibel difference between the two sets of sound pressure levels (●). Corresponding data for the standard 2-cc coupler are shown (× — × —). (Reprinted with permission from Sachs and Burkhard, 1972.)

ears. . . . The mean pressure in real ears and in the Zwislocki coupler differ by no more than 3 dB up to 7.5 kHz." Unfortunately no comparable data seem to be available for supraaural phones.

The comparative data given here should constitute sufficient proof that it is possible to make an ear simulator with a simple and exactly specificable geometry that closely approximates the acoustic characteristics of a median human ear. The simulator can be built entirely out of metal for stability. Note that I discussed the first attempt at making such a simulator; improvements are to be expected in the future. Some have already been proposed by Diestel (1974), Burkhard (1978), and Shaw (1975). The ear simulators open new possibilities for the development and testing of earphones, especially hearing aid earphones. They should eliminate almost entirely the need for time-consuming and inherently inaccurate psychophysical tests in such work. In particular, they should make it possible to compare directly the performance characteristics of earphones of different types and to compare the psychophysical results obtained with them.

Nevertheless, ear simulators will always have one fundamental limitation. They can at best simulate the median or average characteristics of real ears. They will never obviate the need for individual tests.

REFERENCES

Békésy, G.v., and W.A. Rosenblith. 1951. The mechanical properties of the ear. In S.S. Stevens (ed.), Handbook of Experimental Psychology, pp. 1075–1115. John Wiley & Sons, New York.

Beranek, L.L. 1949. Acoustic Measurements. John Wiley & Sons, New York.

Burkhard, M.D. 1978. Ear simulators, designs, stability, etc. In M.D. Burkhard (ed.), Proceedings of a Conference on Manikin Measurements, pp. 69–74. Industrial Research Products, Elk Grove Village, Ill.

Delany, M.E. 1964. The acoustical impedance of human ears. J. Sound Vibr. 1: 455–467.

Delany, M.E., L.S. Whittle, J.P. Cook, and V. Scott. 1967. Performance studies on a new artificial ear. Acustica 18:231–237.

Diestel, H.G. 1974. Measurements on new couplers for insert earphones. Paper presented at Eighth International Congress on Acoustics, July, London.

Inglis, A.H., C.H.G. Gray, and R.T. Jenkins. 1932. A voice and ear for telephone measurements. Bell Syst. Tech. J. 11:293–317.

Nichols, R.H., Jr., R.J. Marquis, W.G. Wiklund, A.S. Filler, D.B. Feer, and P.S. Veneklasen. 1945. Electro-acoustic characteristics of hearing aids. OSRD Report 4666, Electro-Acoustic Laboratory, Harvard University, Cambridge.

Rabinowitz, W.M. 1977. Acoustic-Reflex Effects on the Input Admittance and Transfer Characteristics of the Human Middle Ear. Doctoral dissertation, Massachusetts Institute of Technology, Cambridge.

Sachs, R.M., and M.D. Burkhard. 1972. Earphone Pressure Response in Ears and Couplers. Project 20021 for Knowles Electronics, Franklin Park, Ill.

Shaw, E.A.G. 1974. The external ear. In W.D. Keidel and W.D. Neff (eds.), Handbook of Sensory Physiology, Vol. V (1), pp. 455–490. Springer-Verlag, Berlin.

Shaw, E.A.G. 1975. The external ear: New knowledge. Earmolds and associated problems. Scand. Audiol. 5(suppl.):24–50.

Teranishi, R., and E.A.G. Shaw. 1968. External-ear acoustic models with simple geometry. J. Acoust. Soc. Am. 44:257–263.

Wiener, F.M., and A.S. Filler. 1945. The response of certain earphones on the ear and on closed couplers. Report PNR-2, December 1, Psycho-Acoustic and Electro-Acoustic Laboratories, Harvard University, Cambridge.

Wiener, F.M., and D.A. Ross. 1946. The pressure distribution in the auditory canal in a progressive sound field. J. Acoust. Soc. Am. 18:401–408.

Zwislocki, J.J. 1963. An acoustic method for clinical examination of the ear. J. Speech Hear. Res. 6:303–314.

Zwislocki, J.J. 1967. Couplers for Calibration of Earphones. Report W.G. 48, National Academy of Sciences—National Research Council, Committee on Hearing, Bioacoustics, and Biomechanics, Washington, D.C.

Zwislocki, J.J. 1970. An Acoustic Coupler for Earphone Calibration. Report LSC-S-7, Institute for Sensory Research, Syracuse University, Syracuse, N.Y.

Zwislocki, J.J. 1971. An Ear-like Coupler for Earphone Calibration. Report LSC-S-9, Institute for Sensory Research, Syracuse University, Syracuse, N.Y.

Zwislocki, J.J., and A.S. Feldman. 1970. Acoustic Impedance of Pathological Ears. ASHA Monographs No. 5. American Speech and Hearing Association, Washington, D.C.

CHAPTER 8

CORFIG: COUPLER RESPONSE FOR FLAT INSERTION GAIN

Mead C. Killion and Edward L. Monser, IV

CONTENTS

The acronym CORFIG, which literally stands for "Coupler Response for Flat Insertion Gain," can be thought of as the answer to this question: What would the coupler frequency response of a hearing aid look like if the hearing aid produced a perfectly flat frequency response for the average user?

Another interpretation for CORFIG might be CORrection FIGure, since knowledge of the CORFIG corresponding to a given hearing aid design allows one to estimate the insertion gain that will be provided to a user of that hearing aid. The use of such "coupler correction curves" has a long history.

The relationship between the coupler response of a hearing aid and its in situ response can be reasonably well specified on the average, but such a specification must include the following information:

1. The type of sound field
2. The orientation of the listener in the sound field
3. The location of the microphone on the listener's head

4. The exact construction of the earmold used to measure the in situ response
5. The exact construction of the earmold used to measure the coupler response
6. The type of coupler used to obtain the coupler response

Each of these factors is discussed in conjunction with the presentation of recently obtained average CORFIG data.

RECENT CORFIG DATA

Definitions

The *insertion gain* of a hearing aid on a given user is equal to the ratio of the eardrum pressure produced by the hearing aid to the unaided eardrum pressure that would have been produced by the same sound field. Unless otherwise stated, the sound field is generally assumed to be produced by a point source located directly in front of the listener at a distance 1 m from a hypothetical line drawn between the two ear canal openings.

The term *functional gain* was introduced by Pascoe (1975) to describe the subjective threshold-difference method of determining the in situ gain of a hearing aid. A hearing aid that improved a subject's threshold by 20 dB at 1 kHz, for example, was said to have a functional gain of 20 dB at 1 kHz. To avoid standing wave problems in his test room, Pascoe used ⅓-octave bands of noise as stimuli.

In most instances, the objective and subjective measures might be expected to give the same estimate of the in situ gain of the hearing aid, but a variety of experimental pitfalls can plague either type of measurement. Background noise (Walden and Kasten, 1976), hearing aid-microphone noise (Killion, 1976a), standing waves in the test room with pure tones or insufficiently steep filter slopes with narrowband noise stimuli (Orchik and Mosher, 1975), inadequate specification of sound field incidence, variations in earmold construction and fit, and hearing aid overload (a particular problem at 4 kHz with subjects having noise-induced hearing loss with a "4-kHz notch") have been implicated in producing unusual results during functional gain determinations. Most of these problems, plus problems with probe-tube location and probe-tube microphone calibration (Burkhard and Sachs, 1977), can plague attempts to obtain accurate insertion gain measurements. When everything goes well, however, it is possible to obtain good agreement between functional gain and insertion gain measurements. Causey and Beck (1977), for example, found less than a 2-dB difference at most frequencies between the insertion gain of an over-the-ear (OTE) hearing aid measured on a KEMAR manikin and the average functional gain measured on a group of subjects with sensorineural hearing loss.

The informal phrase *coupler response of a hearing aid* is used in this chapter as a shorthand for the more elaborate phrase *the response of a hearing aid whose input SPL is held constant and whose output SPL is measured into a specified coupler.* The *coupler response* as used here is identical to the *frequency response* of a hearing aid as specified in ANSI S3.22 (1976) when the coupler and earmold are chosen appropriately.

Measurement Techniques

There are several ways to obtain the CORFIG curves. The simplest intuitively would be to construct a hearing aid that produced an exactly flat insertion gain frequency-response curve on the KEMAR manikin and then measure its coupler response.

Less intuitive but more simple in practice, the required CORFIG curve can be predicted from direct measurements of the sort described by Kuhn (Chapter 4, this volume) and Shaw (Chapter 6, this volume). As discussed by Knowles (1959) and others, the CORFIG curve corresponding to a given hearing aid and given sound field condition can be predicted from:

1. The unaided sound pressure level normally produced at the eardrum by that sound field
2. The sound pressure level produced at the in situ hearing aid microphone inlet by the same sound field
3. The difference between the eardrum and coupler response of the earphone-earmold combination used with the hearing aid

The difference (3) should be close to zero on the average when the coupler response of the hearing aid is measured into a Zwislocki coupler and the same earmold construction is used for both coupler response and insertion gain response measurements. Thus a good estimate of the CORFIG can be obtained with the aid of a KEMAR manikin by subtracting the sound pressure level available to the in situ hearing aid microphone from the sound pressure level developed at the eardrum-position microphone of the unaided manikin. The result is the difference that the hearing aid must make up in order to compensate for the loss of external ear resonance and the out-of-ear microphone location. (In a facetious sense, all one has to do to obtain Zwislocki coupler CORFIG curves is subtract everything in Kuhn's chapter from everything in Shaw's chapter.)

This procedure is illustrated in Figure 1, which is based on measurements made on a KEMAR manikin several years ago in the anechoic chamber at Industrial Research Products Inc. (IRPI). The curve on the left shows the pressure developed at the eardrum-position microphone on an

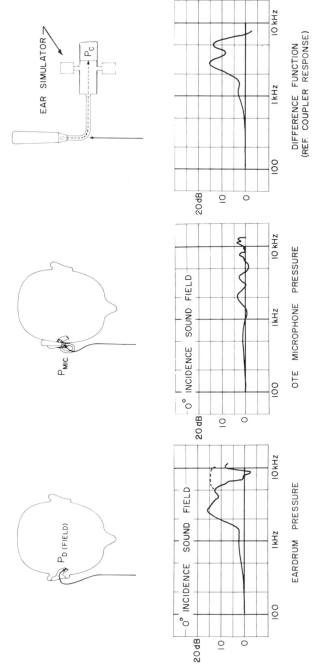

Figure 1. Derivation of coupler response required for unity insertion gain (CORFIG) for an over-the-ear (OTE) hearing aid. (Reprinted with permission from Killion, 1976b.)

unaided KEMAR manikin in a 0° incidence sound field, taken from Burkhard and Sachs (1975). The curve in the middle shows the pressure developed at the side of the head at a location corresponding to the microphone inlet of an OTE hearing aid. This curve was one of the 23 curves obtained by Madaffari (1974) at possible hearing aid microphone locations on the side of the head. The curve on the right shows the difference between the first two curves, and represents an early estimate of the coupler response required of an OTE hearing aid if it is to produce a flat insertion gain frequency-response curve. In other words, the curve on the right is what we now call the CORFIG curve, or more precisely, the Zwislocki coupler CORFIG curve.

A different CORFIG curve results each time the microphone location, sound source location, type of sound field, earmold, or type of coupler is changed.

Effect of Sound Field

The effect of different sound field measurement conditions is illustrated in Figure 2 for three sound field conditions. These curves were obtained with an OTE hearing aid utilizing a "forward-looking" microphone whose inlet was directly over the pinna and slightly forward of the vertical midline of the earcanal. The 0° and 90° incidence CORFIG curves are similar to the (inverted) correction curves shown by Burkhard (1978), although refinements in our measurement techniques (use of more nearly a "point source" loudspeaker) have resulted in minor changes at high frequencies.

A Problem with 0° Incidence Measurements   The 8-kHz notch in the 0° incidence CORFIG curve of Figure 2 is a result of a sharp concha antiresonance, which is discussed by Shaw (Chapter 6, this volume). This antiresonance produces a substantial null in eardrum pressure at approximately 8 kHz. This antiresonance disappears when the concha is filled with an earmold.

In order for a broadband OTE hearing aid to have a flat insertion gain in the 5- to 10-kHz region for 0° incidence sound, either the microphone inlet must be located in front of the ear canal entrance in an unoccluded concha, or an 8-kHz notch filter must be included in the hearing aid in order to duplicate the effect of the concha antiresonance. Viewed another way, an OTE hearing aid that has a smooth coupler response in the 5- to 10-kHz region will have an insertion gain curve that shows a sharp peak at about 8 kHz when the insertion gain is measured in a 0° incidence sound field on a KEMAR manikin. On real subjects, one would obtain a similar peak from either objective insertion gain or subjective functional gain measurements in a 0° incidence field, but the exact frequency and magnitude of the peak would vary among subjects (see discussion below and Chapter 6).

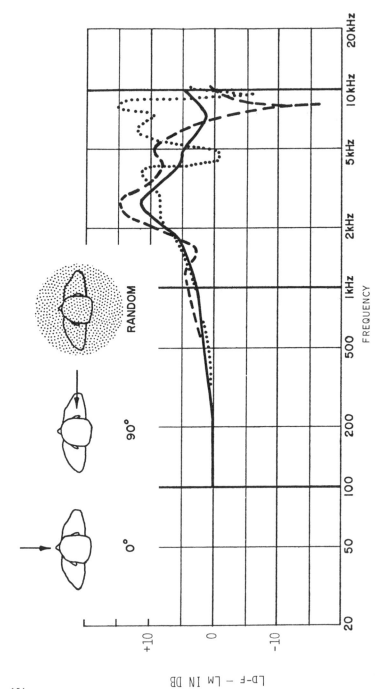

Figure 2. Comparison between 0° incidence (broken line), 90° incidence (dotted line), and random incidence (solid line) Zwislocki coupler CORFIGs for an OTE hearing aid.

154

The 8-kHz CORFIG notch (insertion gain peak) presents a dilemma to the designer of a broadband hearing aid. Insertion gain is now commonly measured at 0° incidence in an anechoic chamber, as discussed in *Manikin Measurements—Conference Proceedings* (Burkhard, 1978) or the 1978 V.A. *Handbook of Hearing Aid Measurement* (Beck, 1978), but listening is seldom done in an anechoic chamber. In most real-life situations, sufficient reflected energy arrives at the listener's ear to effectively fill in the notch caused by the concha antiresonance. Even in close face-to-face conversations, 10% to 20% of the energy arriving at the listener's ears is typically reflected energy. When listening to music, the sound sources are generally located much farther than a critical distance away from the listener, so the majority of the energy arriving at the listener's ear is reflected energy. With high fidelity loudspeakers in a typical living room, for example, the curve given by Olson (1967, p. 285) indicates that 90% of the energy arriving at the listener's ear can be reflected energy. The sound arriving after reflections from the walls, ceiling, and floor more nearly represents a diffuse or random-incidence sound field than a 0° incidence sound wave (which, after all, can only be achieved under anechoic conditions).

These comments should not be misinterpreted to mean the notch caused by the concha antiresonance has little significance. The perceptual importance of the concha antiresonance in vertical localization has been studied quite extensively (see, for example, Butler and Belendivk, 1977). For judgments of overall spectral balance in the perceived sound, however, the experiments performed by Schulein (1975) indicate that random-incidence sound plays a dominant role.

As a result of these considerations, we undertook to obtain random-incidence CORFIG curves.

Random-Incidence Measurements   The estimate of the random-incidence CORFIG shown as the solid curve in Figure 2 was obtained in the IRPI reverberation chamber (width, length, and height of 2.42 mm, 3.58 mm, and 3.64 mm, respectively) using a warble tone with ± 50 Hz deviation and a 10-Hz triangular wave modulation (modulation index of 10). As with the 0° and 90° data, the random-incidence curve represents the difference between the sound pressure level (SPL) measured first at the unaided manikin eardrum location and then at the in situ hearing aid microphone inlet position. (For the latter measurement, the hearing aid was placed on the manikin and attached to an earmold filling the manikin concha.) The inlet pressure was measured with a probe microphone located 1–2 mm from the blocked microphone inlet. The random-incidence curve shown in Figure 2 represents the average of several measurements made with various locations of the manikin in the reverberation room. In some cases, the pressure at the eardrum location was measured with the same probe microphone used to measure microphone inlet pressure, while in others a Brüel & Kjaer 4134 ½-inch condenser microphone was employed. (Both measurements gave essentially similar results after corrections for the pressure frequency response of each microphone.)

Since there appears to be few random-incidence data available in the literature, we were pleased at the close agreement between the eardrum pressure data we obtained and the data calculated by Shaw (1976) from a 52-source position experiment using a KEMAR ear mounted in a small baffle. That comparison is shown in Figure 3. The 3-dB difference shown at low frequencies is the result of a small baffle effect in Shaw's experiment.

Choosing the Reference Sound Field   The choice facing the hearing aid designer may not be as important in practice as it first appears. Note that all the OTE CORFIG curves in Figure 2 are similar everywhere below 4 kHz. When the hearing impairment is severe enough so that maximizing speech intelligibility in face-to-face listening situations is the dominant consideration, a suitable hearing aid may well have a limited enough bandwidth so that it will make little difference which sound field condition is chosen for reference.

When the hearing impairment is mild enough so that sound quality is the major consideration, the random-incidence sound field might be a more suitable reference. The main problem with random-incidence measurements is that they require either a reverberation chamber or the averaging of a large number of anechoic chamber measurements. An alternate possibility might be to average the 0° and 90° incidence anechoic chamber data for hearing aids whose response extends into the 5- to 10-kHz region. This would provide a rough approximation to random-incidence measurements without their complications.

Effect of Microphone Location

If the hearing aid microphone is located in the ear instead of over the ear, the location of the sound source becomes much less important. This is illustrated in Figure 4, where representative CORFIG curves are shown for three different microphone locations: over-the-ear (OTE), in-the-ear (ITE), and in-the-concha (ITC). The OTE curves in Figure 4 are identical to those shown in Figure 2, and are included here for comparison.

The ITE microphone location corresponds to a microphone approximately centered in the face of an ITE aid that fills the concha, so that the microphone is located approximately flush with the plane of the external pinna.

The ITC microphone location is similar to that discussed by Berland and Nielsen (1969), in which the microphone inlet of an OTE hearing aid was extended down into the unoccluded concha in front of a "phantom"

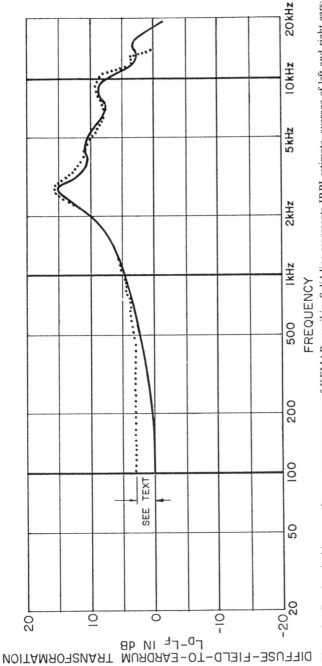

Figure 3. Random-incidence eardrum pressure response of KEMAR manikin. Solid line represents IRPI estimate, average of left and right ears; dotted line represents Shaw estimate, KEMAR ear mounted in small baffle. (Reprinted with permission from Killion, 1979.)

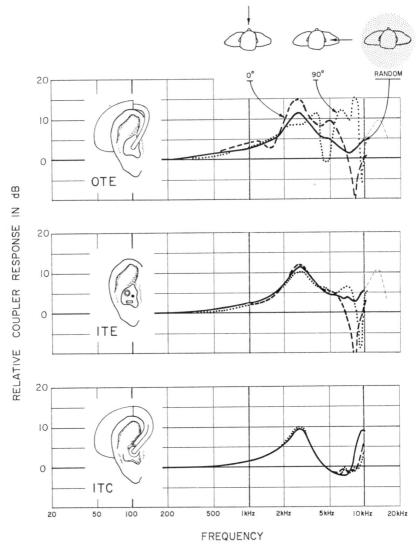

Figure 4. CORFIG curves for three hearing aid types and three sound field conditions.

or "canal-lock" earmold. In this case, nearly all of the directional properties of the ear are obtained, as discussed by Shaw (Chapter 6, this volume).

In a sense, the curves in Figure 4 simply illustrate that one loses more and more of the directional effects of the external ear as the microphone is moved farther from the ear canal entrance, a point that has been made frequently in recent advertising literature for ITE hearing aids. Conversely, the closer the hearing aid microphone is located to the ear canal entrance, the less one has to worry about which sound field condition is used to measure insertion gain.

Effect of the Coupler

So far we have discussed only Zwislocki coupler CORFIG curves. The greater pressure developed by a hearing aid earphone-earmold combination in real ears compared with the 2-cc coupler was shown by Sachs and Burkhard (1972b). As a rough rule of thumb, this amounts to 3.5 dB at low frequencies, 5 dB at 1 kHz, 10 dB at 3 kHz, and 15 dB at 6 kHz, and is nearly independent of the earphone-earmold combination as long as a good earmold seal is obtained. As has been often observed, the error introduced by the 2-cc coupler tends to compensate for the error introduced by ignoring head diffraction and loss of external ear resonance. The comparison shown in Figure 5 illustrates this relationship. The 0° incidence CORFIG for an OTE aid—in other words, the increased response required of the hearing aid to compensate for the loss of external ear resonance and the out-of-ear microphone location—is shown by the solid curve in Figure 5. The Sachs and Burkhard data on the increase in actual eardrum pressure over the 2-cc coupler pressure developed by hearing aid earphones is shown by the broken curve, which was drawn to have the same low frequency asymptote as the solid curve. The net result of these compensating errors is that the traditional 2-cc coupler frequency response curves would not have been too badly in error (Lybarger, 1978) except for yet another (earmold-related) error, which is discussed in the next section.

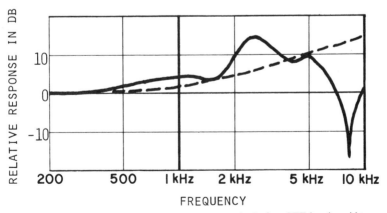

Figure 5.   Comparison between increased response required of an OTE hearing aid to compensate for the loss of external ear resonance and the out-of-ear microphone location. Solid line represents 0° Zwislocki coupler CORFIG; broken line represents difference between real ears and the 2-cc coupler, drawn to the same low frequency asymptote.

The difference between the two curves in Figure 5 represents the 2-cc coupler CORFIG for an OTE hearing aid in a 0° incidence sound field. That difference is shown as a solid curve in Figure 6, a curve sensibly equal to that shown by Cole (1975), Knowles and Burkhard (1975), and Lybarger (1978), except for a 3.5-dB level shift (resulting from our equating the low frequency asymptotes in Figure 5) to emphasize the shape of the difference curve. An OTE hearing aid whose frequency response matched the solid curve of Figure 6 when measured in accordance with ANSI-S3.22 (1976) or IEC-118 (1959) should produce a flat insertion gain frequency-response curve when measured on a KEMAR manikin in a 0° incidence sound field, provided that: 1) the same earmold configuration is used on the KEMAR manikin as is used for the 2-cc coupler measurements, and 2) the OTE hearing aid has the same microphone location used to obtain these COR-FIG curves. The former is the more important provision, and is discussed in the following section.

Effect of Earmold

The earmold simulator specified for OTE hearing aids in the ANSI and IEC standards comprises 25 mm of 2-mm diameter tubing followed by 18 mm of a 3-mm diameter hole leading into the 2-cc coupler volume. This particular earmold-coupler combination is actually one of several so-called "2-cc couplers." In the United States, it is more properly called an "HA-2 earphone coupler with entrance through a rigid tube" (ANSI-S3.7, 1973); in Europe, it is called a "2 cm³ coupler with earmold substitute" (IEC-126, 1973).

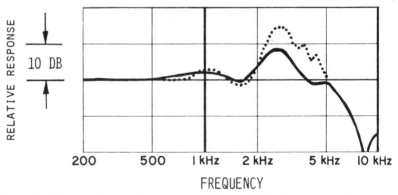

Figure 6.   Difference between the two curves of Figure 5. Solid line represents 2-cc coupler CORFIG for an OTE hearing aid in a 0° incidence field. The dotted line represents the additional high frequency gain required to offset the use of a "conventional" earmold rather than the "HA-2" configuration for one hearing aid; namely, the one illustrated in Figure 5b of Beck (1978).

The typical earmold supplied to a hearing aid purchaser, however, has the flexible tubing extending nearly to the tip of the earmold. The loss of high frequency response resulting from such a constant bore earmold construction was shown by Lybarger (1972) and by Dalsgaard and Jensen (1976). The dotted curve in Figure 6 illustrates, for one particular hearing aid, the additional high frequency gain required to offset the use of a conventional earmold (1.9 mm tubing extending nearly to the tip of the earmold) rather than the "HA-2" configuration. The amount of additional gain required depends on the particular hearing aid design (compare, for example, Figures 5a and 5b of Beck, 1978). Nonetheless, the dotted curve of Figure 6 is similar to the difference between functional gain and coupler gain obtained by Pascoe (1975, Figure 9), who measured functional gain with the subject's custom earmolds and 2-cc coupler gain with the HA-2 coupler, and to the differences between insertion gain and coupler gain obtained by Dalsgaard and Jensen (1976, Figures 8 and 9) under similar measuring conditions.

With broadband earphones and/or the use of earmold venting, the variation in hearing aid frequency response brought about by changes in earmold construction can be much larger than the net correction shown by the solid curve in Figure 6. Stated differently, the error incurred by ignoring the CORFIG corrections may be small compared to the error incurred by ignoring variations in earmold constructions. (Indeed, it is possible to use a large-tube earmold whose acoustics provide a rough "built-in correction" to the solid CORFIG curve of Figure 6; i.e., whose transfer characteristic, when compared with an HA-2 earmold, provides the required boost at 2.7 kHz). These comments are not meant to minimize the importance of the CORFIG determinations, but to reemphasize the large error that can occur when earmold construction is not well defined and controlled. (Extensive discussions of earmold acoustics are found in other chapters of this volume.)

Effect of Individual Differences

All the data discussed so far are average data as represented by the acoustic characteristics of the KEMAR manikin measurements. If a separate hearing aid were to be designed for each user, it would presumably be possible to take into account individual eccentricities in external ear ("ear canal") resonances and eardrum impedances. To be economically practical, however, most hearing aid designs must be based on average data. Under those circumstances, individual variations in external ear resonance and eardrum impedance may cause the (insertion) gain and the (insertion gain) frequency response of a hearing aid to deviate substantially, for a given individual, from the design nominals. This comes about for two reasons: individual differences in external ear resonances and individual differences in ear canal and eardrum impedances.

External Ear Resonances   An estimate of the individual differences in external ear resonance was provided in the data of Filler, Ross, and Wiener (1945). In that report, individual sound field-to-eardrum pressure curves for 12 male and 2 female subjects were given, which can be compared with the overall average curves for the same subjects shown by Wiener and Ross (1946). The standard deviation (from the average curve) of the individual curves ranged from 1 to 2 dB below 1800 Hz up to from 4 to 7 dB in the 5–8 kHz region, with peaks at 2.1 and 3.3 kHz. The peak deviations occurred mostly because individual external ear resonance frequencies were lower or higher than the average. No individual eardrum pressure curve deviated more than 7.5 dB from the average curve below 5 kHz, but the majority deviated by at least 5 dB at some frequency below 5 kHz.

Eardrum Impedance   As part of a study leading to a validation of the modified Zwislocki coupler, Sachs and Burkhard (1972a) reported the probe-tube microphone measurement of the sound pressures developed in 11 occluded ears (6 male and 5 female) by subminiature hearing aid earphones. The standard deviation of the pressure developed ranged from approximately 1 dB at 1 kHz to 5 dB in the 6- to 8-kHz region. Greater pressures were developed in female ears (by 3 to 5 dB at the higher frequencies).

The Net Result   Although the variations in outer ear resonance and eardrum impedance are only partially independent variables, it is clear that no hearing aid designed for the average ear can be expected to produce an insertion gain, in the majority of individual cases, that does not exhibit one deviation of perhaps 7 dB at some frequency. The subjective importance of such deviations to a long-term hearing aid wearer is not known, but even larger deviations in unaided frequency response can occur because of the accumulation of earwax in the canal, deviations that occur with such a gradual onset that they often go unnoticed by the sufferer until the canal becomes almost completely blocked. Thus, in most cases, it seems reasonable to assume that satisfactory adaptation to a slightly inaccurate insertion gain frequency response makes it unnecessary to provide modification for individual eccentricities. Experimental evidence one way or the other is lacking.

HISTORY

Coupler Corrections

The 2-cc coupler was first described by Romanow (1942), who emphasized that it was not a real-ear simulator, but simply a convenient, easily fabricated coupler into which readily reproducible hearing aid measurements could be made. Romanow provided tentative correction curves, to be applied to the 2-cc coupler response curves for body-worn hearing aids, in order to estimate the field-referenced response of the hearing aid.

In 1944, LeBel reported a series of studies in which he obtained 2-cc coupler correction curves similar to those described by Romanow. To illustrate the importance of the corrections, LeBel showed a comparison between two frequency responses: one uncorrected, which looked fairly flat, and one corrected, which showed a 15-dB high frequency loss.

Knowledgeable hearing aid designers have been using these corrections for years. Lybarger described one such hearing aid design in 1947, for example.

With head-worn hearing aids a totally new set of corrections was required, since the hearing aid microphone was now located on the head instead of the chest. Thus in 1959 Knowles described a tentative set of coupler correction curves, based on the data provided by Wiener and Ross (1946) on the (unaided) eardrum pressure produced by a progressive sound field. The refined estimate of the difference between normal eardrum pressure and the pressure available to the microphone of an ITE hearing aid, as given by Knowles in 1967, is essentially similar to the best estimates of today. It was reproduced by Killion and Carlson (1970).

The remaining question was the difference between the eardrum pressures produced in real ears and in the 2-cc coupler. Although several laboratories had reported ear canal plus eardrum impedance values substantially higher than those presented by the 2-cc coupler, no alternate coupler had gained general acceptance until Zwislocki (1970) reviewed the problem and developed a realistic ear simulator. The suitability of this ear simulator (coupler) was verified by Sachs and Burkhard (1972a), who determined the eardrum pressure developed by insert receivers on 11 real ears. A slightly modified version of the original Zwislocki coupler (IRPI DB-100) was described by Sachs and Burkhard (1972b). Most of the modifications were for fabrication convenience, although a small change from Zwislocki's original design was made in one of the four acoustic branches, a change which smoothed the transfer impedance in the 500- to 1000-Hz region by approximately 1 dB.

In 1972 the KEMAR manikin (Burkhard and Sachs, 1975) was introduced, and it became practical to obtain a physical measurement of the insertion gain of hearing aids on a routine basis.

Hearing Aid Research

Before leaving the history of coupler correction curves, it is interesting to note that although those individuals involved with the standardization of hearing aids were acutely aware of the need for such corrections to obtain the true frequency response of a hearing aid as perceived by a wearer, these corrections were almost universally ignored in audiologic research. This was true even when the effect of the frequency response of a hearing aid on aided performance was the experimental question. Figure 7, for example, shows the effective or "orthotelephonic" frequency response calculated by Fletcher (1953) for the master hearing aid used in the Harvard study (Davis et al., 1947). It is evident that the frequency response that was labeled "flat" was not what one would normally consider a flat frequency response. Similarly, it is perhaps not surprising that the frequency response labeled "high-pass 6" gave almost universally better results. (The "high-pass 6" response was not flat either; but except for the peak at 7 kHz it looks like a perfectly sensible real-ear response for a hearing aid.) Lybarger (1978) gives an expanded discussion of these issues.

The first systematic attempt to relate the speech discrimination obtained with hearing aids to their true in situ frequency response was reported by Fournier (1965), who used free field Békésy audiometry to obtain what we now call the functional gain of the hearing aids under evaluation. The functional gain of the hearing aid was obtained from the difference between the aided and unaided threshold tracings for each subject. (This same technique was used a few years later in the United States by Green (1969) to obtain the functional gain frequency response provided by open canal hearing aid fittings.)

Not until the early 1970s did comprehensive research studies begin to appear in which the experimental design took into account the real-ear frequency response of the hearing aid. Excellent studies were conducted by Villchur (1973), Pascoe (1975), Skinner (1976), and Lippman (1978).

An Aside on Terminology

In 1975, Mahlon Burkhard and I became dissatisfied with the term *orthotelephonic* for reasons discussed by Burkhard (1978, p. 17). We were looking for a term that could be unambiguously defined to mean the objective measurement (with a probe-tube microphone at the eardrum, for example) of the in situ gain of a hearing aid. Professor R.V. Schoder, a classics scholar at Loyola University, coined the term *etymotic* (pronounced et-im-oh'-tik; literally, real ear), which he assured us met all the rules for coining "new ancient Greek words."

At about the same time, Dalsgaard (1974) had begun using the term *insertion gain,* a well-defined engineering term that appears to have been first applied to hearing aids by Ayers (1953). Both terms have the same meaning.

The term *insertion gain*—which is self-explanatory—appears to be gaining favor over the term *etymotic* in the United States. Both terms are used in Europe.

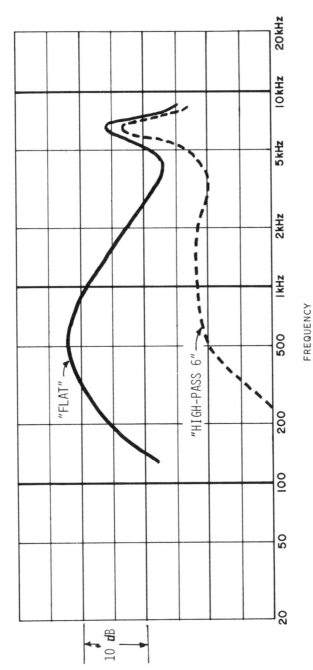

FREQUENCY

Figure 7. Real-ear (orthotelephonic) response of master hearing aid used in the Harvard study. (After Fletcher, 1953.)

165

SUMMARY

As stated earlier, the relationship between the coupler response of a hearing aid and its in situ response can be reasonably well specified on the average, but such a specification must include the following information:

1. The type of sound field
2. The orientation of the listener in the sound field
3. The location of the microphone on the listener's head (or body)
4. The exact construction of the earmold used to measure the in situ response
5. The exact construction of the earmold used to measure the coupler response
6. The type of coupler used to obtain the coupler response

When the bandwidth of the hearing aid is limited, the first three factors may be safely ignored for most practical cases, but they become increasingly important as the hearing aid response is extended above 4 kHz. In the latter case, the use of random-incidence sound field measurements has some appeal.

The large number of variables that must be specified before reproducible CORFIG determinations can be made indicates that the choice of standardized coupler corrections may be a complicated business.

REFERENCES

American National Standards Institute. 1973. Method for Coupler Calibration of Earphones, ANSI-S3.7-1973. American National Standards Institute, New York.
American National Standards Institute. 1976. Specification of Hearing Aid Characteristics, ANSI-S3.22-1976. American National Standards Institute, New York.
Ayers, E.W. 1953. A discussion of some problems involved in deriving objective performance criteria for a wearable aid from clinical measurements with laboratory apparatus. Proceedings of the First ICA Congress, Delft, pp. 141–143. W.D. Meinema, Delft.
Beck, L.B. (ed.). 1978. Handbook of Hearing Aid Measurement 1978. Veterans Administration, Washington, D.C.
Berland, O., and E. Nielsen. 1969. Sound pressure generated in the human external ear by a free sound field. Audecibel 18:103–109.
Burkhard, M.D. (ed.). 1978. Manikin Measurements—Conference Proceedings. Industrial Research Products, Elk Grove Village, Ill.
Burkhard, M.D., and R.M. Sachs. 1975. Anthropometric manikin for acoustic research. J. Acoust. Soc. Am. 58:214–222.
Burkhard, M.D., and R.M. Sachs. 1977. Sound pressure in insert earphone couplers and real ears. J. Speech Hear. Res. 20:799–807.
Butler, R.A., and K. Belendivk. 1977. Spectral cues utilized in the localization of sound in the median sagittal plane. J. Acoust. Soc. Am. 61:1264–1269.
Causey, G.D., and L.B. Beck. 1977. The relation of measurements on KEMAR to

behavioral performance. Paper presented at the annual convention of the American Speech and Hearing Association, November 2–5, Chicago.

Cole, W.A. 1975. Hearing aid gain—A functional approach. Hear. Instr. 26:22–24.

Dalsgaard, S.C. 1971. On the frequency response of hearing aids. In Proceedings of the 7th International Congress on Acoustics, pp. 477–480. Akademiai Kiado, Budapest.

Dalsgaard, S.C. 1974. Measurement of insertion gain of hearing aids. Proceedings of the Eighth International Congress on Acoustics, London, p. 205. Goldcrest Press, Trowbridge.

Dalsgaard, S.C., and O.D. Jensen. 1976. Measurement of the insertion gain of hearing aids. J. Audiol. Tech. 15:170–183.

Davis, H., S.S. Stevens, R.H. Nichols, C.V. Hudgins, R.J. Marquis, G.E. Peterson, and D.A. Ross. 1947. Hearing Aids: An Experimental Study of Design Objectives. Harvard University Press, Cambridge.

Filler, A.S., D.A. Ross, and F.M. Wiener. 1945. The pressure distribution in the auditory canal in a progressive sound field. Harvard University Report PNR-5 (Contract N5ori-76) to Office of Research and Inventions, U.S. Navy.

Fletcher, H. 1953. Speech and Hearing in Communication. D. Van Nostrand Co., New York.

Fournier, J.E. 1965. La correction prothétique des curdités; problèmes techniques et medico-sociaux. Pract. Oto-Rhino-Laryngol. 27:180–198.

Fournier, J.E. 1968. Hearing aid evaluation by pure tone automatic audiometry. Audecibel 17:99–108.

Green, D.A. 1969. Non-occluding earmolds with CROS and IROS hearing aids. Arch Otolaryngol. 89:96–106.

IEC-118. 1959. Recommended methods for measurements of the electro-acoustical characteristics of hearing aids. Bureau Centre de la Communications Electrotechnques Internationales, Geneva, Switzerland.

IEC-126. 1973. IEC reference coupler for the measurement of hearing aids using earphones coupled to the ear by means of ear inserts. Bureau Centre de la Communications Electrotechniques Internationales, Geneva, Switzerland.

Killion, M.C. 1976a. Noise of ears and microphones. J. Acoust. Soc. Am. 59: 424–433.

Killion, M.C. 1976b. Experimental wide-band hearing aid. J. Acoust. Soc. Am. 59 (suppl. 1):S62(A). (Available from Knowles Electronics, Franklin Park, Ill.)

Killion, M.C. 1979. Equalization filter for eardrum-pressure recording using a KEMAR manikin. J. Audio Eng. Soc. 27:13–16.

Killion, M.C., and E.V. Carlson. 1970. A wideband miniature microphone. J. Audio Eng. Soc. 18:631–635.

Knowles, H.S. 1959. Physical aspects of hearing aids. Paper presented at Allerton House, University of Illinois, Urbana.

Knowles, H.S. 1967. Some problems in head-worn aid response measurements. Paper presented at the 33rd Convention of the Audio Engineering Society, October 16, New York.

Knowles, H.S., and M.D. Burkhard. 1975. Hearing aids on KEMAR. Hear. Instr. 26:19–21.

LeBel, C.J. 1944. Pressure and field response of the ear in hearing aid performance determination. J. Acoust. Soc. Am. 16:63–67.

Lippman, R.P. 1978. The effect of amplitude compression on the intelligibility of speech for persons with sensorineural hearing loss. Doctoral thesis, Massachusetts Institute of Technology, Cambridge. (Available from University Microfilms, Ann Arbor, Mich.)

Lybarger, S.F. 1947. Development of a new hearing aid with magnetic microphone. Electric. Manufact. 40:104–108, 162.

Lybarger, S.F. 1972. Ear molds. In J. Katz (ed.), Handbook of Clinical Audiology, pp. 602–623. The Williams & Wilkins Co., Baltimore.

Lybarger, S.F. 1978. Selective amplification—A review and evaluation. J. Am. Audiol. Soc. 3:258–266.

Madaffari, P.L. 1974. Pressure variation about the ear. J. Acoust. Soc. Am. 56:S3(A). (Available from Knowles Electronics, Franklin Park, Ill.)

Olson, H.F. 1967. Music, Physics and Engineering. Dover Publications, New York.

Orchik, D.J., and N.L. Mosher. 1975. Narrow band noise audiometry: The effect of filter slope. J. Am. Audiol. Soc. 1:50–53.

Pascoe, D.P. 1975. Frequency responses of hearing aids and their effects on the speech perception of hearing impaired subjects. Ann. Otol. Rhinol. Laryngol. 84 (suppl. 23):1–40.

Romanow, F.F. 1942. Methods for measuring the performance of hearing aids. J. Acoust. Soc. Am. 13:294–304.

Sachs, R.M., and M.D. Burkhard. 1972a. Earphone pressure response in ears and couplers. J. Acoust. Soc. Am. 51:140(A). (Available from Knowles Electronics, Franklin Park, Ill.)

Sachs, R.M., and M.D. Burkhard. 1972b. Zwislocki coupler evaluation with insert earphones. Industrial Research Products Report No. 20022-1 to Knowles Electronics, Franklin Park, Ill.

Schulein, R.B. 1975. In situ measurement and equalization of sound reproduction systems. J. Audio Eng. Soc. 23:178–186.

Shaw, E.A.G. 1976. Diffuse field sensitivity of external ear based on reciprocity principles. J. Acoust. Soc. Am. 60:S102(A).

Skinner, M.W. 1976. Speech intelligibility in noise-induced hearing loss: Effects of high-frequency compensation. Doctoral thesis, Washington University, St. Louis. (Available from University Microfilms, Ann Arbor, Mich.)

Villchur, E. 1973. Signal processing to improve speech intelligibility in perceptive deafness. J. Acoust. Soc. Am. 53:1646–1657.

Walden, B.E., and R.N. Kasten. 1976. Threshold improvement and acoustic gain with hearing aids. Audiology 15:413–420.

Wiener, F.M., and D.A. Ross. 1946. The pressure distribution in the auditory canal in a progressive sound field. J. Acoust. Soc. Am. 18:401–408.

Zwislocki, J.J. 1970. An acoustic coupler for earphone calibration. Report LSC-S-7, Laboratory of Sensory Communication, Syracuse University, Syracuse, N.Y.

CHAPTER 9

PROBLEMS IN THE RECORDING AND REPRODUCTION OF HEARING AID-PROCESSED SIGNALS

Robyn M. Cox and Gerald A. Studebaker

CONTENTS

Since the late 1950s, a large portion of research concerning hearing aid performance on human subjects has been carried out using a technique that is commonly referred to as *hearing aid processing*. Mainly because of our interest in alternative techniques for the selection of hearing aids, we undertook a detailed analysis of hearing aid processing with the aim of developing a protocol that was as valid and accurate as possible.

The essential features of hearing aid processing schemes are that the test signals are first presented to and processed by the hearing aid(s) under investigation and the hearing aid's output is recorded and stored. Subsequently, this recorded signal is presented to the test subjects. The technique has the advantage of each test subject receiving nearly the same signal, unaffected by variations in position and orientation of the subject's head, hearing aid control settings, battery voltage, or the possible deterioration of the hearing aid's performance over time. Also, use of prerecorded signals greatly facilitates the implementation of techniques like the paired comparison technique explored by Zerlin in 1962 and more recently by other

The work reported in this chapter was supported in part by Public Health Services Grants #NS 13514 and NS 12588 from the National Institute of Neurological and Communicative Disorders and Stroke.

investigators (Studebaker, White, and Hoffnung, 1978; Witter and Gold-stein, 1971). Hearing aid processing was essential to the development of a paired comparison hearing aid selection procedure in which we were interested.

A review of previously used processing techniques revealed that the procedure was often carried out as illustrated in Figure 1. The hearing aid was placed alone before a loudspeaker in some environment, such as an anechoic room (Zerlin, 1962) or a hearing aid test box. The output of the hearing aid was directed to a standard HA-2 coupler, the acoustic signal was transduced by a microphone and recorded on magnetic tape. Subsequently, the recorded signals were played back to each subject, typically through TDH-39 earphones mounted in MX-41/AR supraaural cushions (Jerger, Speaks, and Malmquist, 1966; Zerlin, 1962). Some investigators used earphones mounted in circumaural cushions (Witter and Goldstein, 1971).

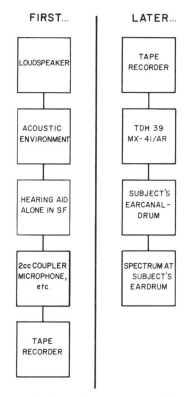

Figure 1.   Essential features of a frequently used procedure for production of hearing aid-processed signals. Each block represents an element that potentially modifies the signal spectrum, which finally reaches the subject's eardrum. (S.F. = sound field.)

In the context of the emerging awareness of the importance and nature of the acoustic factors affecting hearing aid performance, it was apparent that these recording and playback techniques were inadequate. First, head shadow and baffle effects were not allowed to occur. Second, the HA-2 coupler has a frequency response different from that of the real ear at higher frequencies and has less acoustic damping than the real ear. Third, playback via a TDH-39 earphone resulted in signal modification by the frequency response of the earphone. Fourth, use of a supraaural earphone cushion resulted in significant acoustic leakage in the low frequencies, which was highly variable across subjects, and even within a subject, depending on the tension of the headband. Finally, and most important, the resonant behavior of the ear canal and auricle enclosed under an earphone has been shown to be very significantly different from the analogous behavior of the same ear canal occluded by an earmold (Larson, Studebaker, and Cox, 1977; Shaw, 1966; Villchur, 1969). An investigation of these effects was published in 1977 (Cox and Studebaker, 1977).

Of course, not all of these factors necessarily influenced the validity of past research utilizing this hearing aid processing procedure. Since most of the effects were held relatively constant across conditions within a given investigation, the result was a relatively constant modification in the spectrum of the test signal. In studies like those on the effects of distortion or on comparing the results of quality judgments with speech discrimination performance, these spectral modifications would probably have little effect on the relative position of the conditions under investigation. However, the substantial inter- and intrasubject variability associated with low frequency acoustic leakage under supraaural cushions might be expected to blur the data, particularly in cases where hearing aid-processed signals were used to study effects of frequency response. Thus, while the drawbacks of this hearing aid processing scheme do not necessarily invalidate the research of the past, they probably made it more difficult to see clearly the effects of the factors under investigation and, in some cases at least, may have had a significant effect on the main effects under investigation as well.

Much current work suggests that frequency response is one of the most important attributes of hearing aid performance. Since the frequency response of the processed signal is substantially affected by the record-playback system utilized, we believed it was necessary to gain much better control over every phase of the procedure than had been heretofore achieved.

THE GOAL

The goal established for the system was to present to the individual subject's eardrum a signal as spectrally similar as possible to the signal that would have occurred at his eardrum had he actually worn the hearing aid in the same environment in which it was recorded. In Cox and Studebaker (1979), the problem was conceptualized as shown in Figures 2 and 3. Figure 2 shows the essential features of the situation in which an individual wears a hearing aid. (This is referred to as the "direct" condition.) A signal source (such as a loudspeaker) produces an acoustic signal, which is modified by the environment. The hearing aid is located on a subject's head and the signal at the hearing aid microphone is further modified by the presence of the head. After processing by the hearing aid's electroacoustic system, the signal is delivered to the subject's eardrum via a length of tubing, an earmold, and the subject's ear canal. The signal is further modified by each of these elements and by the eardrum's impedance.

Figure 3 shows our conceptualization of one possible hearing aid-processing paradigm. In this case, the hearing aid is placed on a manikin's head in the selected environment. The hearing aid's output in a coupler is recorded and subsequently played back to the subject via a transducer. We used a transducer that permitted the subject to use an individually fabricated standard, vented, or open earmold. Note the dashed rectangle on the

Figure 2.   Essential features of a procedure in which each hearing aid is actually worn by the subject. Each block represents an element that potentially modifies the signal spectrum, which finally reaches the subject's eardrum. (Reprinted with permission from R.M. Cox and G.A. Studebaker, 1979, Hearing-aid–processed signals: A new approach, *Audiology* 18:53–71.

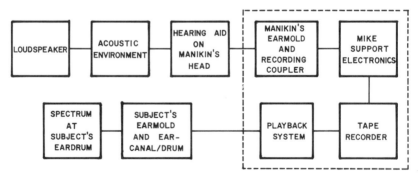

Figure 3.   Essential features conceptualized for a new procedure for production of hearing aid-processed signals. Each block represents an element that potentially modifies the signal spectrum, that finally reaches the subject's eardrum. (Reprinted with permission from R.M. Cox and G.A. Studebaker, 1979, Hearing-aid–processed signals: A new approach, *Audiology* 18:53–71.

right side of the figure. If the goal is to present a signal to the subject that has a frequency response essentially equal to the signal the subject would have received had he actually worn the hearing aid in that environment, then the elements within the dashed rectangle must have no net effect on the frequency response.

IMPLEMENTATION

In order to implement a valid hearing aid processing paradigm, most of the elements shown in Figure 3 require individual and careful consideration. The issues and problems associated with each element are discussed approximately in the order in which they appear in the figure.

The Source

The signal source to date has been a 12-inch coaxial loudspeaker mounted at a height equal to the height of the manikin's ears. However, it should be noted that a loudspeaker of this size is a signal source with dispersion characteristics different from those of a human voice emerging from a human mouth. It is uncertain at this time whether this difference in dispersion characteristic would have a measurable effect on hearing aid performance, but the matter should be investigated. A device like the Field Artificial Voice proposed by Olsen (1972) would appear to be a suitable starting point in that it is relatively simple and is reported to have the desired dispersion characteristics.

The Environment

The nature of the environment in which the recordings are made is another important consideration. Nábělek (Chapter 2, this volume) and Berkley (Chapter 1, this volume) indicated that the normal range of room acoustic effects probably has little influence on the intelligibility of speech heard by normal hearers, but there may be significant effects in the case of hearing-impaired persons listening to speech through hearing aids. Our own work indicates that the environment has a significant effect, at least on the directivity of hearing aids with directional microphones. Although the directivity of nondirectional head-worn aids appears to be much less affected by environment, there are differences in head diffraction effects observed in an anechoic room compared with other, more reverberant rooms. These results were interpreted by us as indicating that hearing aid-processed signals should not be recorded in anechoic rooms when realistic hearing aid performance is an important goal.

The task of defining a "realistic" environment in which to record hearing aid-processed signals is a difficult one. It is uncertain whether the goal should be to simulate a very desirable but realistic listening room or an average listening room, even though that room may be less desirable from a performance viewpoint. At this time there is no entirely satisfactory way to specify the quality of a room for the purposes of communication by the hearing impaired or for the purposes of recording hearing aid-processed signals. However, it seems likely that such a room could be specified using only a few measures. Recent work on the modulation transfer function by Houtgast and Steeneken (1978) looks particularly promising as a functional measure of room acoustics. Measures like reverberation time and the physical dimensions and surface features of the space are obvious possibilities. Other calculation techniques, such as the geometrical acoustic methods of Lochner and Burger (1960, 1961) and the ray and imaging techniques reviewed by Santon in 1976, may also be applied in this area.

Finally, it seems necessary for the recording environment to have an ambient noise level no greater than the equivalent input noise of the typical hearing aid. Most prefabricated sound-treated rooms can meet this requirement within the hearing aid's passband. However, these rooms do not provide a listening situation that is at all realistic in other regards.

Location of the Hearing Aid

The location of the hearing aid should be the same as it would be in actual use. If a manikin is used and if realistic hearing aid performance for the individual subject is our goal, then two factors must be considered. First, does the manikin adequately represent the average head shadow and baffle effects and, second, what is the range of individual variations in these effects? In other words, how well does the manikin represent the range of individual subjects? If intersubject variation is large, then even though the manikin might represent the average person quite well, individual subjects might not be well represented. If, on the other hand, the intersubject variability is small or the effects are not functionally significant, the manikin can be used to represent the heads and bodies of all subjects without further compensation or correction.

In this recording scheme the manikin KEMAR was utilized. After making many comparisons between the results from groups of real human subjects and those from KEMAR, we concluded that KEMAR served the purpose very well under all of the conditions we tested. Figure 4 shows an example of these results.

The data shown in Figure 4 were obtained using a small microphone placed in a forward-facing position inside the case of an ear-level hearing aid. The hearing aid case was placed on KEMAR in the normal way and a broadband signal was presented via a loudspeaker at 0° azimuth. The output of the microphone was amplified and analyzed spectrally. The proce-

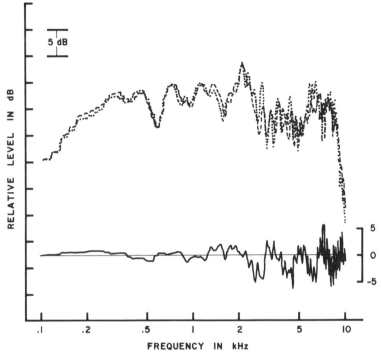

Figure 4. Signal spectrum occurring at a front-facing hearing aid microphone, with the hearing aid worn by KEMAR (dotted line) compared with the mean signal observed with the hearing aid worn by eight adult subjects (broken line). The solid line shows the eight-subject mean data plotted relative to the KEMAR data.

dure was repeated with eight adult subjects. The figure shows the result measured on KEMAR (dotted line) and the mean result for the eight subjects (broken line). The solid line at the bottom shows the difference between the two top curves. This line reveals that the results from KEMAR and the mean of eight subjects did not differ by more than about 3 dB, except in a few narrow frequency regions, up to 5 kHz. Slightly greater differences are shown between 5 and 10 kHz. Since hearing aid output is generally confined to frequencies below 5 kHz, it was concluded from these and other similar data that KEMAR serves very well as an average subject. A similar result was obtained with several complete hearing aids at each of the four cardinal azimuths.

Figure 5 shows the range of the results for the eight individual subjects. These data reflect on intersubject variability. The individual curves were so irregular and so close together that it was impossible to depict them separately. The largest total range below 5 kHz was 12 dB at 4400 Hz. Below

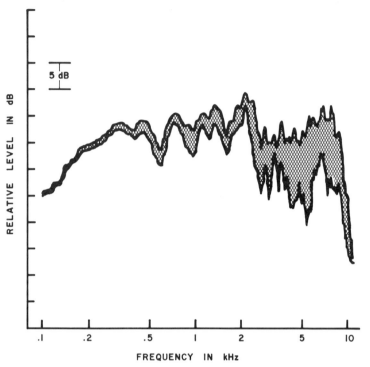

Figure 5. Range of results for the eight adult subjects for whom the mean is shown in Figure 4.

2000 Hz, the largest total range was 4.5 dB. These data indicate that head shadow and baffle effects for the individual subject were not more than a few dB different from the analogous effects on KEMAR at frequencies below 2000 Hz. Between 2 and 5 kHz somewhat larger differences were observed, although most subjects fell within ± 5 dB. Of course, one might anticipate that an occasional subject would show more extreme differences, but no extreme differences were observed in this group. The functional significance, if any, of differences of this magnitude has yet to be determined, but it seems unlikely that they would have major consequences.

Record-Playback

Consider the elements enclosed in the dashed rectangle in Figure 3. The fundamental items here are a recording coupler and associated microphone, a tape recorder, and a playback system. Recall that it is necessary for these items to have no net effect on the signal. Ordinarily, any properly performing tape recorder of reasonable quality will impose no frequency-response shaping on the signal. Hence, the recording coupler and the playback system were the items of concern.

Weighting   There are a number of approaches to this problem, but all involve the use of one or more weighting networks designed to produce a frequency response that is the reciprocal of the net effect of the components located in the rectangle. It is not necessary to physically locate the weighting networks adjacent to the components in the dashed rectangle, provided that the net effect is the same. For example, in some applications it may be useful to place a weighting network before the acoustic signal source (loudspeaker) to offset the frequency response effects of elements that follow (Killion, 1979). However, more often the weighting networks will be located in the recording or in the playback circuits, or both.

Although it is of no theoretical significance whether the weighting precedes or follows the device(s) for which compensation is desired, there are practical reasons related to equipment performance that must be considered. For example, an important practical consideration in locating the weighting network is the performance of the recording device, typically a tape recorder. Tape recorders, because of their high frequency preemphasis, which serves to improve the signal-to-noise ratio in the high frequencies, often have less "head room" at these frequencies, and therefore weighting that increases the relative level of the high frequencies should not be placed before the recorder. If, on the other hand, one were to employ a playback system using earphones in supra- or circumaural cushions, a net decrease in high frequency levels might be required to offset the external ear resonances that are excited in the playback condition but that did not occur in the direct condition. In this case the weighting should precede the recorder, either in the record circuit or possibly in some applications, or even as early as the signal source circuit as suggested by Killion (1979).

Coupling System   The signals from the hearing aid were delivered to the recording coupler located within KEMAR's head via a tygon tube that had a length and diameter similar to what would normally be used with that hearing aid. The tubing extended from the aid in its normal position to the medial end of KEMAR's earmold. KEMAR's earmolds, which were made from Adcomold material, had canal sections that terminated at the entrance to the recording coupler. The earmolds were made to fill KEMAR's concha in a manner similar to the way in which a standard earmold fills the concha of a real person, thus providing an acoustic situation similar to that of an actual person wearing a standard earmold.

Recording Coupler   We have used two different recording couplers: a specially made, simple, cylindrical 1.5 cm^3 coupler and a Zwislocki coupler. Both work satisfactorily and each has its advantages and disadvantages. The main issues in choosing a coupler fall into three categories: cost, damping, and system accuracy and complexity.

We originally chose the 1.5-cm^3 cavity as our record coupler. At that time we conceived the problem as one in which a coupler with a flat transfer function over the frequency range of significant hearing aid output was required. Thus, all coupler dimensions were held to a minimum in order to obtain a flat response to as high a frequency as possible. We used the 1.5-cm^3 coupler successfully over a broad range of conditions.

Subsequently, in order to investigate a question concerning coupler damping, which is discussed later, an alternative system was developed using the Zwislocki coupler as the record coupler in association with a weighting network that removed the Zwislocki coupler's frequency response at the time of playback. This resulted in minimal additional complication because a weighting network was needed in any case to compensate for the frequency response of the playback system. In practice, all the necessary frequency compensation was achieved using a single network that corrected for the net effect of the record and playback systems.

Playback System   Playback systems can be divided into two basic types: those using earphones in supra- or circumaural cushions and those using hearing aid receivers and earmolds. Historically, the system using earphones has been used almost exclusively. There are several problems with this approach. Some of these can be dealt with: the effect of leaks on low frequency response can be eliminated through the use of properly designed and fitted earphones; the introduction of external ear resonances that are not present when an actual hearing aid is used may be at least partially compensated for by a properly designed weighting network. However, the intersubject variability of these effects may be large; thus compensation by a network representing the "average" effect may result in considerable inaccuracy for many individuals.

However, even with these solutions to the earphone playback problems, there remains one aspect that cannot be dealt with by earphone playback in any form at this time. This is the effect of individual variations in earmold configuration and ear canal/eardrum impedance when the earmold is worn. We believed that the acoustic characteristics of the earmold that couples the hearing aid to the ear canal are of very considerable importance in determining the spectrum of the signal that impinges upon the eardrum. Hence we wanted a playback system that would be influenced by these factors as a real hearing aid would be influenced by them. Therefore, we designed a playback system that included a hearing aid receiver, tubing, and an individually fabricated earmold.

A primary consideration in the development of the playback system was the selection of an appropriate playback transducer. Originally we considered using an external button-type receiver because of the relative ease of handling, high power output, and broad frequency response offered by this receiver configuration. In particular, the Western Electric-Audivox 9C receiver seemed promising because of its very wide frequency response. However, we discovered that these button-type receivers, including the 9C, interacted with earmold vents quite differently from the internal receivers found in ear-level hearing aids. Therefore, we decided to try using an internal-type receiver as the playback transducer. We discovered that the state of the art in internal receivers is such that an adequately wide frequency response can be obtained without too much difficulty (see Killion, Chapter 11, this volume). In addition, data obtained with a variety of internal-type receivers of various ages and designs revealed that when they were loaded by a typical length and diameter of tygon tubing, there were virtually no interactions between receivers and earmold vents or other earmold modifications.

An Example   Figure 6 shows the general design of the acoustic portions of one playback system. The internal-type transducer we selected was a Knowles Electronics 1712 receiver, which has a broad frequency response. The output of the receiver is fed into a two-section length of damped tygon tubing. One example of tubing lengths, diameters, and damping elements is shown here. Other arrangements have been used with equal success. The goal in adjusting this tubing arrangement is to use the damping elements to make the output of the tube as smooth and broad as is possible. Then, with the tube coupled to the cavity, which will subsequently be used as the recording coupler, the weighting network is adjusted to produce a flat spectrum in that coupler. This procedure provides the desired result whether the recording coupler is a simple cavity, or a Zwislocki coupler, or any other design.

A weighting network that has proved to be most useful and is relatively easy to use is a form of the bridged "T." An example that we have used is shown in Figure 7. This one is a bandpass design, which produced a peak in the area of 4500 Hz. Killion (1979) used a band reject version of the bridged "T" to produce the inverse of the diffuse field-to-eardrum transfer

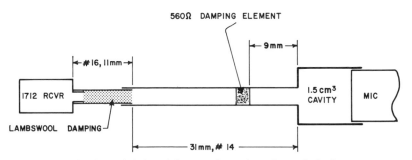

Figure 6.   General design of the acoustic portions of one playback system.

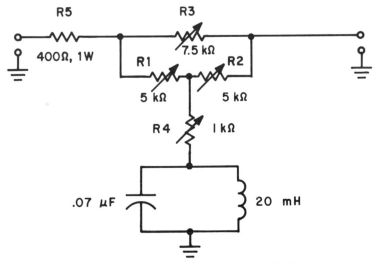

Figure 7.   Schematic drawing of a bridged "T" weighting network, which was used with the acoustic system shown in Figure 6.

function. The bridged "T" design is very flexible and useful in work of this nature.

Figure 8 shows the frequency response of the record/playback system that used the 1.5-cm³ cavity as the recording coupler and the bridged "T" shown in Figure 7. This result is flat ±1 dB from 150 Hz to 5200 Hz. Figure 9 shows the frequency response of a system using a Zwislocki coupler as the recording coupler. In this case the response is flat within ±1.5 dB from below 100 Hz to about 7800 Hz.

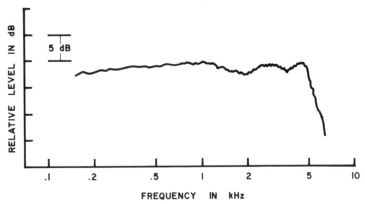

Figure 8.   Frequency response of a record-playback system utilizing the 1.5-cm³ cavity as the recording coupler.

TESTS OF THE SYSTEM

Sealed Earmolds

We have collected considerable data that bear upon how well these systems meet the goal of presenting to the subject's eardrum a signal that is very similar to the one he would have received if he had actually worn the recorded hearing aid under the stated conditions. All the data were obtained using a broadband, thermal noise as the test signal and a spectrum analyzer to derive the signal spectrum at the measurement point.

Our earliest data were obtained using KEMAR as the ever-present and cooperative subject and standard, unvented earmolds. In this case, the "direct" condition is one in which the hearing aid was placed on KEMAR, and its output was fed to a Zwislocki coupler that served as KEMAR's ear canal/eardrum. In the record-playback condition, the signal is prerecorded using one of the systems described earlier, and then played back to the same Zwislocki coupler in KEMAR.

Figure 10 shows a typical example of the data obtained with the record-playback system incorporating the 1.5-cm^3 cavity. The solid line is the spectrum measured at KEMAR's "eardrum" in the direct condition, and the dotted line shows the analogous spectrum measured in the record-playback condition. The two curves are within ± 2 dB below 5000 Hz and the played-back signal corresponds to the direct signal

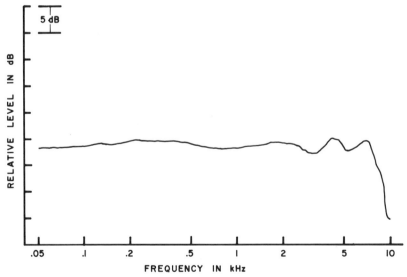

Figure 9.   Frequency response of a record-playback system utilizing a Zwislocki coupler as the recording coupler.

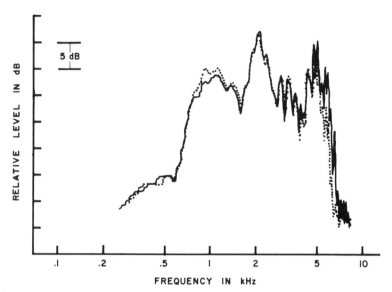

Figure 10.   Hearing aid output spectra obtained at KEMAR's "eardrum" position in the direct condition (solid line) and via a record-playback system using a 1.5-cm³ recording coupler (dotted line) with a standard earmold fitting.

within ± 1 dB in all frequency regions, except at the peak around 1000 Hz. The irregular nature of the frequency response at the high frequencies was produced by standing waves in the environment, which in this case was an audiometric test room. The played-back signal reproduced these effects exactly. Above 5000 Hz the level of the played-back signal drops off rapidly relative to the direct signal. This particular hearing aid was unusual in that it had a significant output above 5000 Hz. For such a hearing aid a record-playback system that does not extend above 5000 Hz is clearly inadequate.

The discrepancy between the two lines in the region of 1000 Hz seen in Figure 10 was seen also in preliminary work using hearing aid receivers alone instead of complete hearing aids. This discrepancy showed up persistently in the region of the hearing aid receiver's so-called "primary peak" —a peak that is produced principally as a reactance resonance between the compliance of the receiver's diaphragm and the inertance of the diaphragm and the tubing system. It was hypothesized that the discrepancy occurred because this resonance reached a higher level in the undamped 1.5-cm³ cavity than it did in the more realistically damped Zwislocki coupler. Thus, a larger peak was produced in the 1.5-cm³ cavity and preserved in the record-playback process than was observed in the direct condition in the Zwislocki coupler.

In order to test this hypothesis, the record-playback system using the Zwislocki coupler as the recording coupler was developed. A typical result using the Zwislocki coupler-based system is shown in Figure 11. Note the pair of lines at the bottom of the figure, which represents the hearing aid's output spectrum in the direct and the processed conditions. The correspondence between the two lines is within ± 1.5 dB across the entire frequency range from 200 to 7000 Hz. There are two regions of improved agreement with this system, as compared with that in Figure 10. First, there is improved agreement in the high frequencies, reflecting the extended high frequency response of this record-playback system, which extends to about 7800 Hz (see Figure 9). Second, the discrepancy noted in the region of the receiver's primary peak in Figure 10 is not seen in Figure 11. The difference between the direct and processed signals is shown as the solid line at the

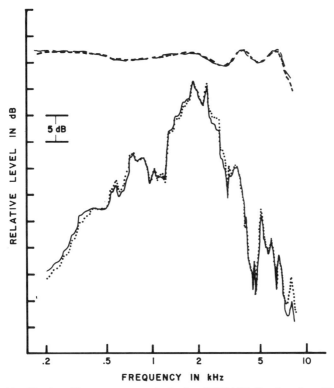

Figure 11.   Hearing aid output spectra obtained at KEMAR's "eardrum" position in the direct condition (dotted line) and via a record-playback system using a Zwislocki coupler as the recording coupler (solid line) with a standard earmold fitting. The two data lines at the top of the figure show the frequency response of the record-playback system (– – –) compared with the difference between the two lower lines (———).

top of the figure. Also shown, superimposed on this line, is the frequency response of the record-playback system (from Figure 9). The close agreement between these two results means that the differences between the direct and processed output curves at the bottom of the figure are entirely attributable to the frequency response of the record-playback system. Thus, damping difference effects do not appear to be playing a role with this Zwislocki coupler-based record-playback system.

Logically, the next step was to obtain similar data using real subjects instead of KEMAR. If the location of the subject in the sound field can be controlled adequately, and if the record-playback system is accurate, then the differences between the results observed on real people and those observed on KEMAR should be attributable to the effects of intersubject variability in head shadow and baffle. For the most part the data support this notion. Figure 12 shows a typical example of the agreement usually observed between direct and played-back signals in real ears. The solid line shows the output spectrum of a hearing aid measured in a subject's ear canal. The dotted line shows the output spectrum of the played-back version of the same hearing aid in the same subject's ear canal. The two curves are within a few dB of each other up to 2500 Hz. At higher frequencies the discrepancies are slightly larger, but differences greater than 5 dB are rarely seen.

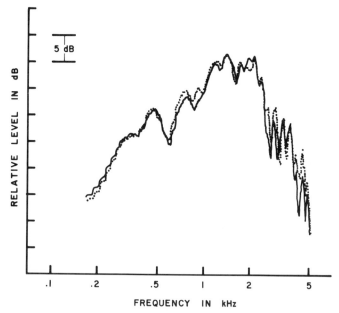

Figure 12. Typical example of results observed in real ear canals using hearing aids fitted with standard earmolds. Output spectrum observed in the direct condition (solid line) is compared with the analogous spectrum observed via the record-playback system (dotted line).

As one would expect when dealing with real people, there are occasional subjects on whom the agreement is markedly poorer than shown in Figure 12. In observations on eight subjects (a different eight from those reported in Figures 4 and 5), seven showed agreement similar to that seen in Figure 12, but one subject was quite different. Figure 13 shows a result from this subject. It is the poorest agreement seen between direct and processed signals in 32 observations using four different hearing aids. Even in this case the agreement is quite acceptable up to 1200 Hz; however, above this frequency, differences as great as 10 dB occur in several places.

Although differences between direct and processed signals have been noted in the last two figures, it should be emphasized that in the great majority of cases the record-playback system achieved very well our stated goal of presenting to the subject a processed signal that was very like the one he would have received if he had actually worn the hearing aid. Figure 14 further illustrates this point. This figure shows two sets of direct and processed hearing aid spectra. The discrepancies between the dotted and solid curves are, in each case, more or less typical. The two sets of data were obtained using the same hearing aid on two different subjects. The shape of the directly measured output curve for this hearing aid is quite different when loaded by the different earmolds and the ear canals of these two subjects. Nevertheless, the processed signal faithfully reproduced these dif-

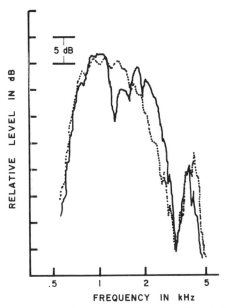

Figure 13.   Poorest agreement seen in real ear canals between direct (solid line) and processed (dotted line) hearing aid output spectra with a standard earmold fitting.

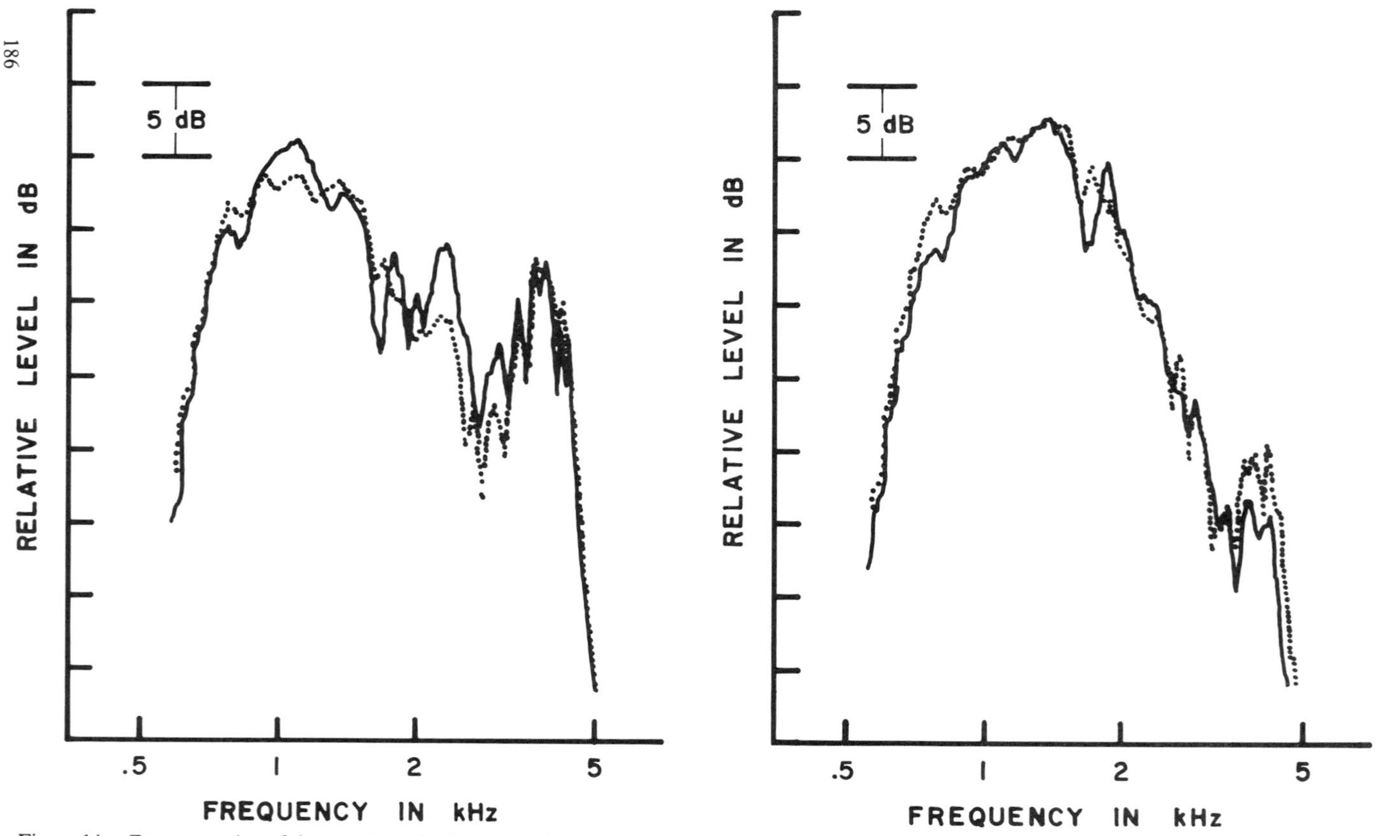

Figure 14. Demonstration of the extent to which the record-playback system preserved intersubject differences. Two pairs of direct (solid line) and processed (dotted line) data...

186

ferences in each case, resulting in a signal at the eardrum of each subject that was very similar to the direct signal.

Unsealed Earmolds

Unintentional Leaks   Although considerable progress has been made toward the goal of prediction of the effects of earmold vents of various configurations (Studebaker and Cox, 1977), an important aspect that remains elusive is the prediction of the magnitude of the effects of unintentional acoustic leaks. Several years of work with earmolds indicates that unintentional acoustic leaks exist with almost all earmolds. However, the magnitude of the effect is quite variable across earmolds, and even leaks of quite small magnitude can have large effects under some circumstances.

Figure 15 shows an example of this. The upper two curves show the spectrum of a signal measured in a real subject's ear canal with a particular

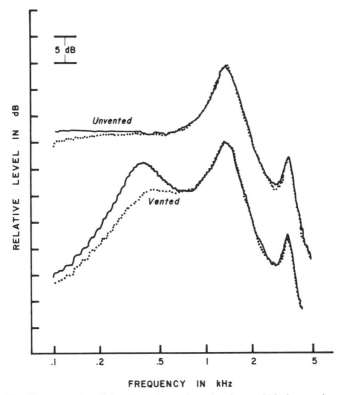

Figure 15.   Demonstration of the effect of an unintentional earmold leak on real-ear vented earmold data. Upper curves show ear canal spectra with standard earmold unsealed (dotted line) and sealed (solid line). Lower curves show earmold now incorporating a parallel vent, unsealed (dotted line) and sealed (solid line).

standard, unvented earmold. The dotted line shows the spectrum with the earmold worn as it normally would be worn. No particular effort was made to prevent acoustic leakage. The solid line shows the spectrum after considerable efforts were made to seal all acoustic leaks. A comparison of these two curves reveals that the acoustic leakage displayed by this earmold in its normal-use condition was quite small, amounting to a 1-dB reduction in level at 200 Hz—quite an insignificant effect as far as hearing aid use is concerned. Now observe the two lower curves. These curves show the spectra measured in the ear canal under the same conditions as for the upper curves: the dotted line shows the normal-use, or unsealed, condition and the solid line shows the sealed condition. In addition, a parallel vent has been added to the earmold—it is the same vent for both curves. Comparison of these two curves reveals a difference of 6 dB at 400 Hz—a difference attributable to the presence or absence of an unintentional acoustic leak, which by itself had no effect at all at 400 Hz. A difference of this magnitude, in this frequency region, may be quite significant and should be accounted for if realistic hearing aid performance is to be achieved using a hearing aid-processed signal. It would be impossible to predict the effect of this vent in this earmold without knowing the magnitude of the unintentional leaks that are present. Work on this problem is in progress, but for the present it seems inevitable that the actual earmold to be used with the hearing aid should be incorporated into the playback system for hearing aid-processed signals.

Modified Earmolds   As noted earlier, a primary motivation for developing a system utilizing an earmold was to permit the characteristics of modified earmolds to exert their normal influence on the processed signal. A basic requirement for the success of the procedure with modified earmolds is that the effect of earmold modifications and variations should be independent of the normally encountered differences in the driving system (receiver and tubing). If they are independent, a hearing aid's output can be recorded in a closed cavity and then played back through a different receiver and a modified earmold, with a result at the eardrum equal to that which would have been obtained had the recorded hearing aid been used with that modified earmold.

Our data testing this issue were all obtained using earmolds with parallel vents. KEMAR was used as the subject. The data obtained using both vented and open earmolds indicate that the correspondence between hearing aid output spectra in the direct and processed conditions is sometimes not as good as that observed when a standard closed earmold is used. The system is subject to three types of errors.

Resonance Shifts   The first type of error occurs because the effects of earmold modifications, such as vents, are not, in fact, entirely independent of the variations in the driving system. Normally, when a hearing aid receiver is used in association with a vented earmold, the presence of the reactance resonance associated with the vent has some slight effect on the location of resonances generated in the acoustic input system, which are within about ± 1 octave of the vent resonance. A resonance that is particularly affected is the primary resonance of the receiver. As a result, when the difference between vented and unvented results is plotted, it may contain one or more peaks and valleys, which is attributable to this slight shift in resonance location.

When the recorded signal is presented to the ear canal or coupler via the playback system, this effect is minimal because all the resonances generated in the playback system are heavily damped. As a result, when the vented response is plotted relative to the unvented response, the curve is very smooth and tends to bisect the peaks and valleys seen in the vented re unvented curve derived for the direct condition.

Figure 16 shows an example of this effect. The two upper curves show the hearing aid's output spectra measured on KEMAR, with a vented earmold in both the direct and processed conditions. The agreement between these two curves is ± 3 dB. Even better agreement than this was often

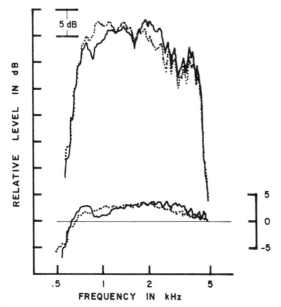

Figure 16.   The upper curves show the hearing aid's output spectrum at KEMAR's "eardrum" position in the direct (solid line) and the processed (dotted line) conditions, with the hearing aid fitted using a vented earmold and a relatively low gain setting. The lower curves show the effect of the vent on the direct (solid line) and on the processed (dotted line) signals.

seen with vented molds on KEMAR. The two lower lines show the effect of the vent on both the direct and the played-back signal. For each line, the spectrum for the vented condition is plotted relative to the spectrum for the unvented condition with the same input signal. Observe that the curve showing the effect of the vent in the direct condition weaves above and below the curve showing the effect of the same vent in the processed condition. Although this effect represents a real difference in the influence of earmold vents on the direct and played-back signals, it is usually of rather small magnitude, and probably is not a functionally significant source of error.

Acoustic Feedback   A second source of error is probably much more significant. The effects of acoustic feedback may become significant considerably before the familiar squeal that we recognize as acoustic feedback becomes audible. As the gain of the hearing aid is increased, a certain amount of the acoustic signal escapes either around or through the earmold. Some of this energy appears at the microphone of the hearing aid and combines with the acoustic energy already present in a way that is dependent upon the level and phase relationships between the direct and the fed-back signal. The result is that the spectrum of the signal appearing at the microphone of the hearing aid begins to take on a much more irregular shape, developing peaks and valleys. This change in the hearing aid's input signal is reflected in a change in the hearing aid's output signal. The spectrum of the signal becomes progressively more and more irregular as the gain of the hearing aid is increased.

As long as a well-fitting, unvented earmold is used, the level of the fed-back signal is normally low enough that the effects on the hearing aid's output are minimal. However, if a vented or open earmold is used, this subaudible acoustic feedback may become a significant factor in determining the hearing aid's output spectrum.

However, when a hearing aid-processed signal is presented via a vented earmold, the acoustic feedback does not have an opportunity to modify the input signal, because the hearing aid's microphone is not physically present in this condition. Consequently, increasing the gain of the playback signal does not result in any change in the spectrum of the signal appearing in the ear canal.

The data shown in Figure 16 were obtained with the hearing aid at a low gain setting. There are no obvious effects attributable to subaudible acoustic feedback in the direct condition. Figure 17 shows the same hearing aid, the same vent, and the same input signal, with the gain of the hearing aid increased, but still below the level of audible acoustic feedback. The upper two curves show the hearing aid's output spectrum for the direct and

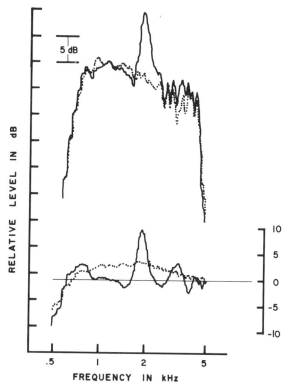

Figure 17.   The upper curves show the hearing aid's output spectrum at KEMAR's "ear-drum" position in the direct (solid line) and the processed (dotted line) conditions, with the hearing aid fitted using a vented earmold and a relatively high gain setting. The lower curves show the effect of the vent on the direct (solid line) and the processed (dotted line) signals.

the record-playback condition. Note that there is a large peak in the spectrum of the direct signal, which is completely absent in the played-back signal. This peak, and a smaller one at 3300 Hz, are both attributable to the effects of subaudible acoustic feedback on the hearing aid's input signal in the direct condition.

The two lower curves show the effect of the vent on the hearing aid's output spectrum with this higher gain setting. As seen in Figure 16, these curves show the difference between the spectrum observed with the vented earmold and the analogous spectrum observed with the unvented earmold in each condition. It is apparent that the two curves are now quite different. In the direct condition, the low frequency cut and the reactance resonance associated with the event are unchanged from those seen in Figure 16. However, at frequencies above 1000 Hz, the effect of the vent is confounded with the effects of acoustic feedback.

These data indicate that when the circumstances are such that a significant amount of acoustic feedback is allowed to occur in the direct condition, the correspondence between the direct and the played-back signal will break down.

Direct Acoustic Input   A third source of error lies in the effects of sound entering the ear canal directly through the vent or opening in the earmold. This problem arises mainly with open earmolds and is generally negligible with vented earmolds. In the direct condition, frequencies within the passband of the hearing aid that enter the ear canal directly through the open earmold combine there with the amplified version of the same signal. Cancellation and addition occur between them according to their relative levels and phase relationships, resulting in some changes in the spectrum of the signal received at the eardrum. In addition, frequencies above and below the hearing aid's passband may also enter the ear canal directly and result in a broader spectrum at the eardrum. These effects, of course, do not occur in the record-playback condition, since there is no opportunity for signal to enter the ear canal directly through the opening in the earmold.

Figure 18 illustrates the result. There are three data lines in this figure: the spectrum at the "eardrum" in the direct condition with the hearing aid turned on, the spectrum at the "eardrum" in the direct condition with the hearing aid turned off, and the spectrum at the "eardrum" in the processed condition. Comparing the direct hearing aid-on result with the processed

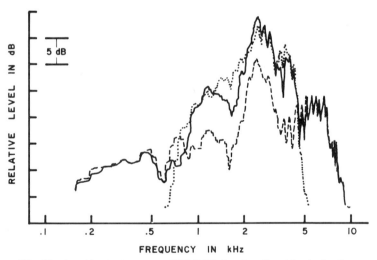

Figure 18.   Hearing aid output spectra at KEMAR's "eardrum" position in the direct condition, hearing aid on (solid line); the direct condition, hearing aid off (broken line); and the processed condition (dotted line). The hearing aid was fitted using an open earmold and a relatively low gain setting.

result shows that the agreement is relatively good within the passband of the hearing aid, although the effects of cancellation in the ear canal can be seen at roughly 800, 1200, and 3600 Hz. Both above and below the passband of the hearing aid, the level in the direct, hearing aid-on condition is higher than the level in the processed condition. Comparison of the hearing aid-on and hearing aid-off lines shows clearly that this is caused by admittance to the ear canal of unamplified signal through the open earmold. In this case, the unamplified signal is a significant proportion of the total signal in the direct condition. Obviously, the hearing aid was being operated at a relatively low gain setting when these data were obtained.

Figure 19 shows the result when the gain of the hearing aid was increased by about 15 dB. The same three data curves are shown in this figure as in Figure 18. The effect of the direct unamplified signal was much less when the hearing aid was operated at a higher gain setting. The general agreement between the spectra of the direct and played-back signals was also better. However, the effects of subaudible acoustic feedback are seen as peaks in the direct signal at 2300 Hz and 3500 Hz.

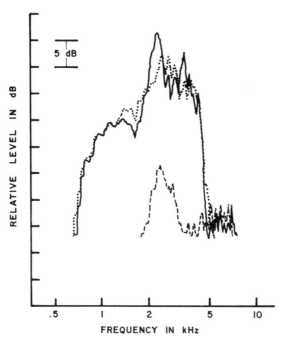

Figure 19.   Hearing aid output spectra at KEMAR's "eardrum" position in the direct condition, hearing aid on (solid line); the direct condition, hearing aid off (broken line); and the processed condition (dotted line). The hearing aid was fitted using an open earmold and a relatively high gain setting.

SUMMARY

We have concluded that the record-playback system, in its present form, usually delivers a hearing aid-processed signal that corresponds very closely to the signal the subject would have received if he had actually worn the hearing aid. The following qualifications must be made, however: 1) the accuracy of the played-back signal is limited by the difference between the head diffraction effects of the individual subject and KEMAR; 2) when used with a vented or open earmold fitting, the accuracy of the played-back signal is decreased when the hearing aid is operated at high gain setting, resulting in high levels of acoustic feedback in the direct condition; 3) when used with an open earmold fitting, the accuracy of the played-back signal is also decreased when the hearing aid is operated at a very low gain setting wherein the unamplified signal admitted to the ear canal via the open earmold constitutes a significant proportion of the total, directly received signal.

Finally, there are some types of hearing aid fittings that may not be amenable for use with the hearing aid-processing procedure in its present form. For example, an earmold incorporating a side-branch vent would appear to be unsuitable, because the tubing portion of the present playback system must be extended completely through the earmold to its medial end and would therefore obstruct any vent that opened into the earmold's main bore. However, preliminary data indicate that the minor modifications necessary to incorporate a side-branch–vented mold into the playback system do not markedly reduce the accuracy of the played-back signal.

In addition, the procedure would require considerable modification before it could be utilized with hearing aids fitted using the "earmold plumbing" techniques described by Killion (1976), because with this technique the tubing connecting the hearing aid to the earmold could not be represented by a simple length of undamped tygon tubing in the recording process.

In spite of these problems there remains a wide range of application for this technique. It has been shown to perform very well in experimental hearing aid selection procedures with closed earmold coupling and with modified earmolds when the hearing aid is operated at a moderate gain level. In addition, hearing aid-processed signals are very convenient for many research applications. Even the discrepancies that exist under certain conditions between directly amplified signals and their hearing aid-processed analogs provide a basis for study of acoustic factors, such as feedback, that are of considerable importance in hearing aid use.

REFERENCES

Cox, R.M., and G.A. Studebaker. 1977. Spectral changes produced by earphone-cushion reproduction of hearing-aid–processed signals. J. Am. Audiol. Soc. 3: 26–33.

Cox, R.M., and G.A. Studebaker. 1979. Hearing-aid–processed signals: A new approach. Audiology 18:53–71.

Houtgast, T., and H.J.M. Steeneken. 1978. The modulation transfer function as a link between room acoustics and speech intelligibility. Paper presented at the 95th meeting of Acoustical Society of America, May, Providence, R.I.

Jerger, J., C. Speaks, and C. Malmquist. 1966. Hearing aid performance and hearing aid selection. J. Speech Hear. Res. 9:136–149.

Killion, M.C. 1976. Earmold plumbing for wide band hearing aids. Paper presented at 91st meeting of Acoustical Society of America, April, Washington, D.C.

Killion, M.C. 1979. Equalization filter for eardrum-pressure recording using a KEMAR manikin. J. Audio Eng. Soc. 27:13–16.

Larson, V., G.A. Studebaker, and R. Cox. 1977. Sound levels in a 2cc cavity, a Zwislocki coupler, and occluded earcanals. J. Am. Aud. Soc. 3:63–70.

Lochner J.P.A., and J.F. Burger. 1960. Optimum reverberation time for speech rooms based on hearing characteristics. Acustica 10:394–399.

Lochner, J.P.A., and J.F. Burger. 1961. The intelligibility of speech under reverberant conditions. Acustica 11:195–200.

Olsen, H.F. 1972. Field type artificial voice. J. Audio Eng. Soc. 20:446–451.

Santon, F. 1976. Numerical production of echograms and of the intelligibility of speech in rooms. J. Acoust. Soc. Am. 59:1399–1405.

Shaw, E.A.G. 1966. Earcanal pressure generated by circumaural and supra-aural earphones. J. Acoust. Soc. Am. 39:471–479.

Studebaker, G.A., and R.M. Cox. 1977. Sidebranch and parallel vent effects in real ears and in acoustical and electrical models. J. Am. Audiol. Soc. 3:108–117.

Studebaker, G.A., R.E.C. White, and S. Hoffnung. 1978. Evaluation of a paired comparison technique for selecting hearing aid characteristics. Paper presented at the annual convention of the American Speech and Hearing Association, November, San Francisco.

Villchur, E. 1969. Free field calibration of earphones. J. Acoust. Soc. Am. 46: 1527–1534.

Witter, H.L., and D.P. Goldstein. 1971. Quality judgements of hearing and transduced speech. J. Speech Hear. Res. 14:312–322.

Zerlin, S. 1962. A new approach to hearing aid selection. J. Speech Hear. Res. 5:370–376.

CHAPTER 10

EARMOLD VENTING AS AN ACOUSTIC CONTROL FACTOR

Samuel F. Lybarger

CONTENTS

Over the past several years, a more orderly understanding of the acoustic factors in earmold design and modification has developed. Although this discussion is limited to some observations on vented and open canal earmold situations, and to acoustic features of earmolds, it must be remembered that the mechanical features, such as the choice of material, the quality of reproduction of the shape of the ear canal and concha, and the smoothness of polish, are also of great importance. These mechanical features are vital to earmold success and are well understood by competent workers in earmold laboratories.

Since the earmold is the last element in the hearing aid system, any changes in performance effected by earmold modification are reflected as changes in gain, frequency response, and output. That saturation output is changed along with gain is often overlooked, yet this can be one of the major advantages of earmold changes as opposed to electronic response changes.

197

A thorough study of the engineering data for a hearing aid may indicate that response, gain, and/or output changes on a hearing aid can be made more easily by electronic controls on the aid in combination with a simple, closed earmold of straightforward construction. This would avoid the feedback problems that might exist with venting, for example. Earmold modification is not always indicated, but rather should be done for a defined reason. Earmolds may be vented for a number of purposes, several of which are considered next.

VENTING

A vent is simply an opening leading from the ear canal, usually from a point at the tip of the earmold, or from a point in the delivery tube near the tip of the earmold, to the outside air. The opening can be large or small, short or long. An example is shown in Figure 1.

Electrical Analogs of Vents

First let us look at the nature of venting by examining a simplified electrical analogy of the receiver-earmold-ear system, as shown in Figure 2. (More detailed analogies may be found in Egolf, Tree, and Feth, 1978; Lybarger, 1972; Studebaker and Cox, 1977.)

In Figure 2, the transduction mechanism of the receiver produces a vibratory force per unit area of diaphragm of F/S. The mechanical impedance of the vibratory system, reduced to an acoustic equivalent, is indicated by Z_m/S^2. The acoustic impedance of the system, including any tubing and damping elements, is indicated by Z_a. An acoustic vol-

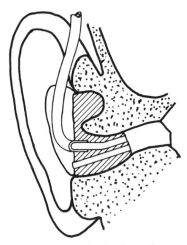

Figure 1.   Example of a vented earmold showing delivery tube and parallel vent.

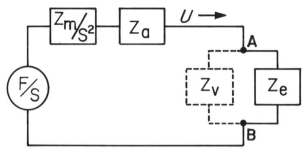

Figure 2. Simplified electrical analogy of the receiver-earmold-ear system, showing vent impedance as being parallel to ear canal impedance.

ume current U is delivered at the end of the tubing system, where it enters the ear canal, whose impedance is shown as Z_e. With no vent present, the sound pressure in the ear canal near the end of the tubing system is $U \cdot Z_e$. The effect of a parallel vent—one that goes from the *tip* of the earmold to the outside—is to add the impedance Z_v in shunt with the ear canal impedance Z_e.

Carlson (1974) has pointed out that in small hearing aid receivers and typical tubing systems, the source impedance is much higher than that of the ear canal or a coupler. This means that at any frequency the volume current U at the end of the tubing is largely a function of the source impedances and the electrical input to the transducer only and is nearly independent of the coupler or vent characteristics.

If we determine the response of the system with the vent closed and then with it open, we can define the *vent response* as the response with the vent open minus the response with the vent closed for the same input to the receiver. The vent response tells us how much the gain and output of the hearing aid are changed by the presence of the vent as opposed to the closed condition. With a very high source impedance, the vent response is dependent only on the acoustic properties of the ear canal and the vent opening itself and is nearly independent of receiver type, tubing diameter and length, or of anything toward the receiver from the point the tubing opens into the ear canal cavity. The vent response then can be expressed simply as the ratio of the complex ear canal impedance and the parallel vent impedance to that of the ear canal alone:

$$\text{Vent response (dB)} = 20 \log_{10} \left| \frac{\dfrac{Z_e Z_v}{Z_e + Z_v}}{|Z_e|} \right| = 20 \log_{10} \left| \frac{Z_v}{Z_e + Z_v} \right|$$

Measurements made with simulated earmolds on a Zwislocki-type coupler (Industrial Research Products Model DB100), discussed below, confirm that the vent response of a particular vent is reasonably independent of receiver and tubing systems that are found in typical hearing aids. This simplifies the prediction of the response with a particular vent from a closed coupler measurement to merely adding the vent response to the measured closed coupler response.

Venting for Static Pressure Equalization

Very few earmolds provide a truly tight seal, in spite of claims to the contrary. However, although a "natural" leak may exist, it is not always dependable, and a drilled hole may be provided to give barometric equalization in the ear canal and to relieve a possible feeling of "pressure." For a medium length earmold tip, a parallel vent with a diameter of 0.6 mm (0.025", #72 drill) will be effective. A vent of 1 mm (0.040", #60 drill) in diameter may be too large in that it can change the hearing aid response noticeably. An equalizing vent should be drilled straight through so that it can be cleaned with a wire.

Venting for Low Frequency Enhancement

The electrical analog of a parallel vent is shown at the left of Figure 3. When the vent has inertance, this is a parallel turned circuit. If the inertance of the vent is large compared with its resistance, the impedance of the vent and ear canal in parallel can be substantially larger than that of the ear canal alone. Thus the sound pressure level in the ear canal can be higher at the vent resonance than with no vent present.

Figure 4 shows the arrangement used for making the various vent response curves that will follow. An extension was made for the DB100 coupler that has the same internal diameter as the coupler canal (7.5

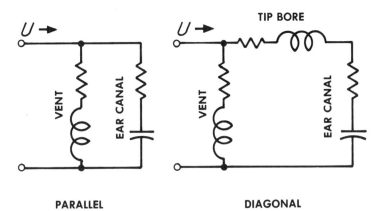

PARALLEL   **DIAGONAL**

Figure 3.   Electrical analogies of parallel and diagonal vents. (Reprinted with permission from Lybarger, 1979, in *Auditory and Hearing Prosthetics Research.*)

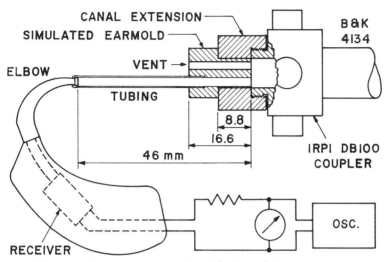

Figure 4.   Apparatus arrangement used for determining vent response.

mm) and whose canal length (8.8 mm = 0.346″) is the same as the length of the portion of the ear canal in KEMAR beyond the coupler. The outer face of this extension thus corresponds in location to the bottom of the concha in KEMAR. The simulated earmolds were made from brass. They have a portion that fits into the coupler extension and a portion of larger diameter to make the vent length correspond to that observed in a number of actual tubing-type earmolds. The fit of the simulated earmold into the coupler extension provides a tight acoustic seal. The signal was delivered from a BK receiver in a hearing aid housing with internal "plumbing" and an earhook (elbow). Attached to the earhook were 46 mm of #13 tubing (ID = 1.93 mm) for the medium-size earmold. To control vent lengths, four "standard" tip lengths were established: medium, long, short, and short-hollowed.

Figure 5 shows the vent response for a medium-length-tip simulated earmold as measured on the equipment shown in Figure 4. The parallel vent length was 16.6 mm (0.65"). The 1-mm diameter vent (#61 drill) shows slight amplification in the 250–300 Hz region, but has little effect above 500 Hz. The 2-mm diameter vent shows substantial amplification, peaking at about 500 Hz, but has little effect above 700 Hz. Above 1500 Hz the vent response of relatively small vents is negligible. An increasing reduction in response occurs below 400 Hz, typical of the general pattern of vent response below the resonant frequency. The 3-mm diameter vent in the medium-length earmold tip produces little amplification, but lowers the response considerably below 550 Hz.

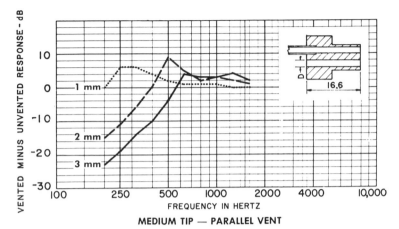

MEDIUM TIP — PARALLEL VENT

Figure 5.   Vent responses for different parallel vent diameters in a simulated medium-tip-length earmold. (Reprinted with permission from Lybarger, 1979, in *Auditory and Hearing Prosthetics Research.*)

For a medium-length tip, it is seen that vent diameters in the range of from, for example, 1.5 mm (0.059″) to 2.5 mm (0.098″) would be effective in enhancing the response in the general region of 500 Hz.

Figure 6 shows similar vent responses for a long tip, Figure 7 for a short tip, and Figure 8 for a short-hollowed tip. The vent resonance peaks move upward in frequency as the length of the vent decreases.

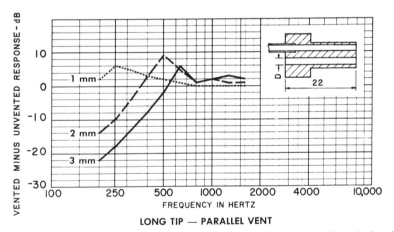

LONG TIP — PARALLEL VENT

Figure 6.   Vent responses for different parallel vent diameters in a simulated long-tip-length earmold. (Reprinted with permission from Lybarger, 1979, in *Auditory and Hearing Prosthetics Research.*)

Venting for Moderate Low Frequency Reduction

With a 2-mm (0.079″) diameter vent, considerable low frequency reduction is seen below 350 to 500 Hz, depending on the tip length. Undesired amplification may also occur at a medium-low frequency. The amplification may be eliminated without losing the low frequency reduction by placing a small amount of acoustic damping over the vent. Figure 9 shows the effect of placing light cloth damping over the outer end of a 2-mm vent in a

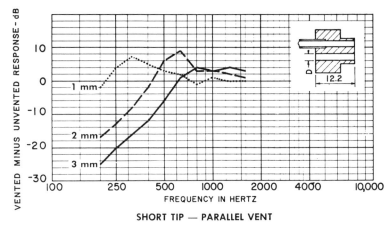

SHORT TIP — PARALLEL VENT

Figure 7.   Vent responses for different parallel vent diameters in a simulated short-tip-length earmold. (Reprinted with permission from Lybarger, 1979, in *Auditory and Hearing Prosthetics Research.*)

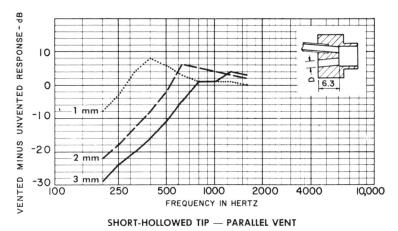

SHORT-HOLLOWED TIP — PARALLEL VENT

Figure 8.   Vent responses for different parallel vent diameters in a simulated short-hollowed-tip-length earmold. (Reprinted with permission from Lybarger, 1979, in *Auditory and Hearing Prosthetics Research.*)

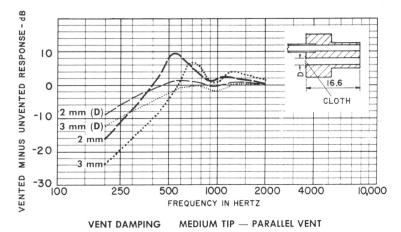

VENT DAMPING   MEDIUM TIP — PARALLEL VENT

Figure 9.   Effects of vent damping on vent response for two parallel vent diameters in a simulated medium-tip-length earmold. (Reprinted with permission from Lybarger, 1979, in *Auditory and Hearing Prosthetics Research.*)

medium-length earmold. The low frequency response is decreased without the vent-caused amplification. To obtain as much low frequency reduction as in an undamped vent, the vent diameter can be increased as shown in the figure. The well-known "modifier" earmold uses the concept of a damped vent.

Venting for Strong Low Frequency Reduction

For strong low frequency reduction, a short vent, large in diameter, must be used. Figure 10 shows the response changes for short-hollowed, short, medium, and long tips when a 3-mm vent is used. For a short-hollowed tip, a reduction of 25 dB at 250 Hz is observed, with the cutoff starting at about 800 Hz. For a long tip, the reduction at 250 Hz is about 18 dB and the cutoff starts at about 500 Hz. In general, a 3-mm or larger vent diameter does not cause much amplification above the cutoff frequency. Some peaking is seen for the long tip with the 3-mm diameter vent.

To obtain more low frequency cutting with a standard type of earmold, vents larger than 3 mm can be used. The shorter the tip used, and therefore the shorter the vent length, the greater will be the effect for a given vent diameter.

Consistency of Vent Response for Different Receivers and Tubing Systems

For the types of venting that have been mentioned so far, it is of interest to examine how constant the vent response remains for different receivers and tubing systems. Figure 11 shows a number of different sound delivery

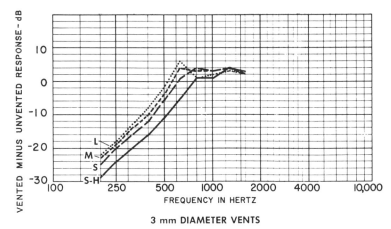

3 mm DIAMETER VENTS

Figure 10. Vent responses for a 3 mm in diameter parallel vent with different simulated earmold tip lengths. (Reprinted with permission from Lybarger, 1979, in *Auditory and Hearing Prosthetics Research.*)

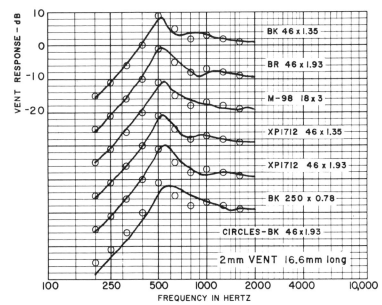

Figure 11. Curves showing relative independence of vent response on earphone type and delivery tube diameter and length (simulated medium-length earmold with a 2 mm in diameter vent).

conditions for a medium-length simulated earmold tip with a 2 mm in diameter by 16.6-mm long vent. The vent response is not greatly affected by a wide range of delivery systems and corresponds adequately for fitting purposes to the points at ⅓-octave intervals for a BK receiver with a 46-mm length of #13 tubing, taken from tables by Lybarger (1978). Tabular values for a wide range of vent diameters and lengths may also be found in Lybarger (1978).

The lowest curve in Figure 11 represents a very unusual tubing system, 250 mm (9.84″) of 0.78 mm (0.031″) tubing. This creates a high impedance damped source, but the difference in the vent response from that in typical tubing systems is small.

Figure 12 shows similar curves for a medium-length tip with a 3-mm (0.118″) diameter vent. Again, for practical purposes, the vent response is not seriously changed by wide changes in the delivery system.

Venting for Extreme Low Frequency Reduction

When one encounters a hearing condition with little or no hearing loss in the low frequencies, but a substantial loss above 500 to 1500 Hz, it is difficult to reduce the low frequency gain and output adequately by venting a standard earmold. To obtain a sufficient reduction in low fre-

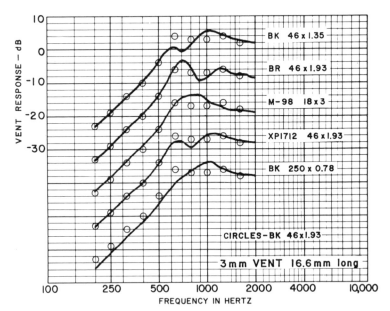

Figure 12. Curves showing relative independence of vent response on earphone type and delivery tube diameter and length (simulated medium-length earmold with a 3 mm in diameter vent).

quency response, a tube fitting like that employed in CROS and IROS provides a good solution. This fitting also allows the ear canal to remain almost fully open, so that low frequency sounds can enter the ear canal with little attenuation.

OPEN CANAL MEASUREMENTS

Apparatus

To obtain the data on tube fittings reported in this chapter, the arrangement shown in Figure 13 was used. An electrical signal was delivered to a hearing aid receiver through a resistor that was large in comparison with the impedance of the receiver, so that constant current conditions were approximated. The receiver was either an older type Knowles BK receiver in a hearing aid case with internal "plumbing" and an elbow, or a Knowles XP1712 broadband receiver with the acoustic equivalent of internal plumbing and an elbow provided by the two-diameter simulator shown. A 46-mm length of tubing was used to deliver the sound to the coupler.

Tests made on KEMAR with two depths of tubing insertion into the ear canal were compared with tests made on the DB100 coupler with the 8.8-mm extension alone. High frequency results were almost identical; in the low frequencies, the simple canal extension gave about 4 dB less response than KEMAR. A conical "concha simulator" was made by a cut-and-try method, as shown in the figure, and produced results for tube fittings almost the same as those obtained with KEMAR, but it made testing much simpler and positioning of the tubing more accurate. This concha simulator arrangement works well when the delivery tube ends

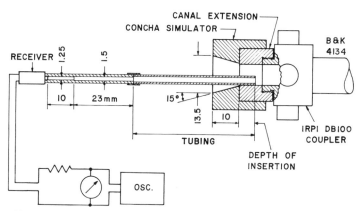

Figure 13. Apparatus arrangement for measuring open canal vent response using a canal extension and concha simulator.

within the canal extension; it cannot be expected to give good results if the tubing ends outside of the cylindrical canal portion.

Again recognizing Carlson's (1974) suggestion that the impedance looking toward the receiver is very high compared with that in the ear canal, one of the major objectives of the tests reported here was to determine to what extent the vent response of the open ear canal was independent of the receiver and tubing system and dependent only on acoustic factors in the ear canal beyond the end of the delivery system. This is a somewhat different approach than that used in the very comprehensive analysis of tube fitting acoustics recently reported by Johansen (1978). Johansen's analysis started at the receiver end of the tubing.

Results

Figure 14 shows the closed coupler response of a Knowles BK receiver and a 1.93-mm diameter by 46-mm long delivery tube (plus internal "plumbing" and earhook). The open ear canal response for the same tubing insertion depth and the vent response are also shown. Notice that the open canal response has all the detailed peak characteristics of the closed coupler response.

The vent response shows a reduction at 200 Hz of about 40 dB, followed by a rising characteristic at about 12 dB/octave, which reaches 0

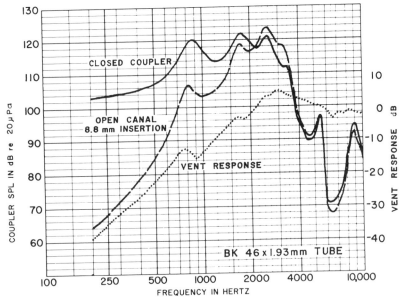

Figure 14.   Closed coupler, open canal, and vent responses for a Knowles BK earphone with elbow and 46-mm long tubing.

dB at about 2200 Hz. This is followed by a very broad peak from 2200 to 5000 Hz, with a maximum of only about 5 dB at 3000 Hz. Above 5000 Hz, the vent response drops slightly to about −2 dB.

Figure 15 shows the same information for a Knowles XP1712 broadband receiver. In spite of the very large difference in the response characteristics of the two receivers, the vent response has not changed significantly, indicating that the vent response is independent of the receiver to a large extent.

It is important to note that open canal fittings do not improve the high frequency performance of a receiver, except for the few decibels obtained in the broad resonance from 2000 to 5000 Hz. If a receiver system cuts off sharply in the closed canal measurement, it will cut off sharply in the open canal measurement.

Tests were then made with different tubing inside diameters to determine whether the vent response would be affected. The data for the XP1712 receiver are shown in Figure 16. Note that the only significant effect on the vent response is a "kink" in the vicinity of the first receiver peak and smaller "kinks" at higher frequency resonances. For practical purposes, it can be said that the vent response is unaffected by the tubing inside diameter for a given depth of insertion. Previously published information indicates that tubing size does offer a good control of low frequency response in tube

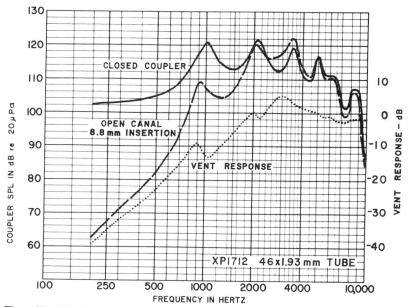

Figure 15.   Closed coupler, open canal, and vent responses for a Knowles XP1712 broadband earphone with tubing system of Figure 13.

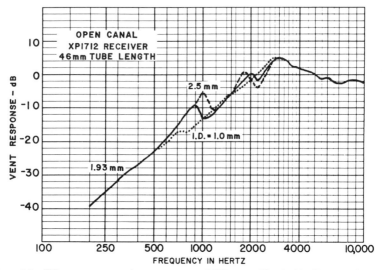

Figure 16.   Effect on open canal vent response of different tubing inside diameters (46-mm length of tubing, 8.8-mm depth of insertion per Figure 13).

fittings (Berland, 1975; Lybarger, 1979); however, it would appear that the changes caused by tubing size occur also in the *closed* response and are not the result of some interaction between the tubing size and the open ear canal.

Depth of insertion, as would be expected, does have a significant effect on vent response, as indicated by Figure 17, which shows smoothed vent responses for 0.76-, 8.8-, and 14.2-mm insertion depths. These curves can be used to estimate the open canal response of a hearing aid if the closed canal response for the same depth of insertion is known. The response is raised pretty much across the frequency band up to about 3 or 4 kHz as the depth of insertion is increased. However, little seems to be gained in level from an insertion depth greater than 10 or 12 mm.

The same information for smoothed vent responses shown in Figure 17 is given in tabular form at ⅓-octave intervals in Table 1. The three sets of data converge above 5000 Hz.

The outside diameter of the tubing could be a factor in changing the vent response, since the acoustic impedance looking outward from the end of the delivery tube would be increased as the available remaining cross-sectional area in the ear canal is diminished. Very little effect was noted up to 4-mm OD tubing, which is seldom exceeded in practice. With a 4.76-mm OD tubing, the response was raised by about 2½ dB in the low and middle frequencies.

The data given on tube fittings are based on the use of tubing alone. When an earmold support is used for the tubing, it should not block the

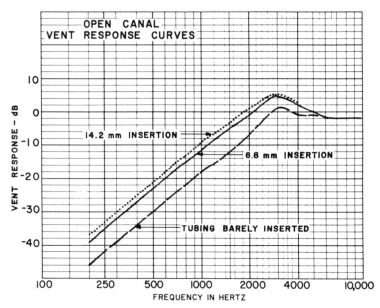

Figure 17.   Smoothed vent response curves for different depths of insertion of a 1.93 × 46 mm delivery tube, using apparatus of Figure 13.

Table 1.   Vent response values for open canal fittings for different depths of tubing insertion (corresponds to smoothed curves of Figure 17)

Frequency (Hz)	Vent response (dB)		
	0.76 mm (0.030″)[a]	8.8 mm (0.346″)[a]	14.2 mm (0.56″)[a]
200	−46	−39	−37
250	−42	−35	−33
316	−38	−31	−29
400	−34	−27	−25
500	−30	−23	−21
630	−26	−19	−17
800	−22	−15	−13
1000	−18	−11	−9
1250	−15	−8	−6
1600	−11	−4	−2
2000	−7	−1	+1
2500	−2	+3	+4
3150	+1	+4	+5
4000	−2	+2	+3
5000	−1	0	−1
6300		−2	
8000		−2	
10,000		−2	

[a] Depth of tubing insertion.

entrance to the ear canal significantly if the maximum reduction in low frequency response is to be achieved. The importance of this has been reported by Hewitt (1977).

Advantage of Damping Element in Open Canal Fittings

It becomes apparent in looking at the closed and open canal responses that all the peaks in response that appear in the closed response also appear in the open mold response, since the vent response is a rather smooth function over the frequency range. The first peak in an open mold fitting is particularly objectionable, because it usually occurs somewhere around 1000 Hz. The tube fitting is used to prevent excessive low frequency amplification, and this peak may produce too much gain at a frequency where the hearing remains quite good. This peak can be greatly lowered by the use of acoustic damping in the delivery system. One solution is to use a Knowles BF-1859 damping element (680 cgs acoustic ohms) in the delivery tube. Figure 18 shows the undamped and damped open canal responses for an XP1712 receiver, with the BF-1859 damping element 30 mm from the open end of the #13 tubing. It seems highly desirable to use acoustic damping routinely with CROS or IROS fittings to prevent excessive gain at the first receiver system resonance.

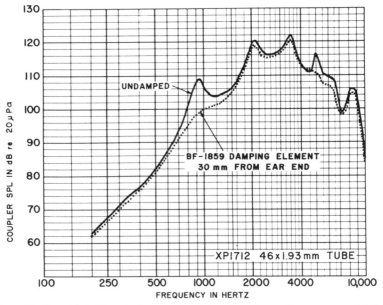

Figure 18. Beneficial effect of using an acoustic damping element in an open canal fitting.

Diagonal Venting

Diagonal venting causes a loss of high frequency response, as has been pointed out by Studebaker and Cox (1977) and Lybarger (1979). This is caused by the remaining inertance in the delivery bore beyond the point at which the vent enters it. Figure 19 shows the vent response for one configuration. Sometimes the size of the ear canal is too small to allow a parallel vent to be drilled beside the main bore. In such cases, a diagonal vent can be used without too much high frequency loss simply by enlarging the main bore as fully as possible from the point where the diagonal vent enters it to the end of the earmold tip.

Venting for Adjustability

When a hearing aid is fitted, it is often difficult to preestablish the amount of venting, if any, that will serve the user best in his actual environment. Both variable and adjustable vents have been developed that allow changes in the venting without remaking the mold.

One popular adjustable vent is the positive venting valve (PVV). A cup-shaped insert with a hole in the bottom fits snugly into a counter-bored hole in the earmold, as seen in the corner illustration of Figure 20. Also shown in Figure 20 are the vent responses for the five sizes (numbered 1–5) of inserts available, as measured on an ear simulator coupler and using a short-hollowed simulated earmold with a short vent channel. The #1, #2 and #3 inserts have little effect compared with no insert, and the #4 and #5 inserts give different amounts of low frequency cut.

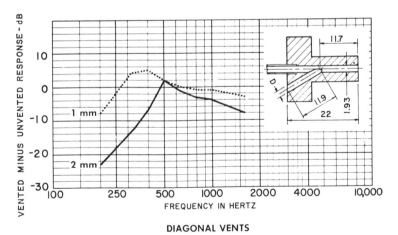

DIAGONAL VENTS

Figure 19.   Vent responses for a simulated earmold with diagonal vent, showing loss of high frequencies not seen in parallel vents. (Reprinted with permission from Lybarger, 1979, in *Auditory and Hearing Prosthetics Research.*)

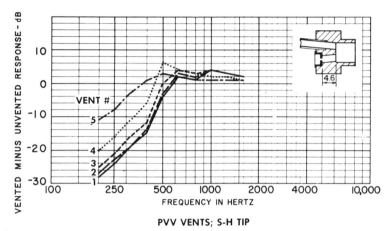

PVV VENTS; S-H TIP

Figure 20.   Vent response curves for "stock" sizes of vent inserts, using a simulated short-hollowed tip earmold (positive venting valve system (PVV)). (Reprinted with permission from Lybarger, 1979, in *Auditory and Hearing Prosthetics Research.*)

Figure 21 shows similar vent response curves for "select-a-vent" inserts. Again, only a couple of the standard inserts give a substantial change in response compared with the open vent hole. Because of the greater length of the vent holes, these inserts give larger peaks in the vented response than do the PVV inserts.

Because of the convenience of changeable vent inserts, it would seem highly advantageous to have the hole sizes graduated more evenly to provide more useful control. Tests were made on the acoustic equivalents of

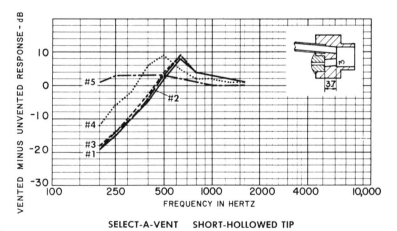

SELECT-A-VENT   SHORT-HOLLOWED TIP

Figure 21.   Vent response curves for "stock" sizes of vent inserts, using a simulated short-hollowed tip earmold—select-a-vent system (SAV). (Reprinted with permission from Lybarger, 1979, in *Auditory and Hearing Prosthetics Research.*)

these two types of vent inserts using a simulated earmold with a short-hollowed tip, and the low frequency reduction in response (at 200 Hz) was plotted against hole diameter, as shown in Figure 22. Inserts were then made to give approximately 5-dB steps of low frequency reduction at 200 Hz. The resulting nicely spaced curves for the positive venting valve are shown in Figure 23 and for the select-a-vent system in Figure 24. Perhaps a fewer evenly spaced sizes would be adequate.

Effect of Vent Channel in Adjustable Vent Devices

In variable and adjustable venting, the fact that the inertance of the vent channel may have a greater effect than the vent insert is often overlooked. Figure 25 shows the response changes first with a short, large-in-diameter

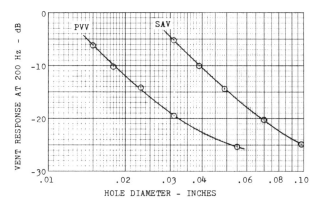

VENT DIAMETER VS. 200 Hz DROP FOR SHORT-HOLLOWED TIP

Figure 22.   Vent insert hole diameters vs. gain loss at 200 Hz for PVV and SAV vent inserts.

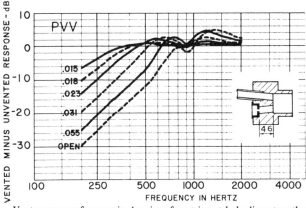

Figure 23.   Vent responses for a revised series of vent insert hole diameters that give better spacing between curves (PVV system).

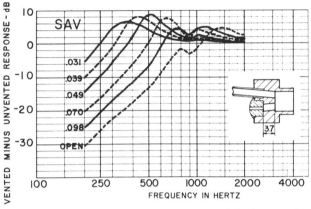

Figure 24.   Vent responses for a revised series of vent insert hole diameters that give better spacing between curves (SAV system).

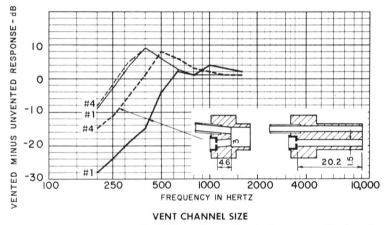

VENT CHANNEL SIZE

Figure 25.   Vent response curves for PVV vent inserts showing importance of a short, large-diameter channel to the vent insert.

vent channel and then with a long, smaller-in-diameter vent channel. It can be seen that for adjustable vents to have a useful and predictable controlling effect, the vent channel must be short and large in diameter; otherwise one could be deceived and believe that something is really happening when inserts are changed.

CONCLUSION

Earmold modifications should be undertaken only when a particular objective is sought and when it appears that electronic controls on the hearing aid will not be adequate. Venting, in particular, can cause feedback problems and has to be limited to moderate or low gain aids. For vents in standard earmolds and for the open canal situation, the vent response is moderately independent of the receiver and tubing system. Therefore, the vented response can be predicted quite well for practical purposes by adding the vent response (as measured with an ear simulator coupler) to the closed coupler response.

REFERENCES

Berland, O. 1975. No-mold fitting of hearing aids, earmolds and associated problems. Scand. Audiol. (suppl. 5).

Carlson, E.V. 1974. Smoothing the hearing aid frequency response. J. Audio Eng. Soc. 22:426–429.

Egolf, D.P., D.R. Tree, and L.L. Feth. 1978. Mathematical predictions of electroacoustic frequency response of in-situ hearing aids. J. Acoust. Soc. Am. 63: 264–271.

Hewitt, C. 1977. New "free-field" earmold. Hear. Aid J. 30(5):10, 32.

Johansen, P.A. 1978. The acoustical properties of a special type of ear mould for hearing aids, tube fittings, evaluated from the acoustical transfer of the mould. J. Audiol. Tech. 17:24–39.

Lybarger, S.F. 1972. Ear molds. In J. Katz (ed.), Handbook of Clinical Audiology, pp. 602–623. The Williams & Wilkins Co., Baltimore.

Lybarger, S.F. 1979. Controlling hearing aid performance by earmold design. In V.D. Larson, D.P. Egolf, R.L. Kirlin, and S.W. Stile (eds.), Auditory and Hearing Prosthetics Research, pp. 101–132. Grune & Stratton, New York.

Lybarger, S.F. 1978. Earmolds. In J. Katz (ed.), Handbook of Clinical Audiology, pp. 508–523. 2nd Ed. The Williams & Wilkins Co., Baltimore.

Studebaker, G.A., and R.M. Cox. 1977. Side branch and parallel vent effects in real ears and in acoustical and electrical models. J. Am. Audiol. Soc. 3:108–117.

CHAPTER 11

PROBLEMS IN THE APPLICATION OF BROADBAND HEARING AID EARPHONES

Mead C. Killion

CONTENTS

This chapter was at least partly prompted by an often-heard statement that starts with the phrase "If only suitable hearing aid transducers were available, ..." The main thrust of this chapter is a demonstration that suitable transducers *are* available. With the exception of hearing aids intended for the severely or profoundly deaf—where extraordinarily high output sound pressures are sometimes required—both transducer and amplifier technolo-

---

An informal listening test demonstration preceded the formal presentation of this paper. A description of that listening test, and a summary of the hearing aid fidelity ratings obtained from 122 (audience) participants, is given in the chapter appendix.

gies have reached the point where the hearing aid designer can produce almost anything one might ask for in a hearing aid, including high fidelity.

A REVIEW OF SOME TRADITIONAL PROBLEM AREAS

Noise in the Hearing Aid

There are two aspects of the noise problem: one involves amplification of noise generated within the hearing aid, and the other involves amplification of external background noise. In a modern hearing aid, the internally generated noise is very low. The primary noise component is caused by the microphone, and modern subminiature microphones have A-weighted noise levels some 5 dB lower than that found in a quiet concert hall and some 15–20 dB lower than that found in a typical residence. The aided threshold determined by the microphone noise level can be within a few dB of normal threshold (Killion, 1976a).

The second aspect of the noise problem is the subjective magnification of background noises. When a hearing aid has (rightly or wrongly) an intentionally restricted bandwidth combined with high frequency emphasis, the initial user reaction is sometimes that of "trying to understand speech in a background of noise." This problem usually disappears after the hearing aid is worn for awhile. There appears to be less adjustment required with broadband hearing aids having smooth frequency response, except when the volume control is turned up too high.

Distortion

Distortion in hearing aids is basically a problem of amplifier design. Surprisingly high peak pressures are frequently presented to the input of the hearing aid; a spoon dropped onto a plate can produce a 110- to 115-dB sound pressure level (SPL) peak. Even commonly encountered speech can produce high SPLs. My daughter's enthusiastic "Hi Dad" at 2½ feet is good for a 114-dB SPL peak at a head-worn hearing aid inlet port. Although commonly available microphones are linear up to 114 dB SPL, that sort of microphone input produces a peak microphone output voltage of 0.15 V. With only a 1.5-V supply available, therefore, anything over 20 dB of *electrical* gain (corresponding to at most a few dB of acoustic gain) will cause amplifier overload unless some sort of low distortion compression limiting is included in the amplifier design.

Earphone Distortion   The earphone itself is seldom the limitation in anything except high-powered class B hearing aid designs. With the high frequency emphasis in most hearing aids, distortion is most likely to occur at high frequencies. Figure 1 shows the maximum undistorted output capability of the Knowles Electronics BP-series of earphones (Carlson, Mostardo, and Diblick, 1976) when used with a well-damped conventional earmold in a Zwislocki-type ear simulator (Zwislocki, 1971). Here we see the direct trade-off available between the maximum undistorted high frequency output and the battery drain of the hearing aid, assuming a constant low frequency output is maintained. The typical earphone response seen in a manufacturer's data sheet is often obtained with the earphone output fed through an undamped tubing to a 2-cc coupler, and thus will have more peaks and valleys than the curves shown here, but is otherwise similar to the broken curve in Figure 1.

The broken curve was obtained under maximum current modulation (MCM) conditions, where a DC current equal to the nominal bias rating of the earphone was applied, and an AC current having a peak value equal to the DC bias current was superimposed. (Thus the net current in the earphone swings from zero to twice the rated bias current and back on each cycle of the AC wave form.) From the earphone manufacturer's viewpoint, such an MCM response curve has the advantage of a single curve applying to all earphones in a model series, regardless of the number of turns wound on the internal coil; i.e., regardless of the electrical impedance of the receiver. From the hearing aid designer's standpoint, however, a family of solid-line curves like those shown in Figure 1 contains more useful information because it tells him the earphone impedance he should order and the battery drain of the amplifier that will be required to produce a given high frequency undistorted sound pressure level.

The family of curves shown in Figure 1 is not unique (a different family is obtained each time the earphone-to-eardrum coupling system is changed), but serves to illustrate three points:

1. The battery drain of the hearing aid is likely to be determined by the high frequency output that is required. (We assume that the low frequency output is to be maintained constant.)
2. If the high frequency sound pressure levels seen on a typical hearing aid earphone data sheet are to be obtained, a fairly low nominal earphone impedance must be selected.
3. The undistorted output capability of the earphone itself substantially exceeds the undistorted output available from a typical single-ended class A hearing aid amplifier.

The Distortion Level in Ears   Another comparison that can be made is that between the distortion produced by nonlinearity in the hearing aid earphone and the apparent distortion produced by nonlinearity in the normal ear itself. (We assume that distortion in the impaired ear is at least as great.) One such comparison is shown in Figure 2, which represents an

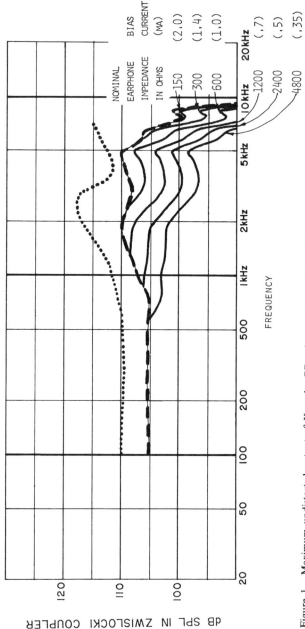

SUPPLY: 1.5V
AMPLIFIER: CLASS A, SINGLE ENDED
BIAS CURRENT: NOMINAL RATED BIAS FOR EACH EARPHONE IMPEDANCE
EARMOLD: 40 MM OF 1.9 MM Ø (#13) TUBING WITH TWO 1500 OHM DAMPERS

Figure 1. Maximum undistorted output of Knowles **BP**-series earphone with well-damped conventional earmold, limited by earphone overload (dotted line), amplifier current clipping (broken line), and amplifier voltage clipping (solid line).

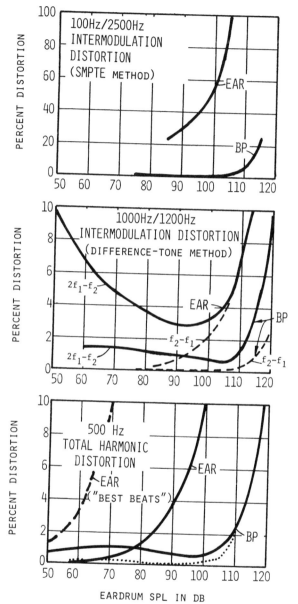

Figure 2. Apparent distortion levels in normal ears and in the BP-series broadband hearing aid earphones.

attempt to pull together some two dozen studies on aural distortion and combination tones.

The top graph in Figure 2 is based on some recent data of Zwicker (1976), who was looking for the psychoacoustic equivalent of the period histogram typically found in single-unit neural firings recorded from nerve VIII. Zwicker found that a brief 2500-Hz tone pip was made completely inaudible when added to an intense 100-Hz tone on one-half of the 100-Hz wave form, but was quite audible on the other half. Such a result corresponds to 100% intermodulation distortion as would be measured on earphones using the so-called SMPTE (low-tone, high-tone) method (Moir, 1958).

The middle graph shows an average of the results of several studies on the level of the two most prominent combination tones: the cubic difference tone at $2f_1 - f_2$ and the simple difference tone at the frequency $f_2 - f_1$. These correspond to the intermodulation distortion products as would be measured on earphones using the so-called CCIF (difference-tone) method (Moir, 1958).

A good estimate of the harmonic distortion levels in the ear is the hardest to come by. For years after it was first used by Wegel and Lane (1924), the "best beats" method was the commonly accepted method for estimating the level of aural harmonics. The broken curve in the bottom graph is derived from several studies in which that method was used. The problem with that method, as pointed out 40 years ago by Trimmer and Firestone (1937), is that no one ever *hears* these harmonic distortion products. A more conservative estimate is shown by the solid curve labeled EAR in the bottom graph, based on a "phase" method used by Clack, Erdreich, and Knighton (1972) and deBoer and Bouwmeester (1975).

The important conclusion that can be drawn from Figure 2 is that the distortion produced by available broadband hearing aid earphones is well below that generated by the nonlinearities in the ear. It would take some 20 dB of pure conductive hearing loss before the earphone distortion became comparable to that of the ear.

The (BP-series) earphone distortion curves shown in Figure 2 are generally "worst-case" figures, since these curves were obtained by using a 6R12 earmold (Knowles and Killion, 1978), whose enhanced high frequency response tends to emphasize the higher frequency distortion products. A further reduction in earphone distortion can be obtained by the use of a low impedance amplifier (as shown by the dotted curve in the bottom graph in Figure 2), rather than the high impedance amplifier used to obtain the solid curve.

Limitations in Frequency Response

Microphones   Until roughly a decade ago, the frequency response available in a practical head-worn hearing aid was limited by the magnetic microphones available at the time. These had been designed intentionally to produce a rising response with a bandpass filter characteristic. Since that time, a wide variety of microphone types, and microphone frequency responses, have become available. Figure 3 shows just a few of the frequency responses that are available today in subminiature microphones. The curves in Figure 3 were obtained on the microphone by itself in each case. When the effect of the various coupling tubes commonly used with these microphones is included, a much wider variety of responses can be obtained.

Although both the ceramic microphones and the electret-condenser microphones are available in flat frequency-response versions, they are most often provided with a peak in response in the neighborhood of the unoccluded external ear resonance. This frequency-response tailoring helps compensate for the loss of external ear resonance when the earmold is inserted; it began in 1968, when the subminiature broadband ceramic microphones were first introduced (Killion and Carlson, 1970).

Earphones   A similar variety of frequency responses is available from subminiature hearing aid earphones. Figure 4 shows a small sampling of the response curves that are available. The original intent was to show the response of a variety of earphone types, mounted in a variety of hearing aid cases, driven from a variety of amplifier source impedances, and coupled to a variety of earmolds. We only got as far as the variety of earmolds. All curves were obtained with a single BP-1712 earphone mounted in an over-the-ear (OTE) hearing aid case (as shown in Figure 16), and driven from a high impedance source (0.01 μF shunted the earphone terminals). The only changes made to obtain the curves shown in Figure 4 were in the earmold constructions. As is obvious from the extensive discussion of earmold venting by Lybarger (Chapter 10, this volume), it would have been possible to add another series of curves to Figure 4 by including a variety of earmold ventings. Only one vented earmold is shown in Figure 4: the near-extreme case of the "acoustic modifier" type of earmold.

It is apparent from the variety shown in Figure 4 that even defining the term *broadband earphone* can be difficult. The curves shown in Figure 5 represent the best comparison we have made to date between conventional and broadband earphones. These response curves were obtained with the earphones mounted in OTE hearing aid cases and coupled with 40 mm of #13 earmold tubing to a Zwislocki coupler. Two 1500-Ω dampers were placed in the earmold tubing to smooth the frequency response, one at the earhook and one 6 mm back from the tubing outlet. The only change

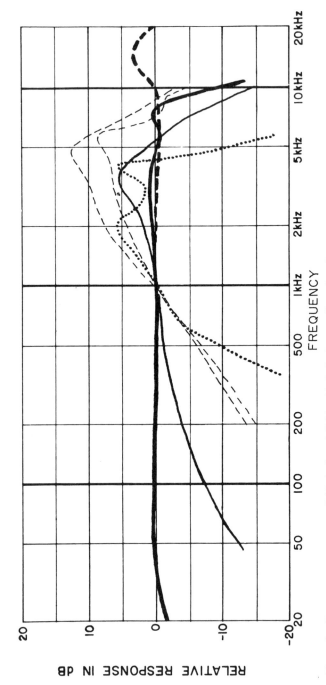

Figure 3. Some response curves available in 6 mm³ (0.01 cubic inches) or smaller microphones.

226

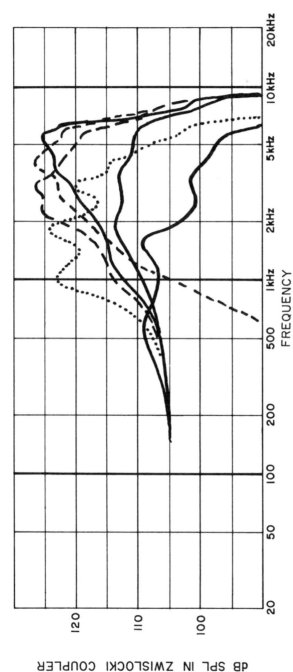

Figure 4. Wide variety of response curves obtained (with a single earphone) by changing the earmold construction.

227

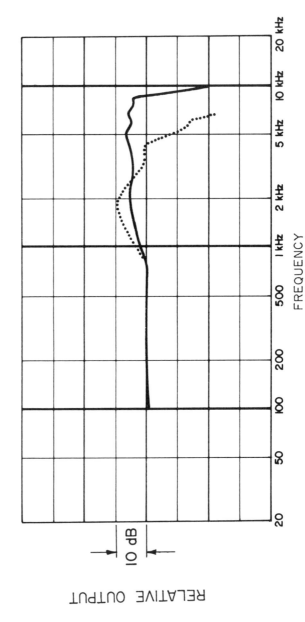

Figure 5. Comparison between broadband BP-series earphone (solid line) and conventional BK-series (dotted line) earphone with one well-damped earmold coupling.

228

between the two curves shown in Figure 5 was a change in the shunt capacitance used to obtain a maximum increase in high frequency output with each earphone.

Before leaving the question of frequency response, one comment is in order: the frequency responses in Figures 1, 3, and 4 have been plotted to the engineering standard scale of 30 dB/decade, instead of the hearing aid standard scale of 50 dB/decade. Figure 6 illustrates the visual effect of the different scales on one completed hearing aid frequency response. Transducers by themselves have traditionally been plotted on the engineering scale, and so most of the frequency response curves used in this chapter are plotted on the 30 dB/decade engineering scale.

However the curves are plotted, it is clear that almost unlimited combinations of frequency responses are available with present hearing aid transducers.

Poor Transient Response

There are two common interpretations for the term *poor transient response.* One is the difference between the output and input wave forms viewed on an oscilloscope when a transient is applied to sound reproduction system. In general, this difference is an inevitable consequence of any frequency-response shaping in the system under test. By this definition, for example, the ear itself has a very poor transient response because of the resonances in the external ear.

To a reasonable first approximation, a hearing aid system can be represented as a minimum-phase network. Under those circumstances, the transient response can be predicted directly from the frequency response. The frequency-response tailoring of a particular hearing aid may or may not be useful, but its effect on "transient response" is inevitable (Mott, 1944). A comprehensive set of frequency-response curves and their corresponding transient-response wave forms was shown by Mott.

The other type of "poor transient response" generally involves an amplifier that exhibits a slow recovery from overload. As mentioned previously, instantaneous peak sound pressure levels of 110 to 115 dB at the hearing aid input are not uncommon. Such peaks can easily cause sufficient amplifier overload to upset the bias levels on the internal coupling capacitors, causing a high amplifier distortion condition lasting much longer than the transient itself (Ingelstam et al., 1971; Killion, Carlson, and Burkhard, 1970). This "blocking distortion" was more often a problem with older amplifier designs, and is much less of a problem now that the majority of hearing aids use some form of fast-acting compression-limiting or AGC system.

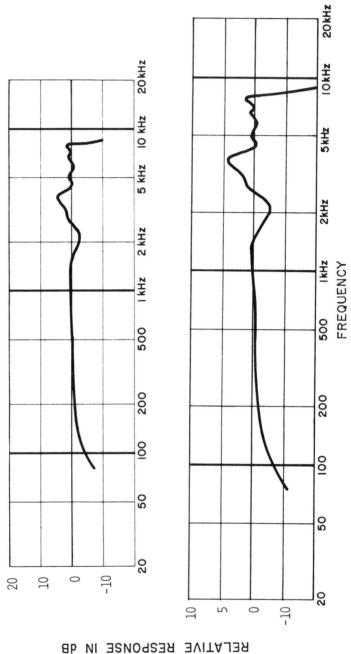

Figure 6. Calculated insertion gain (random incidence) of an OTE experimental hearing aid, plotted on two scales: 50 dB/decade hearing aid standard scale (upper graph), and 30 dB/decade engineering standard scale (lower graph).

Peaks in the Frequency Response

Although many hearing aids are sold that produce peaks in the frequency response, it is not because of a lack of engineering solutions. Figure 7, for example, shows the smooth response obtained in a body-worn aid designed by Lybarger (1949).

The AR series of earphones that Knowles introduced in 1955 were the first earphones small enough to be mounted in a head-worn hearing aid and were specifically designed to be used with a long transmission line, i.e., with a plastic tube coupling the earphone output down to the tip of the earmold. As part of the original design, removable damping elements were provided. These did an excellent job of smoothing the tubing resonances, as shown in Figure 8. At that time, the earphone was large enough that its acoustic impedance was quite low, and thus the tubing resonances could be well damped by the use of an acoustic resistance placed directly at the earphone outlet. These dampers were made available to dispensers through the manufacturer who supplied the hearing aids, but most hearing aids were sold without dampers. A typical reaction was, "Sounds louder without the damper," so the damper was not used.

It would be easy to show a series of frequency responses of hearing aids designed over the years with smooth frequency responses obtained by the proper use of damping; nonetheless, these hearing aids have remained a small factor in the marketplace.

There appear to be several reasons for this. First, the peaks caused by tubing resonances produce a greater hearing aid output for the same battery

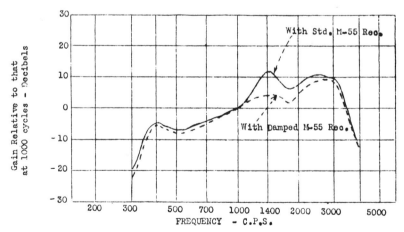

Figure 7.   Well-damped (body-worn) hearing aid response designed in the 1940s. (Reprinted with permission from Lybarger, 1949, Damping elements for M-55 receivers, *Radioear Voice,* Jan.-Feb., pp. T1–T4.)

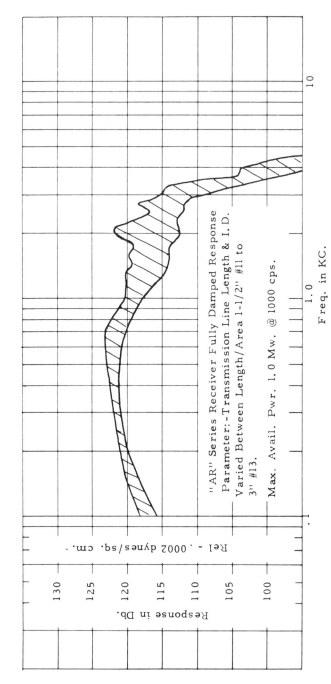

Figure 8. Well-damped (head-worn) hearing aid earphone response available in the 1950s. (Reprinted with permission from Knowles, 1955.)

drain. Producing a truly well-damped earphone-tubing system may lower the high frequency saturation sound pressure level (SSPL) output figure on the data sheet by some 6 dB. Restoring that saturation output will require typically two to four times the battery drain. The first head-worn aids were power limited because of available battery and amplifier technology, so the biggest obstacle to the sale of head-worn aids was their lack of loudness. Removing the damping made the head-worn aids sound much louder, and dealers found that "loudness sells."

Another reason was that the perturbation in the binaural phase-frequency relationships introduced by undamped tubing resonances turned out to be much less of a problem than originally anticipated. Experiments conducted by Knowles in the 1950s showed that only a few days of reeducation were generally required before normal binaural localization was restored for the head-worn hearing aid user.

Probably the most important reason that well-damped hearing aid responses have not dominated the marketplace to date is that even completely undamped earmold-tubing resonances do not sound as bad as might be expected, especially with the stepped diameter coupling used in most modern hearing aids. Figure 9 shows a comparison between two different earphone earmold responses used with an experimental OTE hearing aid to produce some of the prerecorded comparisons used in the listening test demonstration (see appendix). As might be expected, the smooth response obtained with the well-damped earmold produced significantly higher average fidelity ratings (80%) than the peaked response obtained with the undamped earmold (63%). The relatively low fidelity rating obtained with the undamped earmold, however, exceeded the even lower ratings (56%) obtained for a (probably better-than-average) speech audiometer using the TDH-39/MX-41R headphones—a system that has often been characterized as "high fidelity" in the literature.

The results of the informal listening test conducted during the original presentation of this paper are similar to results obtained in more formal listening test evaluations using trained listeners. The loss of bass response because of the well-known cushion leak with the MX-41AR ear cushion appears to be more objectional to most listeners than the peaks introduced by the undamped earmold tubing. Indeed, the insertion gain frequency response of the experimental hearing aid with the undamped earmold (curve 4 in Figure 31) is not much different from the real-ear response of common airline stereo systems, when both are measured with $\frac{1}{3}$-octave bands of noise.

Nonetheless, there are several reasons to suspect that well-damped hearing aid responses will be more common in the future. Most of the listening test participants had normal hearing. Strong peaks in the fre-

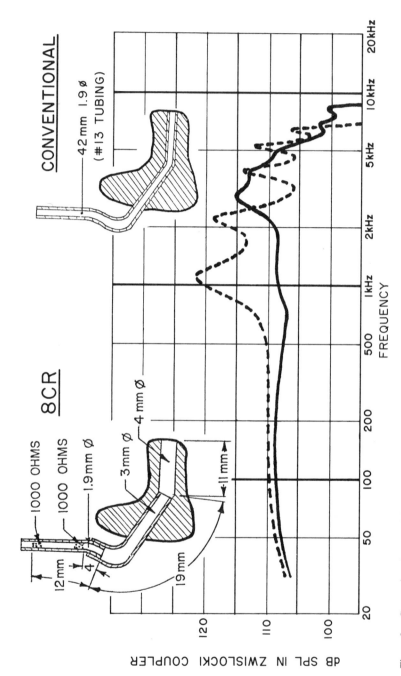

Figure 9. Comparison between 8CR (solid line) and conventional (broken line) earmolds used with broadband earphone. *Note:* BP-1712 earphone is mounted in OTE hearing aid case as shown in Figure 16, and driven from low impedance amplifier (0.7 mA DC bias, 0.56 VAC behind 330 Ω source impedance).

quency response are likely to be much more troublesome to someone with a severely limited dynamic range, because such peaks limit the maximum useful gain a user can employ without experiencing occasional discomfort when an intense vowel formant peak coincides with a peak in the hearing aid response. Moreover, peaks in the transmission characteristic tend to reduce the maximum usable gain before "whistling" because of acoustic feedback (or before changes in the effective frequency response occur as the whistling condition is approached; Lybarger, 1966; Cox and Studebaker, Chapter 9, this volume).

Last, there appears to be an increasing demand for higher quality sound reproduction in hearing aids. This demand seems likely to accelerate as the increasing availability of high fidelity hearing aids makes hearing aid usage more attractive to those for whom hearing aid usage is not a necessity, but simply makes hearing easier (Killion and Carlson, 1970, 1974). Such users generally do not require high sound pressure levels, so that the trade-off between battery drain and SSPL output of the hearing aid may be of little concern.

Fortunately, the usual trade-off between response smoothness and battery drain can be at least partially circumvented by the use of stepped-bore earmolds, as discussed in the next section.

A LOOK AT THE COUPLING PROBLEM

The Problem with Tubing Resonances

As smaller and smaller hearing aid earphones were designed, their acoustic output impedance inevitably went higher and higher. The broadband earphone discussed in this chapter, which was the smallest earphone commercially available when this chapter was written, has an acoustic source impedance that is almost 100 times greater than the acoustic impedance of the load represented by the ear canal and eardrum impedance. Figure 10 shows the well-known tubing resonance peaks that occur when such a high impedance source is connected to a low impedance load through a section of tubing. The curve in this figure was computer generated under the assumption that the earphone itself had a flat frequency response extending well beyond 40 kHz.

The advent of the digital computer made the routine calculation of the frequency response of a transducer tubing system practical. Before then, a typical transducer manufacturer would construct an electrical analog (using physical components) corresponding to each microphone and earphone series, and perhaps a lumped-element approximation to the transmission line formed by the tubing used to couple the earphone to the ear. (A

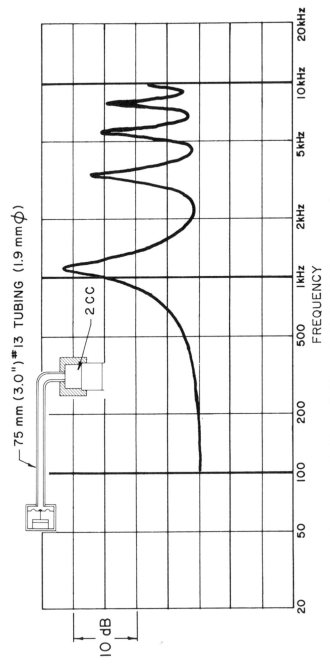

Figure 10.   Calculated tubing resonances for 75 mm of tubing used with ultra-high frequency earphone.

10-section lumped approximation to 95 mm of 1.9 mm in diameter tubing was used at the writer's firm (I.R.P.I.) back in the late 1950s, for example.)

As discussed by White et al. (Chapter 12, this volume), the use of electrical analogs in transducer design has been routine since the Harrison (1929) and Norton (1928) disclosures. Similarly, the use of transmission line equations in treating acoustic coupling tubes has been routine since Mason's papers (1927, 1930) and Stewart's disclosure (1928), in which the continuous and equivalent-T lumped-element representations were discussed. The curve shown in Figure 10 was calculated using Zuercher's (1977) approximations to the full Bessel function solution for acoustic tubes, approximations that allow highly accurate curve generation at a speed equal to that obtainable using automatic analog plotting equipment with physical analogs.

Figure 10 provides a graphic illustration of the problem of delivering a flat frequency response at the other end of a transmission line, a problem that was solved more than 80 years ago. The solution is to terminate the transmission line with a resistance equal to the characteristic impedance of the transmission line (Heaviside, 1899). In the case of an acoustic tube, the characteristic impedance is equal to 41 cgs Ω (ρc for air) divided by the cross-sectional area of the tubing. Approximately 1400 Ω is calculated for #13 (1.9-mm) tubing, for example.

Figure 11 illustrates the effect of terminating resistance on the behavior of an acoustic transmission line formed from the 1.9-mm tubing commonly used with hearing aids. Observe that too low a value of resistance provides insufficient damping of the response peaks, and that too high a value of resistance introduces a new set of peaks (this time at the even multiples of a quarter wavelength), whereas the proper resistance provides an almost perfectly flat frequency response.

Olney, Slaymaker, and Meeker (1945) applied this solution to eliminate the resonance peaks in a probe-tube style of noise-canceling microphone. The effectiveness of this approach when used with a broadband hearing aid earphone is shown in Figure 12. (Note that for consistency the last three figures have been 2-cc coupler curves.)

When applied to a hearing aid, the problem with the classical approach is that terminating the low impedance end of the transmission line with a resistance means placing a damping element at the tip of the earmold—a location where it is likely to become quickly clogged with earwax.

When both ends of the transmission line are connected to a low impedance (as in hearing aids using 1955 vintage large-volume earphones), the damping resistance can be placed at either end of the transmission line. With the high acoustic impedance of a modern broadband earphone, on the other hand, a damping resistor located at the receiver outlet will be safe from earwax but may have little effect on the tubing resonances. This is illustrated in Figure 13.

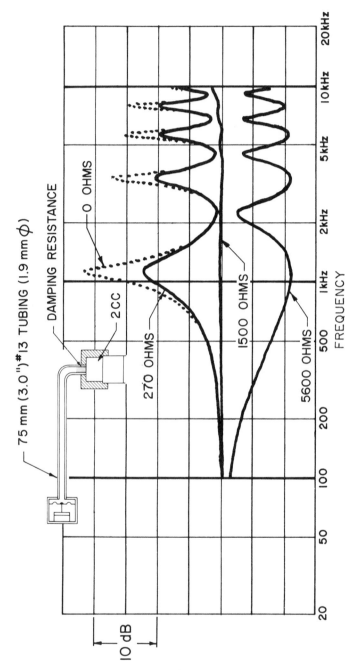

Figure 11. Calculated effect of damping applied to coupler end of 75 mm of tubing used with ultra-high frequency earphone.

238

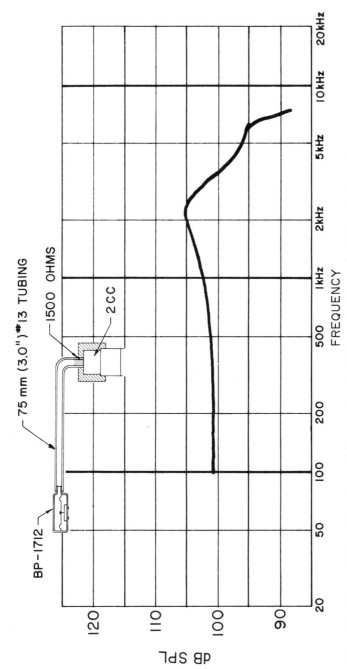

Figure 12. Measured effect of damping applied to coupler end of 75 mm of tubing used with broadband earphone.

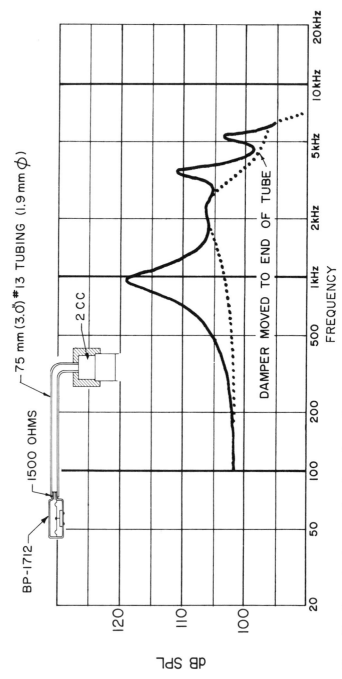

Figure 13. Comparison between effect of damping applied at earphone (solid line) and at coupler end (dotted line) of 75 mm of tubing used with broadband earphone.

240

Carlson's Twin-Tube Solution

A novel and elegant method for damping all tubing resonances by use of dampers located near the earphone was described by Carlson (1974). One mechanical construction using this approach is shown in Figure 14. In essence, the Carlson twin-tube method uses an auxiliary tube that is blocked at the far end in order to provide an impedance conjugate to that of the main coupling tube, and thus in effect cancels the tubing resonances.

The effectiveness of the Carlson twin-tube approach is shown by the curves in Figure 15. Note that almost identical frequency response curves are obtained regardless of the length of the tube pair.

The Carlson solution is being used commercially by at least two hearing aid manufacturers; one uses a simplified lumped-element version. In the lumped-element version, a damped "Helmholtz resonator" is placed in shunt with the receiver output to absorb energy at an otherwise troublesome peak in the tubing transmission (Carlson, 1977).

Stepped-Diameter Tubing

There is another approach to the impedance mismatch problem: to use some form of "horn coupling" between the earphone and the eardrum. As discovered years ago by hearing aid designers, a stepped-diameter coupling system not only improves the high frequency output of the hearing aid, but reduces the peak-to-valley ratio in the output frequency-response curve. Figure 16 shows a typical coupling system, with a small-in-diameter flexible tube connecting the earphone to an earhook, which has a slightly larger

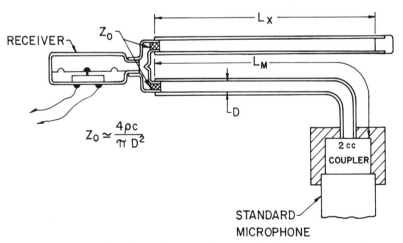

Figure 14.   Example of Carlson twin-tube earphone coupling for smoothing the frequency response. (Reprinted with permission from Carlson, 1974.)

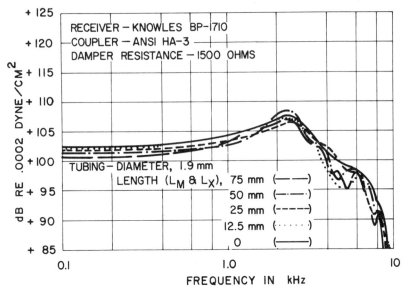

Figure 15. Measured effectiveness of Carlson twin-tube coupling used with various tubing lengths and a broadband earphone.

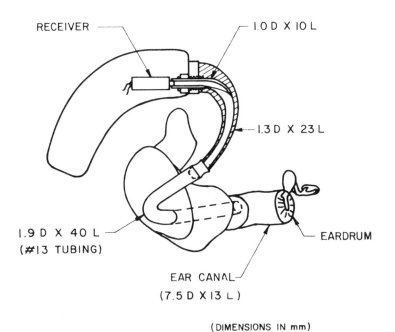

(DIMENSIONS IN mm)

Figure 16. Typical earphone-to-eardrum coupling for OTE hearing aid.

diameter internal bore. The still larger-in-diameter earmold tubing then extends to the tip of the earmold.

The impedance transformation provided by this stepped-diameter coupling makes a damper placed at the earphone outlet much more effective in damping the tubing resonances, as illustrated in Figure 17. The combination of stepped-diameter coupling and a damping resistance placed at the receiver outlet or in the earhook has allowed hearing aid designers to produce hearing aids with smooth frequency responses for nearly two decades.

Stepped-Diameter Earmold Construction   In the 1960s, Knowles suggested that the stepped-diameter approach to the transmission line problem be extended into the earmold itself. This approach was used in a commercial hearing aid design by Lybarger in 1970. Lybarger's earmold recommendation is shown in Figure 18. The increased high frequency response that could be obtained from such a stepped-diameter earmold in a hearing aid using a conventional earphone is shown in Figure 19.

Several years ago, I became interested in the potential improvement in the response smoothness and high frequency output that could be obtained by the use of stepped-diameter earmold bore in conjunction with broadband earphones. Initial experiments resulted in a demonstration that a relatively smooth response extending to 9 kHz was entirely practical using the proper earmold in conjunction with a broadband earphone (Killion and Carlson, 1973). Further experiments led to the development of the 6R10 earmold (Killion, 1976b).

The evolution of the frequency response of the 6R10 earmold is shown in Figure 20. The first thing to note is that placing a capacitor across the terminals of the earphone can produce substantial high frequency response smoothing when the earphone is driven from a high impedance amplifier. If, in addition, the "horn" formed by a stepped-bore earmold construction is added, an additional improvement in response smoothing is obtained, as well as a substantial increase in output above 2 kHz, as shown in Figure 20b. Note also that the stepped-bore earmold has shifted the second response peak at 2200 Hz up to 2700 Hz, the average frequency of the external ear resonance. In order to help compensate for the loss of external ear resonance that accompanies earmold insertion, it is sometimes useful to retain a peak in the earphone coupling system at 2700 Hz while damping all the other peaks, which can be obtained by the proper combination of damping and tubing diameter. For the present purposes, however, note that the application of two damping elements in the earmold tubing produces a smooth response curve even though neither damper is near the tip of the earmold. This two-damper approach is similar to the one described by DiMattia (1958) for use with stethoscope-type headphones.

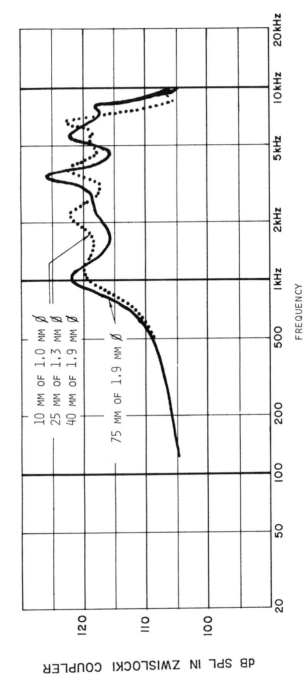

Figure 17. Response smoothing with 1500 Ω located at BP-1712 earphone outlet and 75 mm of constant diameter coupling (solid line) and stepped diameter coupling (dotted line).

244

TO NUB

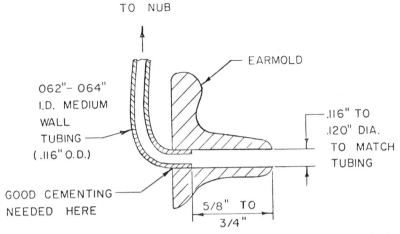

Figure 18.   Dual-bore earmold recommended by Radioear for use with a hearing aid that employs a conventional earphone. (Reprinted with permission from Lybarger, 1970.)

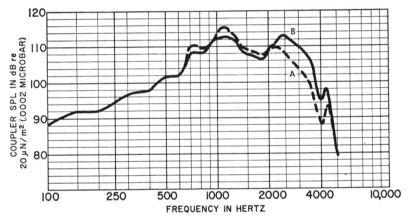

Figure 19.   Increased high frequency output obtainable with conventional hearing aid earphone using dual-bore tubing configuration. (Reprinted with permission from S.F. Lybarger, Ear molds, in J. Katz (ed.), *Handbook of Clinical Audiology*, © 1972, The Williams & Wilkins Co., Baltimore.)

This earmold was originally called a 6R10 earmold because until recently the only suitable damping elements available were of the sintered-metal construction (such as the Knowles BF-1540 series). When dampers having better acoustic characteristics became available, a greater high frequency output with better overall response smoothness became possible, which resulted in the change of the earmold name from 6R10 to 6R12 (meaning the Zwislocki coupler response at 6 kHz is 12 dB higher than the

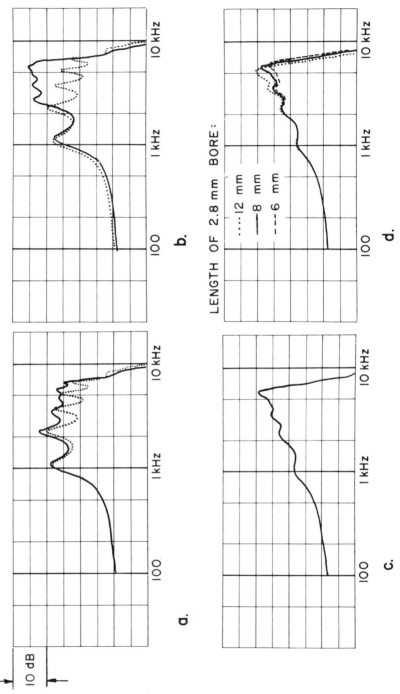

Figure 20. Evolution of 6R10 broadband earmold. a, Shunt capacitor added (0.2μF across receiver; b, "horn" added; c, damping added; d, final tuning.

1-kHz response). The construction and frequency response of the 6R12 earmold are shown in Figure 21 for the earphone mounting and electrical drive conditions described for Figure 4.

An Improved Damping Resistance The improved fused-mesh damping element developed by Carlson and Mostardo (1976) has been made commercially available (Knowles Electronics, BF-1859 series). From a technical standpoint, these damping elements exhibit a nearly pure resistance, so that they introduce only a negligible discontinuity into the transmission line. From a practical standpoint, they allow better smoothing of low frequency tubing resonances with improved high frequency output.

Although these damping elements were primarily designed to fit snugly in 1.9-mm (#13) earmold tubing, they can be used at several other places along the transmission line. Figure 22, for example, illustrates the fact that frequency responses very similar to the 6R12 frequency response can be obtained with a *single* (very carefully placed) damping element located in the earmold tubing, or with two damping elements located inside the hearing aid. In the latter case, the earmold itself retains only the stepped-bore construction and is called an "undamped 6R12."

Exponential Couplers (Horns)

The theory of acoustic horns is well known. The impedance transformation that can be obtained by the use of a horn is equal to the ratio of the outlet area to inlet area. The pressure transformation is therefore equal to the square root of the outlet-to-inlet area ratio, or directly equal to the output-to-inlet *diameter* ratio. This transformation obtains above some minimum cutoff frequency, which for exponential horns is commonly called the flare cutoff frequency, $f_c = mc/4\pi$, where c is the velocity of sound and m is the flare constant determining the increase in area A, with distance x along the horn in accordance with the equation $A = A_0\epsilon^{mx}$ (see, for example, Olson, 1957, pp. 112–114).

Figure 23 shows the relationship between tubing cross-sectional area and tubing length for an exponential horn having two different theoretical flare rate cutoff frequencies: 1 kHz and 2 kHz.

Three comments need to be made about the relationship in Figure 23:

1. When an "exponential horn" is used in a closed system it is more properly called an "exponential coupler."
2. The effective cutoff frequency of an exponential coupler is usually two to three times the theoretical flare cutoff frequency in the configurations of interest in hearing aid design.
3. A "horn" coupler does not have to have a smooth taper in order to achieve effective transfer impedance. In practice, very similar results to

the continuously tapered transmission line can be obtained as long as the distance between the steps is small compared with one-quarter of a wavelength. When the distance between steps is *not* small compared with one-quarter of a wavelength, the impedance discontinuities introduced by the steps can be exploited to produce additional frequency-response control.

It is interesting to see how the steps in a practical stepped-diameter hearing aid coupling system compare with that required of a perfect exponential coupler. One such comparison is included in Figure 23. Keeping in mind that the practical cutoff frequency is likely to be two to three times the calculated flare cutoff frequency, the curves in Figure 23 would indicate that perhaps above 2 to 3 kHz (when mounted in a hearing aid with internal stepped-bore tubing), a high acoustic impedance earphone used with a conventional earmold should exhibit a 6-dB greater high frequency output than if the stepped-diameter coupling system were not used. With the addition of a stepped bore in the earmold, a total of 12 dB greater high frequency output can be expected. In particular, the use of an earmold such as the 6R12 earmold might be expected to produce a net improvement of 6dB in high frequency output compared with a conventional earmold. An approximation of that result obtains in practice, as shown by the curves in Figure 24. These curves show the increase in high frequency response of a broadband hearing aid that can be obtained by progressively boring out the earmold. (It should be noted that the increased high frequency output shown in Figure 24 can only be obtained with an earphone designed to maintain a high acoustic source impedance at high frequency. The results shown here were obtained with a BP-1712 earphone mounted in an OTE hearing aid case, as shown in Figure 16.)

A Delivery Problem

There is a practical obstacle to the design of hearing aids using more sophisticated earmold acoustics: the delivery problem. Many hearing aid dispensers are still not adequately familiar with the predictable low frequency response control (Lybarger, Chapter 10, this volume) that can be obtained through the proper use of earmold venting. The addition of variations in high frequency response caused by earmold design is not likely to improve the situation.

Assume for the moment that someone designs a hearing aid using a variation of the 6R12 earmold, and produces the smooth high frequency emphasis response shown in Figure 25. (Note that Figure 25 shows the *insertion gain* and not the coupler response of the laboratory hearing aid. The latter exhibits a substantial peak at 2700 Hz.) If this hearing aid is

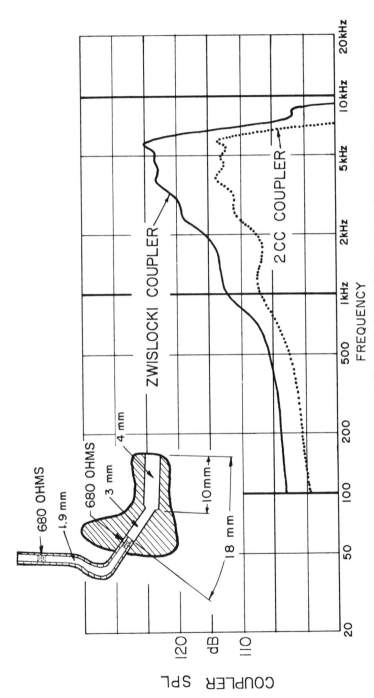

Figure 21. Response of broadband earphone with 6R12 earmold. (Reprinted with permission from Knowles and Killion, 1978.)

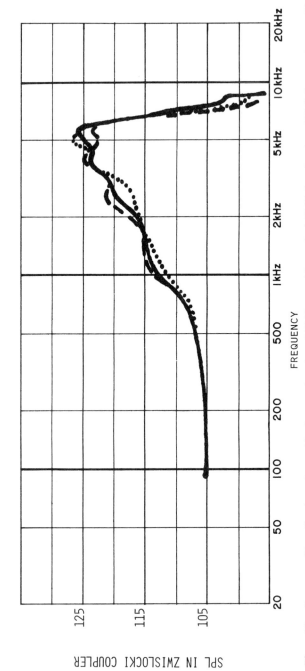

Figure 22. Response of broadband earphone with alternate damping of 6R12 earmold: standard 6R12 damping (two 680-Ω dampers in earmold tubing) (solid line), in-the-aid damping (1500 Ω at earphone outlet and 680 Ω at earhook tip) (broken line), and single-element damping (single 1500 Ω in earmold tubing) (dotted line).

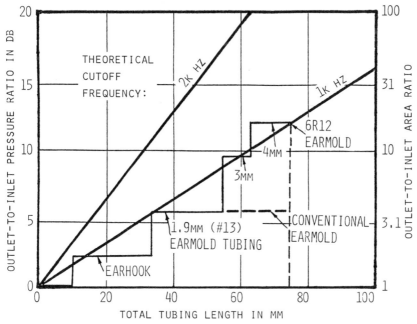

Figure 23. Area vs. length required of "exponential coupler," assuming theoretical flare rate cutoff frequencies shown. The area vs. length of a practical coupling system is shown for comparison.

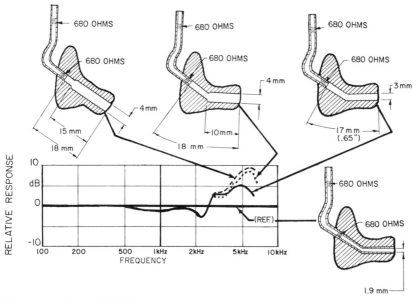

Figure 24. Change in high frequency response of broadband hearing aid caused by boring out earmold.

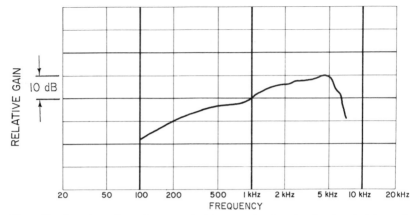

Figure 25.   Insertion gain vs. frequency for laboratory hearing aid. (Reprinted with permission from Knowles and Killion, 1978.)

delivered along with a conventional earmold, the response change shown in Figure 26 is obtained. Worse yet, if it is evaluated in the clinic using a "stock" earmold (as is often done, unfortunately), the response change shown in Figure 27 may be obtained. Instead of a hearing aid with a 6-kHz bandwidth, the hearing aid now has an apparent bandwidth of only 2 to 3 kHz.

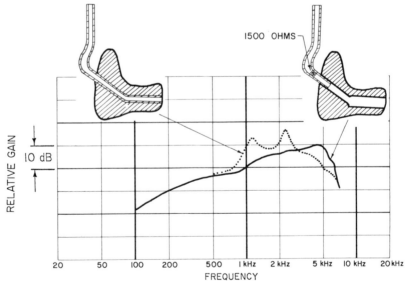

Figure 26.   Insertion gain of laboratory hearing aid with recommended earmold (solid line) and conventional earmold (dotted line). (Reprinted with permission from Knowles and Killion, 1978.)

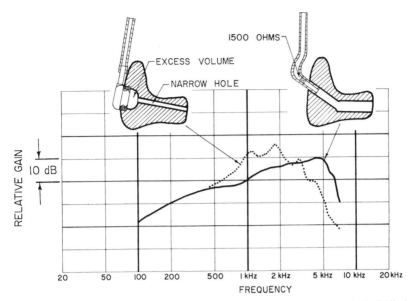

Figure 27.   Insertion gain of laboratory hearing aid with recommended earmold (solid line) and improper snap-in earmold (dotted line). (Reprinted with permission from Knowles and Killion, 1978.)

There may be times when exactly the response change shown in Figure 27 is desired. The increased response in the 1- to 2-kHz region caused by the presence of the cavity in the transmission line might mean that a hearing aid with substantially lower battery drain could be used with some clients. Such an earmold design has been suggested recently by Goldberg (1977, 1978), as shown in Figure 28. This earmold is commercially available through at least one earmold laboratory.

Another Look at the Battery Drain Trade-Off

It is of some interest to look at the reduction in battery drain (class A amplifier) that can be obtained by the use of stepped-bore earmold construction. Figure 29 shows the same family of maximum undistorted output curves shown in Figure 1, except that a 6R12 earmold has been substituted for the well-damped conventional earmold. An undistorted output exceeding 105 dB can be obtained over the entire frequency range from 200 to 6000 Hz, with a battery drain of only 0.35 mA. This compared favorably with the 1 mA that was required with the well-damped conventional earmold. The earmold damping for Figure 1 probably exceeds the level that would be used in a commercial hearing aid, so the comparison may be slightly exaggerated, but the point is not. The use of a stepped-bore earmold may make it practical to employ sufficient damping in the response curve to

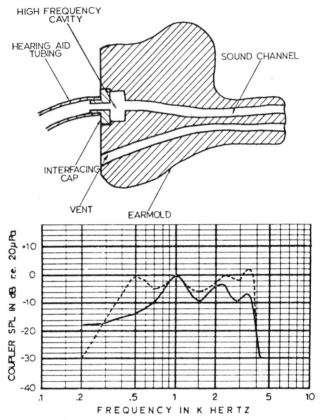

Figure 28.   Example of earmold using lumped elements in transmission line (dotted line) compared with conventional (solid line) earmold. (Reprinted with permission from Goldberg, 1977.)

produce a truly smooth frequency response, without incurring the increased battery drain penalty discussed earlier.

Amplifier Source Impedance

A final response control measure available to the hearing aid designer should be mentioned. Figure 30 shows the change in earphone frequency response caused by changes in the source impedance, i.e., changes in the output impedance of the hearing aid amplifier.

The three curves in Figure 30 have been normalized to the same output level at 100 Hz to illustrate the change in frequency-response curve shape. The *maximum* undistorted output levels obtainable with a single-ended class A amplifier stage and 1.5 V-supply are independent of the amplifier

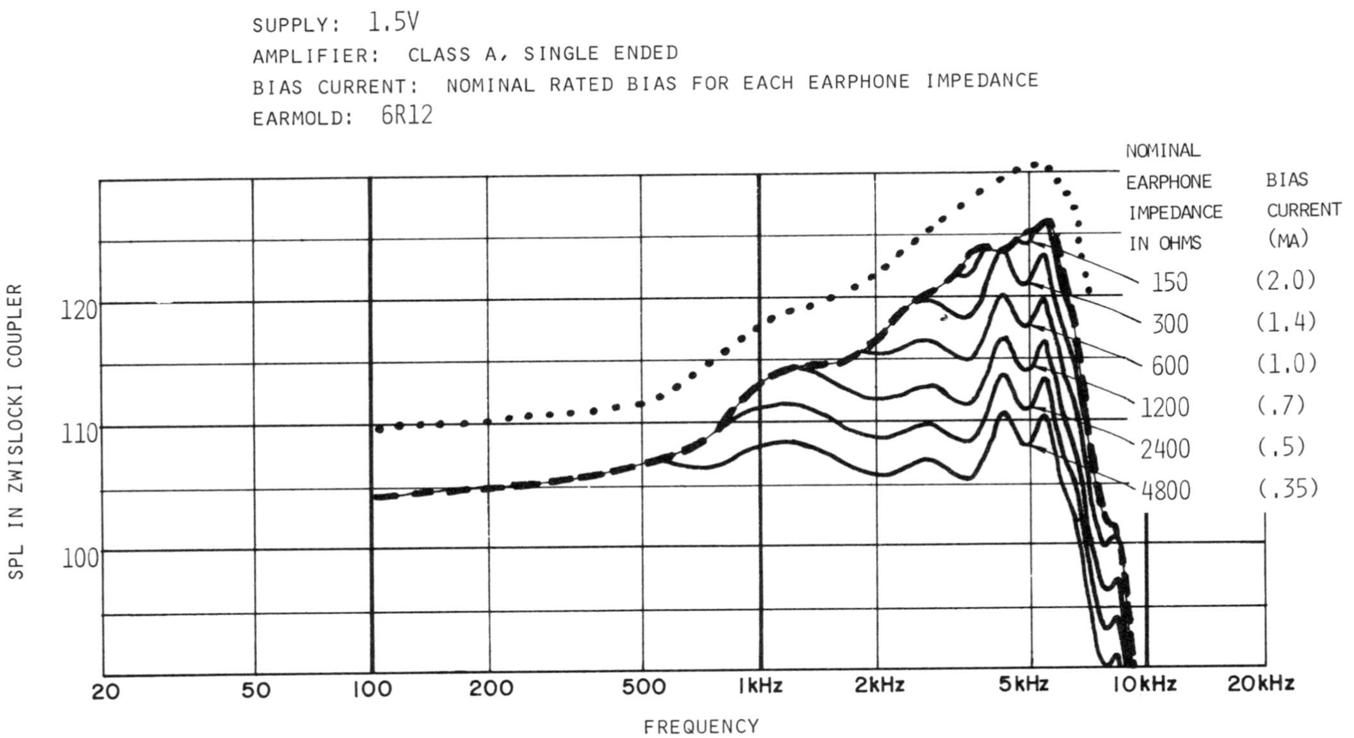

Figure 29. Maximum undistorted output of **BP**-series earphone with well-damped stepped-bore earmold, limited by earphone overload (dotted line), amplifier current clipping (broken line), and amplifier voltage clipping (solid line).

255

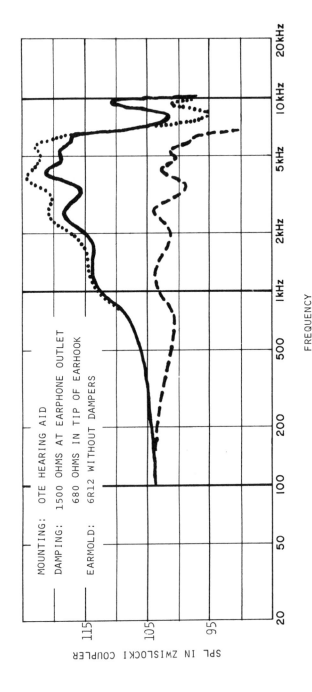

MOUNTING: OTE HEARING AID

DAMPING: 1500 OHMS AT EARPHONE OUTLET
680 OHMS IN TIP OF EARHOOK

EARMOLD: 6R12 WITHOUT DAMPERS

FREQUENCY

Figure 30.   Effect of source impedance on frequency response of BP-series earphone. Response obtained with electrical source of high impedance (constant current) (solid line), high impedance with shunt capacitor (dotted line), and low impedance (constant voltage) (broken line).

output impedance, which is generally determined by the electrical feedback conditions. Thus the *maximum* undistorted high frequency output will be the same for both high impedance and low impedance class A amplifiers (see Figure 29), but the low level frequency response curve will be substantially different (as illustrated in Figure 30).

SUMMARY AND CONCLUSIONS

There are surprisingly few technical limitations to hearing aid performance. Except in those cases where extremely high gains or sound pressure levels are required, the present state of the electronic and transducer art permits almost any conceivable combination of electroacoustic characteristics in head-worn hearing aids, up to and including high fidelity hearing aids. The problem of response peaks introduced by the coupling between the earphone and the earmold tip can be solved in many ways, some of them quite old. (Incidentally, the same conclusion holds for the coupling problem with in-the-ear (ITE) hearing aids, although they were not considered in this chapter.)

Unfortunately, the main problems remain unsolved; we know embarrassingly little about what is required to optimize the electroacoustic characteristics of a hearing aid for a given individual as he goes about his daily life. Indeed, only recently has a consensus appeared in the research literature on the relatively simple question of what electroacoustic characteristics can be used to optimize speech discrimination in a laboratory setting where the task is to repeat fixed presentation level words or syllables. Because hearing aid designers can now produce almost any conceivable combination of characteristics in a hearing aid, they are in an even bigger quandary than before: Which combination of characteristics will provide increased utility for enough hearing aid users to make it economically feasible to produce and properly deliver the improvement to the end user? Providing answers to this question is the challenge facing us. Once it is answered, providing the required hearing aid characteristics will be relatively simple.

APPENDIX: A LISTENING TEST DEMONSTRATION

PROCEDURE

In order to illustrate some of the points made in the formal presentation, a listening test demonstration preceded the presentation itself. Five sound reproduction systems were represented (three of them hearing aids), reproducing first a recording of the Oscar Peterson Trio and then of the New York Philharmonic. The standard of reference for all comparisons was a pair of AR3a loudspeakers spaced along one wall of a 170-m³ (6000 cubic feet) room at the Auditory Research Laboratories at Northwestern University. The sound absorption treatment on the walls of that room was adjusted to provide the 0.3- to 0.5-sec reverberation time typically recommended for recording studios of that volume.

The demonstration used recorded A-B-A comparisons obtained from the output of eardrum-position microphones in a KEMAR manikin after electrical equalization to remove the spectral peak of approximately 15 dB at 2800 Hz produced by the external-ear resonances (a description of the "bridged T" equalization filter is given in Killion, 1979a). In the case of the hearing aid comparisons, the signal obtained with KEMAR "listening" directly to the AR3a loudspeakers (A) was compared with the signal obtained with KEMAR "listening" through the hearing aids (B) to the same loudspeakers. In the other comparisons, the sound reproduction system under test was simply substituted for the reference loudspeakers during the comparison portion.

The audience participants were provided with rating sheets and IBM cards to fill out. They were asked to rate the *similarity* between the sounds heard in segment A and segment B on a 0% to 100% fidelity scale. They were told that if the comparison sound reproduction system did a perfect job of duplicating the sound of the reference sound system, there would be no audible difference between the two segments and a 100% fidelity rating would be expected. A low fidelity rating would be expected if a large difference in sound quality between the reference and comparison systems was heard.

THE SYSTEMS COMPARED

Figure 31 shows the ⅓-octave band frequency response of the five sound reproduction systems used in the prerecorded listening test comparisons.

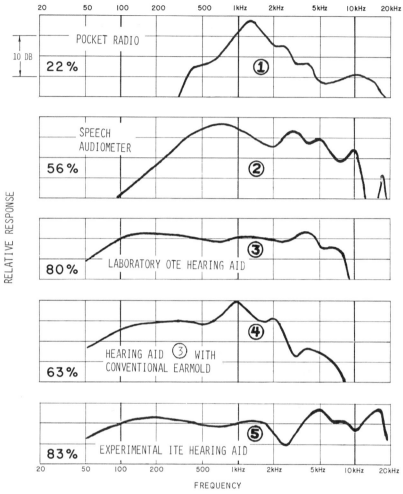

Figure 31.   Frequency response of five sound reproduction systems, as measured with ⅓-octave bands of noise and an equalized KEMAR manikin.

System 1 was an inexpensive ($4.95) pocket radio included as a low fidelity reference.

System 2 represented a high quality speech audiometer assembled from a professional Ampex tape recorder, a Marantz Model 250 high fidelity stereo amplifier, and a pair of TDH-39 earphones—selected to have a typical frequency response—mounted in the standard MX-41 ear cushions. The earphones were carefully taped on a KEMAR manikin to produce a low frequency roll-off (caused by the leak around the ear cushions) equal to the average obtained from probe-tube measurements on real ears as given by Shaw (1966). Between 200 and 10,000 Hz, the resulting "eardrum pressure" response measured on the KEMAR manikin fell to within 2–4 dB of the predicted average real-ear response calculated from Shaw's data.

System 3 was an experimental OTE hearing aid with an 8-kHz bandwidth. It has less than 1.0-mA battery drain, yet it reproduces a fortissimo passage from a full symphony orchestra at concert hall levels without audible distortion. The 8CR earmold used with this aid is shown in Figure 9.

System 4 was included to provide preliminary information on the effect of earmold variations on the fidelity ratings. System 4 used the same 8-kHz OTE hearing aid that was used for System 3, except that a conventional (undamped) earmold was substituted for the 8CR earmold. A frequency-response comparison between these two earmolds is shown in Figure 9.

System 5 was an experimental ITE hearing aid with 16-kHz bandwidth. Like System 3, it reproduces a fortissimo passage at original levels without audible distortion, but requires a battery drain of 5 mA to accomplish it. With the limited capacity of the small cells typically used in ITE hearing aids, such a high drain corresponds to only a few hours of battery life. System 5 was not constructed as an example of a practical hearing aid, but simply as a demonstration that a hearing aid earphone was capable of delivering that combination of bandwidth and output level. (From a more practical standpoint, at least one individual appears to be benefiting from an upward-shifting frequency transposing body-worn hearing aid whose output is fed to the same type of broadband hearing aid earphone used for these demonstrations. The user in question has an unusual hearing loss configuration: a profound loss at the standard audiometric frequencies but near-normal hearing above 10 kHz (Halperin et al., 1977). In that application, a few extra milliwatts of battery drain was considered a small price to pay for usable output in the 12- to 18-kHz region.) Since the time this paper was presented, a new hearing aid earphone type (Knowles ED series) has been released. That earphone appears to make an ITE hearing aid with 16-kHz bandwidth more practical.

RESULTS

The system presentation order for the first set of comparisons (Oscar Peterson Trio) was 1,2,3,4,5. The system order for the second set of comparisons (New York Philharmonic) was 1,5,4,3,2. System 2 (the speech audiometer) for example, was heard in the 2nd and the 10th comparison.

The results of the audience ratings are shown in Table 1. The mean percentage of the two ratings for each system is shown in Figure 31. Because of the large sample size ($N = 122$), the standard error of the individual

Table 1.   Results of listening test demonstrations conducted during conference presentation

Sound system	Fidelity average ratings		Overall average ratings
	Piano trio[a]	Orchestra[b]	
1—Pocket radio	25%	19%	22%
2—Speech audiometer	51%	62%	56%
3—Experimental OTE hearing aid	71%	89%	80%
4—Experimental OTE aid with undamped earmold	61%	66%	63%
5—Experimental ITE hearing aid	80%	85%	83%

[a]Oscar Peterson Trio, sixth chorus of "The Smudge" (cheerful blues).

[b]New York Philharmonic, fortissimo passage from Beethoven's Violin Concerto in D.

means was small, ranging between 1.3% and 1.9%. Thus most of the differences between the hearing aids and the other sound systems exceeded 5σ. These results were subsequently corroborated in more formal listening test comparisons involving additional sound reproduction systems and more comparisons. A description of the more formal tests has been published elsewhere (Killion, 1979b), but the conclusion can be stated here: It is clearly possible to build practical hearing aids whose sound quality will be rated comparably with that of good high fidelity sound reproduction systems.

ACKNOWLEDGMENTS

The longstanding intense interest of H.S. Knowles in "the hearing aid problem" has been directly responsible for most of the developments reported in this chapter. Many of the recent technical breakthroughs have been direct or indirect results of the ingenuity and teaching of E.V. Carlson. Too many others to mention have contributed solutions, direct experimental data, and/or insights.

The listening test comparisons were recorded at the Auditory Research Laboratories of Northwestern University, Evanston, Ill. The audience rating cards were computer processed by Harvey Stromberg of the City University of New York.

REFERENCES

Carlson, E. V. 1974. Smoothing the hearing aid frequency response. J. Audio Eng. Soc. 22:426–429.

Carlson, E. V. 1977. Transducer Coupling System. U.S. Patent #4,006,321.

Carlson, E. V., and A. F. Mostardo.   1976.   Damping Element.   U.S. Patent #3,930,560.

Carlson, E. V., A. F. Mostardo, and A. V. Diblick. 1976. Transducer with Improved Armature and Yoke Construction. U.S. Patent #3,935,398.

Clack, T. D., J. Erdreich, and R. W. Knighton. 1972. Aural harmonics: The monaural phase effects. J. Acoust. Soc. Am. 52:536–541.

deBoer, E., and J. Bouwmeester. 1975. Clinical psychophysics. Audiology 14: 274–299.

DiMattia, A. L. 1958. Headphone Device. U.S. Patent #2,849,533.

Goldberg, H. 1977. Earmold Technology Application Notes. Audiol. Hear. Educ. 3:11–14.

Goldberg, H. 1978. An extended range universal earmold. Hear. Aid J., March: 10.

Halperin, H. R., J. K. Cullen, C. I. Berlin, and M. C. Killion. 1977. Translating hearing aid for 'deaf' patients with nearly normal ultra-audiometric hearing. J. Acoust. Soc. Am. 62:S76(A).

Harrison, H. C. 1929. Acoustic Device. U. S. Patent #1,730,425.

Heaviside, O. 1899. Cases of vanishing or constancy of the reflection coefficients. In Electromagnetic Theory, Vol. II, p. 80. Dover Publications, New York. (Reprinted in 1950.)

Ingelstam, R., B. Johansson, A. Pettersson, and H. Sjögren. 1971. Non-linear distortion in hearing aids. Scand. Audiol. (suppl. 1): 126–134.

Killion, M.C. 1976a. Noise of ears and microphones. J. Acoust. Soc. Am. 59: 424–433.

Killion, M.C. 1976b. Earmold plumbing for wideband hearing aids. J. Acoust. Soc. Am. 59:S62(A). (Available from Knowles Electronics, Franklin Park, Ill.)

Killion, M.C. 1979a. Equalization filter for eardrum-pressure recording using a KEMAR manikin. J. Audio Eng. Soc. 27:13–16.

Killion, M.C. 1979b. Design and evaluation of high-fidelity hearing aids. Doctoral dissertation, Northwestern University, Evanston, Ill. (Available from University Microfilms, Ann Arbor, Mich.)

Killion, M. C., and E. V. Carlson. 1970. A wideband miniature microphone. J. Audio Eng. Soc. 18:631–635.

Killion, M. C., and E. V. Carlson. 1973. A new generation of hearing aid transducers. J. Acoust. Soc. Am. 55:393(A).

Killion, M. C., and E. V. Carlson. 1974. A subminiature electret-condenser microphone of new design. J. Audio Eng. Soc. 22:237–243.

Killion, M. C., E. V. Carlson, and M. D. Burkhard. 1970. Audio Frequency Amplification Circuit. U.S. Patent #3,522,100.

Knowles, H.S. 1955. Application Notes on "AR" Series Receiver. Knowles Electronics Engineering Bulletin #ER-101.

Knowles, H.S., and M.C. Killion. 1978. Frequency characteristics of recent broadband receivers. J. Audiol. Tech. 17:86–99, 136–140.

Lybarger, S.F. 1949. Damping elements for M-55 receivers. Radioear Voice, Jan.–Feb.:T1–T4.

Lybarger, S.F. 1966. A discussion of hearing aid trends. Int. Audiol. 5:376–383.

Lybarger, S.F. 1970. Advance Dealer Information: Radioear Model 1010 Eyeglass Hearing Aid. Radioear, Cannonsburg, Pa.

Lybarger, S.F. 1972. Ear molds. In J. Katz (ed.), Handbook of Clinical Audiology, pp. 602–603. The Williams & Wilkins Co., Baltimore.

Mason, W.P. 1927. A study of the regular combination of acoustic elements, with applications to recurrent acoustic filters, tapered acoustic filters, and horns. Bell Syst. Tech. J. 6:258–294.

Mason, W.P. 1930. The approximate networks of acoustic filters. J. Acoust. Soc. Am. 1:263–272.

Moir, J. 1958. High Quality Sound Reproduction, pp. 50, 225. Macmillan Publishing Co., New York.

Mott, E.E. 1944. Indicial response of telephone receivers. Bell Syst. Tech. J. 23: 135–150.

Norton, E.L. 1928. Wave Filter. U.S. Patent #1,681,554.

Olney, B., F. H. Slaymaker, and W.F. Meeker. 1945. The dipole microphone. J. Acoust. Soc. Am. 16:172–177.

Olson, H.F. 1957. Acoustical Engineering. Van Nostrand Reinhold Co., New York.

Shaw, E.A.G. 1966. Earcanal pressure generated by circumaural and supra-aural earphones. J. Acoust. Soc. Am. 39:471–479.

Stewart, G.W. 1928. Acoustic Wave Filter. U.S. Patent #1,692,317.

Trimmer, J.D., and F.A. Firestone. 1937. Investigation of subjective tones by means of the steady tone phase effects. J. Acoust. Soc. Am. 9:24–29.

Wegel, R.L., and C.E. Lane. 1924. The auditory masking of one pure tone by another and its probable relation to the dynamics of the inner ear. Phys. Rev. 23:266–285.

Zuercher, J.C. 1977. The calculation of isothermal and viscous effects in acoustic tubes. J. Acoust. Soc. Am. 62:S56(A).

Zwicker, E. 1976. Psychoacoustic equivalent of period histograms. J. Acoust. Soc. Am. 59:166–175.

Zwislocki, J.J. 1971. Special Report: An Ear-like Coupler for Ear-phone Calibration. Laboratory of Sensory Communication, Syracuse University, Syracuse, N.Y.

Modeling
Techniques

CHAPTER 12

THE APPLICATION OF MODELING TECHNIQUES TO THE STUDY OF HEARING AID ACOUSTIC SYSTEMS

Richard E.C. White, Gerald A. Studebaker,
Harry Levitt, and Douglas Mook

CONTENTS

Models are used in a variety of ways in speech and hearing sciences, ranging from qualitative descriptions of speech production to quantitative representations of sound propagation in a probe-tube microphone. Their structure depends upon the nature of the system being modeled, the intended application, and the personal preferences and background of the model builder. Typically, the motive for developing a model is a desire to represent a real system in a simplified form, in which essential principles are apparent without being obscured by extraneous details. A useful model is generally more readily understood and more easily manipulated than the system it represents, but it is valid only within certain limits. The art of successful modeling is to gradually introduce refinements that expand these limits and lead to new insights about the real system. The development of such models rarely takes place accord-

Parts of the work described in this chapter were supported by Public Health Services Grants NS-13514 and NS-12588 from the National Institute of Neurological and Communicative Disorders and Stroke.

ing to fixed rules; rather, at each step one is forced to deal with the gap between theory and practice.

Our own work has been directed toward the eventual development of clinically usable models for predicting the real-ear frequency response of a hearing aid. We have taken the view that such techniques cannot be placed on a sound basis unless the mechanisms that control the energy flow through the real system are understood. Acoustic systems are difficult to manipulate. Therefore all of the models described in this chapter rely extensively on analogies between different types of oscillating systems.

The value of formal analogies has been appreciated for a long time, and the history of their development and applications is an interesting one. It is not necessary that an analog model be realized using electrical components. Mechanical or other types of models can be used, and indeed have been. In the nineteenth century, when the understanding of mechanics was highly developed, mechanical analogies were used as aids to understanding newly discovered electrical phenomena. An example of this is Maxwell's (1868) explanation of the oscillatory discharge behavior of the Leyden Jar in terms of the then familiar concepts of inertia, compliance, and viscous loss. Other examples can be found in the work of Campbell (1903), Heaviside (1892), and Pupin (1900).

The development of the telephone and the invention of radio in the late nineteenth century prompted the rapid growth of electrical theory and techniques, and electrical models began to be used more widely than mechanical ones. Webster, in 1919, pointed out the parallels between the acoustic variables of pressure and volume velocity, and the electrical variables of voltage and current. He also noted that each pair could be related analogously by using the concept of impedance. Many important acoustic design techniques were applied in the 1920s. A classic example is Mason's (1927) treatment of acoustic filters involving wave motion.

Perhaps the strongest stimulus for the application of analog modeling methods occurred during the 1920s, when acoustic transducers were needed in increasing numbers by the rapidly developing mass communication industries. Electrical input energy is changed into one or more other forms as it passes through a transducer. (For example, in an electromagnetic loudspeaker, an input current is first used to generate a mechanical force, which in turn gives rise to an acoustic pressure.) Electrical analogs offered the elegance and simplicity of being able to handle these different effects within a unified analytic framework, and thus were quickly seen to be important tools for research and development. Some groundwork had been laid by Kennelly and Pierce's (1912) proposal of the concept of a "motional electrical impedance" caused by the movements of an earphone diaphragm. However, it was the work of Harrison (1927), Norton (1924), and Maxfield and Harrison (1926), among others, that established the practical importance of analog modeling in the design of broadband transducers. Their work formed the basis of methods that are still widely used.

A conclusion that could be drawn from this review is that heavy investment in areas of technology relevant to our present mass communication systems led to the wide use of electrical analog models.[1] It is worth noting, however, that advances in digital computer techniques for signal processing and simulation make it less and less necessary to construct hardware models of any kind. Electrical network analogs are still used conceptually (no doubt because of their familiarity), but the next evolutionary stage may be a return to the practice of 100 years ago: to develop and solve the equations for a system directly.

At this stage some brief clarifying remarks on models of sound propagation may be helpful. In free space, sound waves can travel in three orthogonal directions, and also vary with time. Because of the small diameters of the tubes, in hearing aid systems, it is possible to simplify this situation and to deal only with the time-varying sound propagation along the axis of the tube. This simplification facilitates analogies with electrical energy propagation in cables (transmission lines), which can also be considered unidimensional in space. This view is the basis for the development of transmission line models of sound propagation in tubes. In some special cases a further simplification can be introduced: If the system is small enough, the spatial dimension can be eliminated entirely, which is equivalent to assuming that disturbances propagate instantaneously through regions of the system. This view is the basis for lumped parameter models, in which characteristics of regions of the real system are "lumped" into single circuit elements. Such models can have wide applicability in spite of their simplicity.

The first step in analyzing the behavior of any vibrating system is usually to develop a system of equations that describe its motion. It often turns out that the form of such sets of equations is the same, regardless of whether the original system was an acoustic, a mechanical, or an electrical one. This identity of form makes possible the development of analogies in which electrical circuit elements are used to represent the dynamic behavior of mechanical or acoustic systems. If one remembers that in an electrical analog model, voltages and currents at different points in the model may represent very different physical quantities, one can, because of the strictness of the underlying formal analogous relationships, use the model with

---

[1]Unfortunately, the word *analog* has become ambiguous. It was originally used as a noun, indicating a system exhibiting features parallel or similar to those of another system. It is now also used as an adjective, sometimes replacing *analogical* (as in "analog model") and sometimes opposing *digital* (as in "analog filter").

confidence to predict the behavior of the real system. The advantage of this approach is that the model is usually easier to manipulate than the proto-type system, and its predictions can be fully quantitative.

In the remainder of this chapter we describe three examples of analog models drawn from our own work. The first is a lumped-parameter model, representing the system from the hearing aid earphone to eardrum, which was designed to reveal the effects of earmold venting. The second example is a derivation, from first principles, of the parameters of a transmission line analog of sound propagation in narrow tubes. The third example illustrates the development of a quantitative physical analog of a hearing aid earphone. The examples are presented in a sequence that illustrates how a basic model can stimulate further work.

EARMOLD VENTING: A LUMPED-PARAMETER ANALOG

The first example is a quantitative model for explaining some of the acoustic effects seen when earmolds are vented. Apart from allowing external sounds to enter the ear canal directly, which was not considered in this model, the main effect of an earmold vent was originally thought to be to attenuate the low frequency output of the hearing aid and thus to act as an acoustic high-pass filter. However, it became evident that venting introduces other effects as well, and an analog model seemed an ideal means of rapidly examining the effects of different variables without the need for lengthy real-world trials. By varying the values of the components in an equivalent circuit, it is possible to simulate the effects of many different types and combinations of tubing and vent. After a desired frequency response charac-teristic has been attained, the final component values can be translated into physical measurements, and a practical system can be realized very quickly. By comparing the behavior of the model with that of real acoustic systems, the frequency range over which the model is applicable and the accuracy to be gained by using more elaborate models can be determined.

The goal was to develop a model that could adequately explain the behavior of the system under a variety of conditions. The approach chosen was to develop a lumped-parameter electronic circuit model of the imped-ance analogy type. In this type of analogy: 1) electrical voltage represents acoustic pressure, 2) current represents acoustic volume velocity, 3) induc-tance represents acoustic inertance, and 4) capacitance represents acoustic compliance. Studebaker and Cox (1977), after considering earlier lumped-parameter models of the hearing aid earphone, tubing, and eardrum (e.g., Bentzen, 1975; Dalsgaard, Johansen, and Chisnall, 1966; Lybarger, 1969; Wansdronk, 1962), synthesized the electrical circuit shown in Figure 1. In addition, Figure 1 shows schematic diagrams of the analogous side-branch

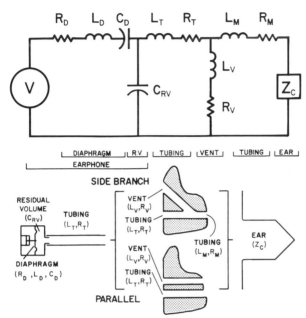

Figure 1. Relationships between a physical acoustic system and an electrical analog. *Bottom:* Schematic drawings of two types of vented earmold. Also shown are a hearing aid earphone, connecting tubing, and the ear canal. Labels followed by symbols in parentheses show relationships between regions of the acoustic system and individual elements in the electrical model above. *Top:* Electrical analog model of a vented earmold configuration. The circle at the left *(V)* represents an acoustic source, and the square at the right *(Z_C)* represents an acoustic load. (*Top* reprinted with permission from Studebaker and Cox, 1977.)

and parallel-vented earmolds. In the analog circuit the earphone diaphragm's resistance, mass, and compliance are represented by R_D, L_D, and C_D. The parallel capacitor C_{RV} represents the small air volume over the diaphragm. The total inertance and resistance of the air in the tubing leading from the earphone to the vent opening are represented by L_T and R_T, respectively. When a vent is present, its inertance and resistance are represented by L_V and R_V. If there is additional tubing between the vent opening and the ear canal itself, as in the case of a side-branch vent, it is represented by L_M and R_M. For parallel vents, which open directly into the ear canal, L_M and R_M are omitted. Z_C is the acoustic load presented by the ear canal and eardrum. In the electrical model we used an electrical analog of the Zwislocki coupler (1970) to represent a typical human ear. We have found that under many conditions this load gives results significantly different from those obtained using a single capacitor, which is the analog of a simple cavity. The values of the electrical components in the model were in all cases obtained independently by physically measuring the sizes of the tubes and vents, calculating their inertances and compliances, and then converting these results into analogous electrical quantities.

Figure 2 shows the results obtained for parallel venting under three conditions. Each of the three curves represents the difference between vented and unvented data. The agreement was very close, the largest differences being about 3 dB. These results were obtained without any "after-the-fact" adjustment of the electrical model.

Figure 3 shows the results obtained with a side-branch vent. Again, the agreement between the electrical model and the acoustic system was quite good. Note that both actual and simulated side-branch vents reduce the pressure levels measured at high frequencies relative to the unvented condition. In contrast, parallel vents had no effect at high frequencies, as shown in Figure 2. In each case, the electrical model replicated the effects seen in the acoustic domain.

This simple lumped-parameter analog model proved to be quite adequate to simulate all earmold modification effects up to at least 1500 Hz, and usually higher. An exception might be those effects that result from acoustic feedback because of radiation of sound from the vent opening, as discussed by Cox and Studebaker (Chapter 9). In any case, the perform-

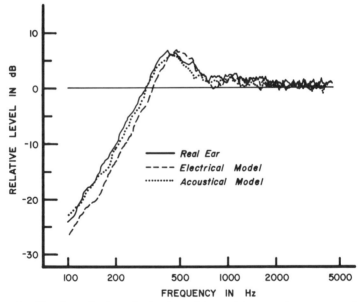

Figure 2.   The effects of real and simulated parallel earmold vents: real-ear response of vented earmold (solid line), Zwislocki coupler response of same earmold (dotted line), and response of analog circuit, measured in electrical analog of Zwislocki coupler (broken line). Response of each system is plotted as pressure developed with the vent open relative to pressure developed with the vent closed. (Reprinted with permission from Studebaker and Cox, 1977.)

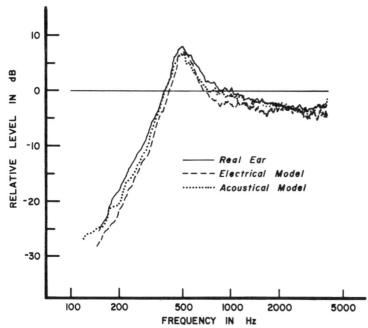

Figure 3.   The effects of real and simulated side-branch earmold vents. The solid, dotted, and broken curves represent the results of real-ear measurements, Zwislocki coupler measurements, and measurements made on the electrical analog model, respectively. (Reprinted with permission from Studebaker and Cox, 1977.)

ance of lumped-parameter models will differ from that of acoustic systems at high frequencies, because the wavelengths of acoustic and electrical signals are very different at a given frequency. Therefore, the electrical model cannot simulate any effects caused by the wavelength of the acoustic signal. Such effects become significant at frequencies where the quarter-wavelength of the acoustic signal approaches the length of a tube or a dimension of a cavity. In normal hearing aid systems such effects are never seen below 1000 Hz and only rarely below 2000 Hz. As an example, a lumped-parameter model will not reveal the resonances that normally occur when the length of the hearing aid's output tube is one-half or one-quarter of the signal wavelength. Under most conditions, the half-wavelength resonance occurs above 3000 Hz, and thus is not a problem in the simulation of venting effects. However, the quarter-wavelength resonance seen with modern miniature earphones occurs at frequencies well below 3000 Hz. This, considered with the recent development of high frequency hearing aids discussed by Killion (Chapter 11), emphasizes the need for more sophisticated models for predicting the behavior of such systems at frequencies above 1500 Hz.

Even with earmold vents, which are normally thought to have only low frequency effects, there can, under some circumstances, be effects at high frequencies that are not predicted by the model in Figure 1. An example is the irregular pattern sometimes seen with side-branch vents in the higher frequency regions when side-branch vent results are plotted relative to unvented results. This pattern appears because side-branch vents produce an impedance discontinuity in the hearing aid's output tube at the point where the vent intersects the main bore. This discontinuity appears to divide the main tube into two shorter ones, thereby raising the half-wavelength resonant frequency and the frequencies of its multiples, producing the typical sharp up-and-down pattern between 3000 and 5000 Hz in relative plots. These effects, as well as those noted previously, arise whenever the physical size of a system becomes comparable with the signal wavelength. The problem of dealing adequately with such wavelength-dependent effects stimulated further work, which is described in the next section.

A TRANSMISSION LINE ANALOG MODEL

In passing from the earphone diaphragm to the eardrum, sound waves pass through a complex series of tubes and cavities, which have a significant influence on their frequency characteristics. The ability of transmission line theory to deal with such influences in a unified way was recognized early (see, for example, Mason, 1927; Meeker and Slaymaker, 1945; Olney, Slaymaker, and Meeker, 1945). Although equations can be developed in acoustic terms, it is usually more convenient to work with electrical transmission line analogs. As Killion (Chapter 11) points out, such models are now widely used for the computation of the overall frequency response of hearing aid systems. For example, Zuercher (1977) described an efficient computational model of sound propagation in tubes. Another example is the work of Egolf (Chapter 13), based on four-pole network theory.

The first stage in the development of a transmission line analog is to formalize the relationships between the equations of motion for the air in the hearing aid tubing system and the parameters of the electrical analog circuit. Because of the analytic and computational difficulty of this problem, it is usual to introduce simplifying assumptions in order to facilitate the solution. One example of such an assumption would be the "small tube" approximation described in many basic acoustic texts. The development to be presented here involves no assumptions other than that of plane-wave propagation in the tube, and thus represents an exact solution of Rayleigh's (1877) original formulation of this problem. Although this work is still incomplete, in that it has not yet been incorporated into a usable practical model of a complete hearing aid system, it is included here because of its basic theoretical importance.

The transmission line analog model is summarized in Figure 4. The figure illustrates a small segment (length Δx) of an infinitely long tube (radius r_i), together with an electrical analog network representing the properties of the air in the tube element. The values of the components in the analog circuit are derived by analyzing the motion of the air in response to an acoustic pressure (P). The characteristics of this motion are specified completely by the relationship between pressure (P) (analogous to voltage, E) and the resulting volume velocity (U) (analogous to current, I).

When an acoustic wave travels along the tube section, it is subject to four types of effects. First, since the air in the tube possesses mass, force is needed to accelerate it. Mass is represented by its electrical impedance analog, inductance (L). Second, when the air in the tube is moving, there are viscous frictional effects at the tube walls that resist its motion; these are represented by resistance (R). Third, the air in the tube can be compressed elastically by the applied pressure. The compliance of the air is represented in the electrical analog by the capacitor (C). Fourth, compression generates heat, which can leak through the tube walls, thereby reducing the "springiness" of the air. This effect is represented by the shunting resistance, which is associated with an electrical conductance (G).

The parameters R and L were obtained by applying Newton's law of motion to the behavior of the air within the small segment of tube shown in Figure 4. From the resulting equations, exact expressions were developed for R and L that depended on no assumptions other than that of plane-wave

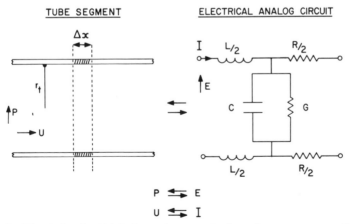

Figure 4.   Electrical analog (right) of acoustic propagation in an elementary segment of a long tube (left). The analogy is indicated by the arrows ($\rightleftarrows$). Voltage (E) is analogous to acoustic pressure (P), and current (I) is analogous to acoustic volume velocity (U).

propagation in the tube. These expressions were solved numerically for a wide range of frequencies and tube dimensions. The expressions for the remaining two elements in the analog *(G* and *C)* were obtained indirectly. It proved easier to develop an expression for the propagation constant of the tube element, and this expression was first solved numerically. The result, together with those obtained for *R* and *L,* allowed exact expressions for the parameters *G* and *C* to be developed. For the details of this solution, the reader is referred to the chapter appendix.

Once the behavior of an elementary tube section has been analyzed in detail, the acoustic propagation in a tube of any length can be modeled by breaking up the tube into short sections, each of which can be represented by an elementary analog section with suitably chosen parameter values. These sections can then be strung together into a "ladder network" to represent the complete tube. In the limit, the tube sections can be made infinitely short, and their number then becomes infinitely large. The resulting wave propagation is described by transmission line equations into which one need merely substitute the values of *L, R, C,* and *G.* Thus, propagation in tubes of any size can be described.

As described in the appendix, results were compared with those obtained by Flanagan (1972), who derived approximate expressions valid for the larger tubes used in models of the human vocal tract. We found that even for tubes as small as 0.5 mm in diameter, Flanagan's values were in error by only about 1% for a wide range of frequencies. Although this work is still in progress, it has been established so far that the values of these constants are accurate for the limiting cases of large and small tubes. The next stage in the development involves designing models of complete systems and the comparison of predictions with actual measurements. We expect significantly better agreement at high frequencies than would be obtained with a lumped-constant model, because wavelength-dependent effects have been included.

The application of transmission line models entails cost as well as provides benefits. The cost is a simplicity that is useful for gaining basic insights. The benefits are precision and an extended range of validity. It is worth noting that the approach typified by the model described here can be a powerful complement to that of the minimal lumped-parameter model presented earlier. One important application of the transmission line model may be as a "benchmark" against which the errors introduced by different degrees of simplification can be assessed, thus providing a basis for selecting the most suitable model for a particular need. Our hope is that combining the two approaches offers the best route to the design of models that are not so difficult to apply that they stifle "game playing," but that are soundly based in theory and sufficiently accurate for their purpose.

A MODEL OF A HEARING AID EARPHONE: A MOBILITY ANALOGY

The transmission line model described in the previous section can represent the propagation of acoustic waves through the tubing and other acoustic networks that couple the output of a hearing aid to the ear, but it cannot, by itself, predict the overall frequency response of the hearing aid. Apart from the characteristics of the hearing aid's microphone and amplifier, it is also necessary to account for the properties of the earphone, which converts the electrical output of the hearing aid amplifier into an acoustic signal. The properties of the earphone are important for two reasons. First, the earphone is not equally sensitive to all frequencies and will modify the frequency content of any signal passing through it. Second, the frequency response of the tubing and earmold themselves are determined partly by the impedance characteristics of the acoustic source. Thus, it is important to specify the characteristics of this source as accurately as possible. The model described here represents a traditional approach to problems of this type (e.g., Beranek, 1954), in which the aim is to design an electrical analog circuit model in which each component is clearly analogous to a specific physical component of the real earphone. Such a model, when used as the signal source for a transmission line model (such as that described previously), provides a basis for precise description of the frequency characteristics of hearing aid output signals.

In the miniature earphone that was studied (Knowles Model 1712), the conversion of electrical input to acoustic output occurs in two stages. In the first stage, an alternating electrical current flowing in a coil induces mechanical oscillations in a small magnetic "reed." The second stage involves a diaphragm that is connected to the reed, and whose movements generate an acoustic pressure at the sound outlet port of the earphone. Our aim was to see whether it was possible, by dismantling the earphone and measuring the structural, mechanical, and electrical characteristics of its component parts, to design an analog equivalent model capable of predicting the relationship between electrical input and acoustic output.

The first step was to examine the magnetic reed and identify the parameters of its response to an electrical input. We found that: 1) the current flowing in the coil is determined partly by electrical factors, including the resistance and inductance of the coil itself, 2) the motion of the reed in response to a given electrical current depends on the size of the force generated by that current, and 3) the effectiveness of a given force in actually moving the reed depends on the reed's resistance to movement; this resistance is determined by the reed's effective mass, as well as by stiffness and damping effects caused by the elasticity and friction in its suspension. The physical constants of each of these effects—except for friction—were measured and the values of the analogous electrical components were derived. The second step was to identify and measure the properties of the diaphragm and its suspension. The acoustic impedance presented to the diaphragm by the different sections of the metal case enclosing the earphone was also estimated.

Figure 5 shows a schematic analog model of the earphone. The terminals at the left represent the electrical input connection, and those at the right represent the sound outlet port. This model was derived using a mobility analogy in which current represents pressure or force (shown by the horizontal or "flow" arrows), and voltage represents velocity or volume velocity (shown by the vertical or "potential" arrows). Another feature of a mobility analogy is that inductance is analogous to compliance and capacitance is analogous to mass.

The lowest line of labels in Figure 5 shows that three types of effects are represented in this model: electrical, mechanical, and acoustic. The next level of labels shows the sources of these effects. The highest level of labels, which indicates individual circuit elements in the diagram, shows the separate parameters that were measured. The transformer connecting the electrical and mechanical sections of the diagram represents the transduction of electrical current into mechanical force and also accounts for the relationship between mechanical velocity and induced EMF. The transformer connecting the mechanical and acoustic sections of the diagram represents

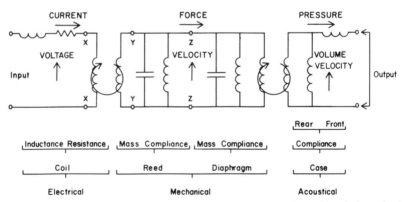

Figure 5.   Mobility analog of a balanced armature hearing aid earphone. Electrical, mechanical, and acoustic factors are represented in three regions of the circuit. Transformers connecting adjacent regions represent transduction of electrical current into mechanical force, and of mechanical force into acoustic pressure. The horizontal arrows indicate analogies between current, force, and acoustic pressure. The vertical arrows indicate analogies between voltage, velocity, and volume velocity. Electrical input is on the left, and the terminals at the right represent the sound outlet port. The short circuit at the sound outlet port, and the points X, Y, and Z are discussed in the text. Labels below individual circuit elements indicate parameters whose values were measured.

the operation of the diaphragm, which converts mechanical force into acoustic pressure and linear velocity into volume velocity.

Consider first the electro-mechanical transformer. Because of the magnetic linkages between its windings, this transformer causes an effective electrical impedance to appear between its primary terminals (X-X), which is proportional to the mechanical impedance at its secondary terminals (Y-Y). Thus the net electrical input impedance of the earphone contains two components: the electrical impedance of the coil and a "referred" component caused by the motion of the reed—hence the term *motional impedance*. A simple extension of this argument shows that the mechanical impedance of the reed is itself modified, via the mechano-acoustic transformer in Figure 5, to include a component due to the impedance of the acoustic load on the diaphragm. Thus, analysis of the electrical input impedance of an active earphone can yield information on the motion of its components.

Two tests of the model are described here. First, the input impedance of an earphone whose reed had been disconnected from its diaphragm was compared with that predicted by the model. (In Figure 5, this corresponds to breaking the circuit at terminals Z-Z.) Second, the input impedance of an intact earphone with a blocked sound outlet port (i.e., no external acoustic load) was compared with that of the model for analogous loading conditions (i.e., output volume velocity = 0, indicated by the output short circuit shown in Figure 5).

Figure 6 shows the results of the first test. The agreement is encouraging, particularly for the magnitude data (solid line and X's), where both calculated and measured results show the characteristic peak caused by the mechanical resonance of the moving reed, as well as the higher frequency dip representing a series resonance determined by the mass-controlled motion of the reed and the inductance of the earphone coil winding. The fairly close prediction of the frequencies of these resonances also implies that the estimated parameter values are reasonably accurate. The agreement is less close for the phase data (broken line and circles), although the forms of the curves are similar. One reason for the difference is the omission of damping from the model, which also explains the excessive height of the resonance peak in the magnitude curve (solid line).

Figure 7 contains the results of the second test of the model, and compares the magnitude of the electrical input impedance of the complete circuit in Figure 6 (curve) with measurements made on a real unit (X's). Again, the agreement is encouraging in that the model correctly predicts the general form of the data. However, there are also discrepancies, some of which are again due to the omission of damping.

As noted elsewhere, one of the benefits of developing a representational model is that insights are gained into how the real device operates. Al-

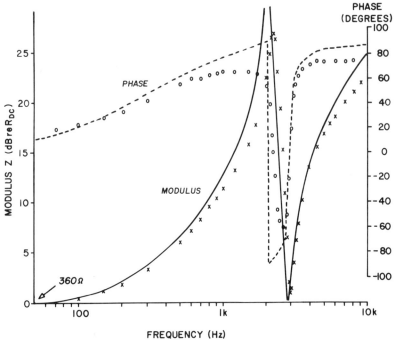

Figure 6.   Evaluation of a model of the magnetic reed portion of a miniature hearing aid earphone, showing modulus (solid line) and phase (broken line) of its electrical input impedance (Z). The plotted points (X, O) are measurements made at the input terminals of a real earphone. Note that modulus data are plotted on a dB scale using the DC resistance of the earphone coil (360Ω) as reference.

though the earphone model described here is still incomplete, these insights have already proved useful in the solution of a practical problem. When a hearing aid earphone is coupled to the ear by tubing, resonances are typically seen in the overall frequency response, some of which are caused by the tubing system and some of which are internal to the earphone. Reference to the earphone model allowed these effects to be separated by indicating which of them were likely to be caused by the earphone itself. The model also suggested how external components could be added to smooth and control the overall frequency response to achieve the desired result.

CONCLUSIONS AND SUMMARY

A model can be defined as an abstract and simplified representation of reality that incorporates the essential principles of a scientific theory in an accessible and usable form. In a sense, a model is the tool by which a theory can be brought to bear on explaining and predicting real effects. Each model

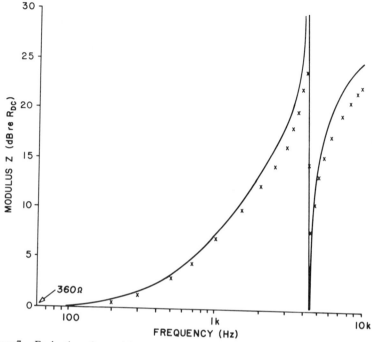

Figure 7.   Evaluation of a model of a miniature hearing aid earphone, showing modulus of the electrical input impedance of an earphone with the sound outlet port blocked (i.e., no external acoustic load) (X's), and input impedance of the model (Figure 5) under analogous output loading conditions (i.e., output short circuit) (solid line).

described in this chapter was developed to explore a different problem in representing the characteristics of hearing aid-processed signals. The vented earmold lumped-parameter model is easy to understand, and is a source of valuable intuitive insights despite its limited accuracy at higher frequencies; the transmission line model is a precise, quantitative analog, and is valuable as a reference system against which simpler systems can be compared; the earphone model, because of the strict analogic method used, is particularly closely related to the physical principles of operation of the real earphone. Although each has its special area of relevance, the three models also illustrate a developmental sequence that is common in modeling. Recall that the high frequency limitations of the minimal lumped-parameter model indicated the need for a new approach based on transmission line theory, and that the earphone model was then developed in order to represent more accurately the acoustic output characteristics of the hearing aid. Each new development adds refinements that allow applications to a wider range of practical situations. However, the process is ongoing. An improved model will no doubt also have its shortcomings, which could be reduced by further refinement at the cost of increased complexity. At each stage it is necessary to ask whether the precision and range of validity of the model are sufficient for the purpose at hand. Superfluous refinement may be counter-productive, and one secret of success is knowing when to stop.

APPENDIX

The acoustic response of the air in a tube is governed by equations analogous to those governing the propagation of an electromagnetic wave along a transmission line. Expressions for the values of the components in a transmission line analog were derived by Flanagan (1972) for the case of tubes of intermediate diameter sizes. In those derivations it was assumed that the tube radius was very large compared with any radial variations in the propagating sound wave. The object of the work described here was to develop an analytical framework that could be used to assess the effects of such approximations and to determine the range of their applicability.

The parameters of the transmission line model are derived by considering the behavior of an incremental segment of a tube (radius r_t and length Δx) similar to that shown in Figure 4. The sound propagation in the tube segment is defined by two variables, the pressure *(P)* and the volume velocity *(U)*. Volume velocity is defined as the integral, over the cross-section of the tube, of the particle velocity of the acoustic wave. Thus, volume velocity has the units of cm^3/S, or Velocity $\times$ Area, and is an appropriate analog to current because it represents a flow that is conserved through nodes. Acoustic pressure is analogous to voltage.

When a source of sound is applied to one end of the tube, the air responds in a number of ways. It accelerates (with acceleration, a) in response to the impressed force, but like any mass it resists acceleration according to Newton's law:

$$f - ma - bV = 0$$

The effective force *(f)* is caused by the acoustic pressure gradient along the axis of the tube section. The mass *(m)* is related to density, and the associated inertial resistance is analogous to inductance. The resistance term *(bV)* results from sliding friction against the tube walls.

Application of pressure also compresses the air in the tube segment. The air resists this compression elastically and tends to return to its original volume when the pressure is relaxed. This springiness of the air is analogous to a shunt capacitance. Furthermore, when the air in the tube is compressed, its temperature is raised slightly. Some of the heat generated by this compression leaks away into the walls of the tube and reduces the tendency of the air to return to its original volume. This effect is analogous to that of a shunt conductance. These effects can be represented by the electrical analog model shown in Figure 8. Each electrical element represents a portion of the differential equations that govern the transmission of sound through the analogous tube segment.

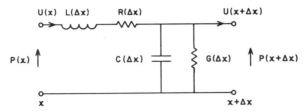

Figure 8.   Analog of acoustic propagation in an elementary segment (Δx) of an infinitely long tube. Acoustic pressures at the beginning and end of the segment are represented by $P(x)$ and $P(x+\Delta x)$, respectively. Similarly, the input and output volume velocities are $U(x)$ and $U(x+\Delta x)$, respectively. Inductance (L) is the analog of acoustic inertance (due partly to the mass of the air and partly to viscous friction). Resistance (R) represents the real part of the viscous drag at the tube wall. Capacitance (C) represents the acoustic compliance of the air, and conductance (G) represents the mechanism of heat loss through the walls of the tube.

To see how the tube segment affects P and U, we write

$$P(x + \Delta x) = P(x) - U(x)[R + j\omega L]\Delta x$$
$$U(x + \Delta x) = U(x) - P(x + \Delta x)[G + j\omega C]\Delta x$$

In the limit, as $\Delta x \to 0$, we have

$$\frac{dP(x)}{dx} = - (R + j\omega L)\, U(x)$$

$$\frac{dU(x)}{dx} = - (G + j\omega C)\, P(x)$$

which together imply

$$\frac{d^2 P(x)}{dx^2} = (R + j\omega L)(G + j\omega C)\, P(x)$$

which leads to the solution

$$P(x) = P_+ e^{-\gamma x} + P_- e^{-\gamma x}$$

where γ, the propagation constant, is defined by

$$\gamma = \sqrt{(R + j\omega L)(G + j\omega C)}$$

and since

$$U(x) = \frac{-1}{(R + j\omega L)} \frac{dP(x)}{dx}$$

we have

$$U(x) = Y_0[P_+ e^{-\gamma x} - P_- e^{-\gamma x}]$$

where

$$Y_0 = \frac{\gamma}{(R + j\omega L)} = \sqrt{\frac{G + j\omega C}{R + j\omega L}}$$

and is called the characteristic admittance. Thus, to characterize sound transmission through the tube we need only find R, L, C, and G. In our application it turned out to be easier to find R, L, and γ, and then to use

$$G + j\omega C = \frac{\gamma^2}{(R + j\omega L)}$$

which implies

$$G = \text{Re}\left\{ \frac{\gamma^2}{R + j\omega L} \right\}$$

$$C = \text{Im}\left\{ \frac{\gamma^2/\omega}{R + j\omega L} \right\}$$

The derivation of the series components R and L used methods quite similar to those used by Flanagan (1972), but with some differences, which will be pointed out in context. The solution for the propagation constant (γ) was based on the method used originally by Rayleigh (1877), but the resulting equations were solved numerically.

INDUCTANCE AND RESISTANCE

The effective acoustic inductance consists of two terms. One is due directly to the inertia of the mass of air, and is equal to ρ/A_0. The derivation of this is clearly explained in Flanagan (1972). The second term is an indirect result of frictional resistance at the tube wall. We derived an expression for this term in a manner similar to Flanagan's but we used cylindrical coordinates and did not assume a large tube radius. The appropriate vector equation for Newton's law is found in Rayleigh (1877, §345):

$$\rho \frac{d\bar{V}}{dt} + \nabla P - \mu \nabla^2 \bar{V} - \tfrac{1}{3} \mu \nabla (\nabla \cdot \bar{V}) = 0$$

We isolated the resistance term by setting the capacitance to zero, meaning that we imagined the air as being incompressible. We also ignored the inertial resistance component of the inductance by imagining the air to be without mass. We next analyzed the effect of moving the tube relative to the air contained in it, and developed an expression for the velocity of the air as a function of radial distance. From this expression it was possible to calculate the energy directed into the wall as a result of sliding friction; this was the resistance term. The expression for the resistance term contained an imaginary component, indicating that an additional term must be added to the inductance. The equations that accomplished this sequence are now sketched.

Setting $\nabla P = 0$ in the vector equation for Newton's law above, writing in cylindrical coordinates, and using

$$\frac{d}{dt} = j\omega$$

$$\frac{\partial}{\partial z} = 0$$

we have

$$\rho j\omega \bar{V}(r) - \mu\nabla^2 \bar{V}(r) - \frac{\mu}{3} \nabla \frac{\overset{= 0}{\cancel{(\partial \bar{V}(r))}}}{\partial z} = 0$$

and hence

$$\rho j\omega V_z(r) - \mu\left[\frac{\partial^2 V_z(r)}{\partial r^2} + \frac{1}{r}\frac{\partial V_z(r)}{\partial r} \right] = 0$$

Writing

$$\Lambda^2 \equiv -j\omega\rho/\mu$$

we have

$$\frac{\partial^2 V_z(r)}{\partial r^2} + \frac{1}{r}\frac{\partial V_z(r)}{\partial r} + \Lambda^2 V_z(r) = 0$$

which is Bessel's equation. This has the solution

$$V_z(r) = \frac{V_0}{J_0(\Lambda r_t)} J_0(\Lambda r)$$

(using $V_z(0) \neq \infty$; $V(r_t) = V_0$)

Now the force at the tube wall is

$$f_z = -\mu(\nabla V_z \cdot \hat{\imath}_r) = -\mu\frac{\partial V_z(r)}{\partial r} = \frac{-\mu V_0 \Lambda J_1(\Lambda r_t)}{J_0(\Lambda r_t)}$$

and the power dissipated at the wall is

$$<P> = |\frac{1}{2} V_0 \operatorname{Re}\{f_z\}| = |\frac{1}{2} V_0^2 \mu\operatorname{Re}\left\{ \frac{\Lambda J_1(\Lambda r_t)}{J_0(\Lambda r_t)} \right\}|$$

which allows us to identify an analogous resistance

$$R = \frac{2\pi r_t}{A_0^2} \mu\operatorname{Re}\left\{ \frac{\Lambda J_1(\Lambda r_t)}{J_0(\Lambda r_t)} \right\}$$

as well as the additional inductance term

$$L_{extra} = \frac{2\pi r_t}{A_0^2} \frac{\mu}{\omega} \operatorname{Im}\left\{ \frac{\Lambda J_1(\Lambda r_t)}{J_0(\Lambda r_t)} \right\}$$

PROPAGATION CONSTANT

Next we derived the expression for γ, taking essentially the same approach as that used by Rayleigh. However, modern computer methods make it possible to obtain numerical solutions without the need for the simplifying approximations that Rayleigh was forced to make. The four basic equations describing the propagation of sound in the tube are:

Conservation of mass:

$$\nabla \cdot \bar{V} + j\omega S = 0 \tag{1}$$

Newton's law:

$$j\omega \bar{V} + \nabla \left\{ \frac{P}{\rho_0} - \frac{1}{3} \frac{\mu}{\rho_0} \nabla \cdot \bar{V} \right\} - \frac{\mu}{\rho_0} \nabla^2 \bar{V} = 0 \tag{2}$$

Equation for generation and flow of heat:

$$j\omega\theta - j\beta\omega S - \nu\nabla^2\theta = 0 \quad \dagger \tag{3}$$

Equation of state relating pressure, condensation, ambient density, and excess temperature:

$$P - P_0 (1 + S + \alpha\, \theta) = 0 \tag{4}$$

Manipulation of these four equations leads to the following result:

$$-\omega^2\theta' - \{a^2 + j\omega\,(\nu + \mu' + \mu'')\}\{\nabla^2\theta' + \frac{\nu}{j\omega}\,\{b^2 + j\omega(\mu' + \mu'')\}\{\nabla\theta' = 0$$

$$(\text{Rayleigh, 1877, § 348})$$

Solving for θ' we obtain

$$\theta' = A_1 Q_1 + A_2 Q_2$$

where θ_i satisfy

$$\nabla^2\theta_i = \lambda_i\theta_i$$

and λ_i are roots of

$$\frac{\nu}{j\omega}\,\{b^2 + j\omega(\mu' + \mu'')\}\lambda^2 - \{a^2 + j\omega(\mu' + \mu'' + \nu)\}\lambda - \omega^2 = 0$$

The expression for θ' can now be substituted into equation 3 to solve for S, and the result substituted into equation 1 to solve for $\bar{V}$. Assuming

---

†Here Flanagan uses

$$\frac{d\theta}{dt} = \frac{\lambda}{K_P\rho}\,\nabla^2\theta$$

and deals only with heat that is generated far from the tube wall. We allow heat generation throughout the tube radius.

$$V(r,z) = V(r)e^{\gamma z}$$

the resulting equations for V and θ' can be written

$$V_z = AJ_0\left(r\sqrt{\gamma^2 - \frac{j\omega}{\mu'}}\right) - A_1\gamma(j\frac{\omega}{\lambda_1} - \nu)J_0(r\sqrt{\gamma^2 - \lambda_1})$$

$$- A_2\gamma(j\frac{\omega}{\lambda_2} - \nu)J_0(r\sqrt{\gamma^2 - \lambda_2})$$

$$V_\phi = -A\,\frac{\gamma}{j\omega/\mu' - \gamma^2}\,\frac{\partial}{\partial r}\,J_0\left(r\sqrt{\gamma^2 - \frac{j\omega}{\mu'}}\right) - A_1(\frac{j\omega}{\lambda_1} - \nu)\,\frac{\partial}{\partial r}\,[J_0$$

$$(r\sqrt{\gamma^2 - \lambda_1})] - A_2(j\frac{\omega}{\lambda_2} - \nu)\,\frac{\partial}{\partial r}\,[J_0\,(r\sqrt{\gamma^2 - \lambda_2})]$$

$$\theta' = A_1J_0(r\sqrt{\gamma^2 - \lambda_1}) + A_2J_0(r\sqrt{\gamma^2 - \lambda_2})$$

Writing these in matrix form we have

$$\begin{vmatrix} V_z \\ V_\phi \\ \theta' \end{vmatrix} = \begin{vmatrix} Q & -\gamma\xi_1Q_1 & -\xi_2Q_2 \\ \dfrac{-\gamma}{j\omega/\mu' - \gamma^2}\dfrac{d\theta}{dt} & -\xi_1\dfrac{dQ_1}{dt} & -\xi_2\dfrac{dQ_2}{dt} \\ 0 & Q_1 & Q_2 \end{vmatrix} \begin{vmatrix} A \\ A_1 \\ A_2 \end{vmatrix}$$

Where

$$Q = J_0(r\sqrt{\gamma^2 - j\omega/\mu'})$$
$$Q_1 = J_0\,(r\sqrt{\gamma^2 - \lambda_1})$$
$$Q_2 = J_0\,(r\sqrt{\gamma^2 - \lambda_2})$$
$$\xi_i = (j\omega/\lambda_i - \nu)$$

At the tube walls we require that

$$\begin{vmatrix} V_z \\ V_\phi \\ \theta' \end{vmatrix} = \begin{vmatrix} 0 \\ 0 \\ 0 \end{vmatrix} \quad \dagger\dagger$$

Thus, the determinant of the matrix must be zero when evaluated at $r = r_t$. This gives an implicit equation, which can be solved for γ, the propagation constant:

$$\gamma^2\,\frac{[\xi_1 - \xi_2]}{\sqrt{\gamma^2 - j\omega/\mu'}}\,\frac{J_1(r_t\sqrt{\gamma^2 - j\omega/\mu'})}{J_0(r_t\sqrt{\gamma^2 - j\omega/\mu'})} + \xi_2\sqrt{\gamma^2 - \lambda_2}\,\frac{J_1\,(r_t\sqrt{\gamma^2 - \lambda_2})}{J_0(r_t\sqrt{\gamma^2 - \lambda_2})}$$

††Note that different boundary conditions could be applied at this point.

$$- \xi_1 \sqrt{\gamma^2 - \lambda_1} \; \frac{J_1(r_t\sqrt{\gamma^2 - \lambda_1})}{J_0\,(r_t\sqrt{\gamma^2 - \lambda_1})} = 0$$

Rayleigh provided two approximate solutions valid for very small tubes, and a third valid for slightly larger tubes. We solved this equation numerically for γ.

Thus we have expressions for all the parameters of the transmission line analog model. They are summarized below:

$$L = \frac{\rho}{A_0} - \frac{2\mu}{\omega\pi r_t^3} \, \mathrm{Im} \left\{ \frac{\Lambda J_1(\Lambda r_t)}{J_0(\Lambda r_t)} \right\}$$

$$R = \frac{2\mu}{\pi r_t^3} \, \mathrm{Re} \left\{ \frac{\Lambda J_1(\Lambda r_t)}{J_0(\Lambda r_t)} \right\}$$

$$G = \mathrm{Re} \left\{ \frac{\gamma^2}{R + j\omega L} \right\}; \; C = \frac{1}{\omega} \, \mathrm{Im} \left\{ \frac{\gamma^2}{R + j\omega L} \right\}$$

$$Y_0 = \gamma/[R + j\omega L]$$

COMPARISON OF NUMERICAL AND APPROXIMATE SOLUTIONS

Once the expressions summarized above had been obtained, numerical solutions were computed for a range of tube sizes and frequencies. These solutions were then compared with the corresponding values obtained from Flanagan's approximate equations. Some of our results are shown in Figures 9–12. Specifically, R, L, G, and C are shown for three radii ($r_t = 0.05$, 0.1, and 0.374 cm). The first two radii correspond to the small tube sizes often used in hearing aids, and the third radius is roughly equal to that of the ear canal. Note that because of the large changes of R, L, and C with tube radius, a different ordinate scale is used for each of the three radii considered.

As can be seen from the diagrams, there is generally close agreement between the numerical and approximate solutions. The largest approximation errors (i.e., differences between the numerical and the approximate solutions) occur with the resistive components, $(R$ and $G)$. The approximation error for R is roughly independent of frequency, but its magnitude increases as tube radius r_t decreases. However, the value of R itself also increases as r_t decreases and, expressed as a proportion of R, the proportional approximation error increases only moderately with decreasing r_t over the range considered. For example, at a frequency of 3900 Hz and for $r_t = 0.374$ cm, the approximation error for R is 3 mΩ, or about 0.5% of R. For $r_t = 0.05$ cm, on the other hand, the approximation error is 9.56 Ω, or roughly 3.8% of R.

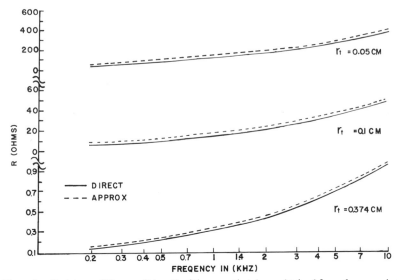

Figure 9.   Resistance *(R)* per unit length of the acoustic tube as obtained from the numerical and approximate solutions. The values of *R* are shown for three tube radii, r_t = 0.05, 0.1, and 0.374 cm. Note the changes in the scale of the ordinate.

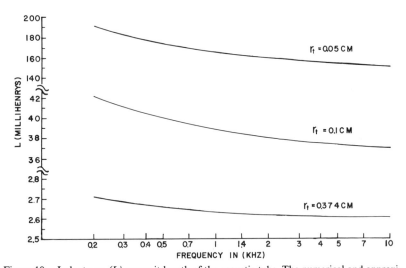

Figure 10.   Inductance *(L)* per unit length of the acoustic tube. The numerical and approximate solutions yielded values of *L* within 0.5% of each other (for the frequency range and tube radii considered), and both cases are represented by a single curve for each of three radii. Note the changes in the scale of the ordinate.

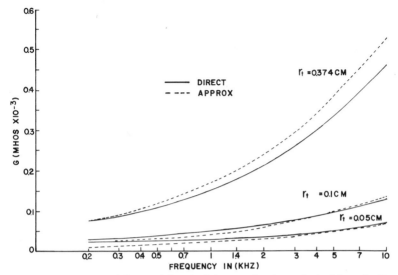

Figure 11.   Conductance *(G)* per unit length of the acoustic tube as obtained from the direct and approximate solutions. The values of G are shown for three tube radii, $r_t = 0.05, 0.1$, and 0.374 cm.

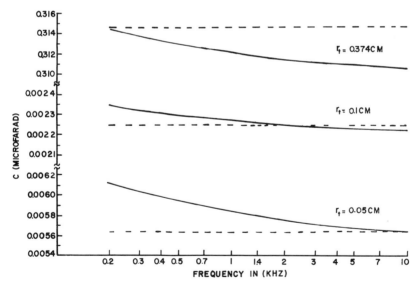

Figure 12.   Capacitance *(C)* per unit length of the acoustic tube, as obtained from the numerical and approximate solutions. The values of C are shown for three tube radii, $r_t = 0.05, 0.1$, and 0.374 cm. Note the changes in the scale of the ordinate. Note also that the middle section of the ordinate scale should cover 0.021 to 0.024 microfarads, and not 0.0021 to 0.0024, as shown.

It should be noted that at lower frequencies the value of the resistive component (R) is relatively small. Thus, the proportional approximation error will appear quite large in the low frequency region. For example, at 200 Hz and $r_t = 0.05$ cm, the approximation error is 10 Ω, which is almost 20% of R itself. This error, however, turns out to have relatively little effect on the transfer function of the network because of the small value of R at this frequency.

The proportional approximation error for the conductance parameter (G) is larger than that for R and it also shows a greater degree of frequency dependence. At low frequencies, the approximate value of G is slightly lower than that obtained numerically, but at high frequencies the situation is reversed. For a tube with a radius of 0.05 cm, the approximation error is roughly zero at 12 kHz. For tubes with radii of 0.1 and 0.374 cm, the corresponding zero-error frequencies are 3 kHz and 235 Hz, respectively.

The approximate solutions for capacitance (C) are extremely accurate for large tubes, but less so for small tubes at low frequencies. For example, at 200 Hz and with $r_t = 0.05$ cm, the approximation error is 500 picofarads, or roughly 10% of C at this frequency. At 10 kHz, on the other hand, the two values of C are within 0.2% of each other.

In contrast to the results obtained for R, G, and C, the approximation errors for L, the inductance, were found to be negligibly small. The numerical and approximate solutions for L were within 0.5% of each other over the entire range of frequencies and tube sizes considered.

Of great practical interest is the effect of the approximation error on the transfer function of the acoustic tube. Figure 13 shows the pressure ratio $/P_{out} / P_{in}/$ as derived from both the approximate and the numerical solutions. The results shown are for two tubes, each 2.25 cm long and with radii of 0.05 and 0.374 cm. Each tube is assumed to terminate in an infinite acoustic impedance (i.e., a rigid closure). The two solutions produce very similar transfer functions that differ primarily in the location of the resonant frequency (for the frequency range considered). For the larger tube, which approximates a typical ear canal, the frequency of resonance obtained from the numerical solution exceeds that obtained from the approximate solution by roughly 20 Hz. The situation is reversed for the smaller tube, where the resonant frequency obtained from the numerical solution is less than that of the approximate solution, also by about 20 Hz. The reason is apparent from Figure 12, which shows that in the vicinity of the resonant frequency the proportional approximation errors for the capacitance term are roughly equal and opposite for these two cases. It should be remembered that the approximation error is negligible for the inductance term, and the resistive terms $(R$ and $G)$ have little effect on the frequency of the resonance. Note

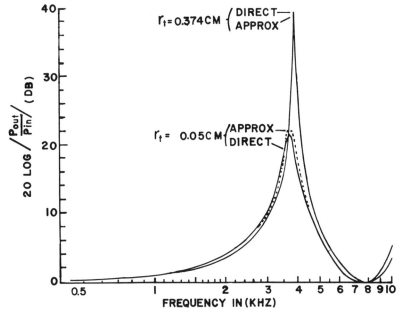

Figure 13.   Transfer function $/P_{out}/P_{in}/$ for an acoustic tube 2.25 cm long. Transfer functions obtained from the numerical and approximate solutions are shown for two tube radii. The larger tube radius is equal to that of a typical human ear. The smaller radius is typical of small gauge tubing often used with hearing aids.

also that the value of $/P_{out}$ / P_{in} / at resonance is substantially greater for the larger tube. This is because, as can be seen from Figure 9, the value of R at the resonant frequency is about 400 times larger for a tube of radius 0.05 cm than for a tube of radius 0.374 cm. The smaller losses in the larger tube account for the increased height of its resonance peak. The differences in the heights of the resonance peaks for numerical and approximate solutions are a result of the approximation errors for R and G.

In summary, a model of sound transmission in tubes has been developed that is based on Rayleigh's original formulation of the problem. A direct numerical solution has been obtained in terms of an equivalent electrical network representing an infinitesimal length of tube. The results of the direct numerical solution have been compared with those of a simpler, approximate solution developed by Flanagan (1972). Excellent agreement was found for the inductance term (L), and good agreement was found for the resistance term (R) except for a small, relatively fixed approximation error inversely related to the tube radius. The agreement was not quite as good for the shunt terms of the model $(G$ and $C)$. For the situations of interest here (sound transmission in the ear canal or in tubing of the size used with hearing aids), the resistive component is of secondary importance except in the region of a resonance. Even under the worst conditions, the difference between the numerical and approximate computed transfer functions is on the order of 5%, either for the ear canal or for a narrow tube of length equal to the ear canal.

SYMBOLS

j = $\sqrt{-1}$

L = Inductance per unit length of tube

R = Resistance per unit length of tube

C = Capacitance per unit length of tube

G = Shunt conductance per unit length of tube

P = Pressure

U = Volume velocity

ω = Radial frequency ($2\pi \times$ frequency in Hz)

γ = Propagation constant

r_t = Tube radius

x = Linear distance along tube

Δx = Incremental distance along tube

P_+ = Amplitude of forward-traveling pressure wave (determined by source and boundary conditions)

P_- = Amplitude of backward-traveling (reflected) pressure wave

Y_0 = Characteristic admittance of tube

$\text{Re}\{z\}$ = Real part of complex number z

$\text{Im}\{z\}$ = Imaginary part of complex number z ($z = \text{Re}\{z\} + j \cdot \text{Im}\{z\}$)

ρ = Density

ρ_0 = Ambient density

$\nabla \cdot \bar{V}$ = Divergence of V. In cylindrical coordinates:

$$\nabla \cdot \bar{V} = \frac{\partial}{\partial x} V_x + \frac{\partial}{\partial y} V_y + \frac{\partial}{\partial z} V_z$$

where $V = V_x \hat{i}_x + V_y \hat{i}_y + V_z \hat{i}_z$

$\nabla^2 \bar{V}$ = Defined in Cartesian coordinates to mean:

$$\left[\frac{\partial^2}{\partial x^2} V_x \right] \hat{i}_x + \left[\frac{\partial^2}{\partial y^2} V_y \right] \hat{i}_y + \left[\frac{\partial^2}{\partial z^2} V_z \right] \hat{i}_z$$

$\nabla \phi$ = $\frac{\partial \phi}{\partial x} \hat{i}_x + \frac{\partial \phi}{\partial y} \hat{i}_y + \frac{\partial \phi}{\partial z} \hat{i}_z$ (ϕ is a scalar, not a vector)

$\hat{i}_r$ = Unit vector in radial direction

A_0 = Cross-sectional area of tube

$J_0(z)$ = Zero order Bessel function of z (z may be complex)

$J_1(z)$ = First order Bessel function of z (z may be complex)

Λ^2 = $-j\omega\rho/\mu$ · Hence: $\Lambda = (-1+j)\sqrt{\omega\rho/2\mu}$

S = Condensation, i.e., $\rho = \rho_0(1+S)$

θ = Excess temperature (as Rayleigh, 1877, § 348)

β = Thermal constant from: $\dfrac{d\theta}{dt} = \beta \dfrac{dS}{dt} + \nu \nabla^2 \theta$

θ' = θ/β

b = Newton's value for the speed of sound, i.e., $\sqrt{P_0/\rho_0}$

a = Laplace's value for the speed of sound, i.e., $b \sqrt{K_p/K_v}$ (Rayleigh, 1877, § 348)

λ = Material parameter (as Flanagan, 1972)

μ = Coefficient of friction

μ' = μ/ρ_0

μ'' = $\mu'/3$

ν = $\rho_0 \lambda / K_p$

K_p = Specific heat of air at constant pressure

K_v = Specific heat of air at constant volume

α = Constant of proportionality (as Rayleigh, 1877, § 247)

ACKNOWLEDGMENTS

We would like to thank Harvey Stromberg and Ronald Slosberg for their help in computer programming. Dr. Robyn M. Cox made some of the acoustic measurements on which part of this chapter was based. Many friends and colleagues read and discussed earlier versions of the manuscript, and thoughtful comments by Professor Katherine S. Harris, Steven Hoffnung, and Dr. Mead C. Killion were particularly helpful. Our greatest debt is to Dr. Sheila J. White, whose incisive and constructive criticisms played a central part in the final shaping of this chapter.

REFERENCES

Bentzen, N. 1975. Computer calculated equivalent circuits for couplers and earphones. Scand. Audiol. (suppl. 5):138–155.

Beranek, L. L. 1954. Acoustics. McGraw-Hill Book Co., New York.

Campbell, G. A. 1903. On loaded lines in telephonic transmission. Phil. Mag. 5 (ser. 6):313–330.

Dalsgaard, S. C., P. A. Johansen, and L. G. Chisnall. 1966. On the frequency response of ear-moulds. J. Audiol. Tech. 5:2–15.

Flanagan, J. L. 1972. Speech Analysis, Synthesis and Perception. Springer-Verlag, New York.

Harrison, H. C. 1927. Acoustic Device. U.S. Patent #1,730,425.

Heaviside, O. 1892. Electrical Papers. MacMillan & Co., Ltd., London.

Kennelly, A. E., and G. W. Pierce. 1912. The impedance of telephone receivers as affected by the motion of their diaphragms. Proc. Am. Acad. Arts Sci. 48: 113–151.

Lybarger, S. F. 1969. The earmold from an acoustical standpoint. Paper presented at Audiology Lecture Program, Mayo Clinic, March 8, Rochester, Minn.

Mason, W. P. 1927. A study of the regular combination of acoustic elements with applications to recurrent acoustic filters, tapered acoustic filters, and horns. Bell Syst. Tech. J. 6:258–294.

Maxfield, J. P., and H. C. Harrison. 1926. Methods of high quality recording and reproducing of music and speech based on telephone research. Bell Syst. Tech. J. 5:493–523.

Maxwell, J. C. 1868. On Mr. Grove's experiment in magneto-electric induction. Phil. Mag. 35:360–363.

Meeker, W. F., and F. H. Slaymaker. 1945. A wide range adjustable acoustic impedance. J. Acoust. Soc. Am. 16(3):178–182.

Norton, E. L. 1924. Wave Filter. U.S. Patent #1,681,554.

Olney, B., F. H. Slaymaker, and W. F. Meeker. 1945. The dipole microphone. J. Acoust. Soc. Am. 16(3):172–177.

Pupin, M. I. 1900. Wave transmission over non uniform cables and long distance air lines. Trans. Am. Inst. Elec. Eng. 17:445–507.

Rayleigh, Lord. 1877. The Theory of Sound. Macmillan & Co., Ltd., London. (Reprinted in 1945 by Dover Publications, New York.)

Studebaker, G. A., and R. M. Cox. 1977. Side branch and parallel vent effects in real ears and in acoustical and electrical models. J. Am. Audiol. Soc. 3:108–117.

Wansdronk, C. 1962. On the mechanism of hearing. Philips Res. Rep. 62:1–140.

Webster, A. G. 1919. Acoustical impedance and the theory of horns and of the phonograph. Proc. Nat. Acad. Sci. 5:275–282.

Zwislocki, J. J. 1970. An Acoustic Coupler for Earphone Calibration. Report No. LSC-S-7, Laboratory of Sensory Communication, Syracuse University, Syracuse, N.Y.

Zuercher, J.C. 1977. The calculation of isothermal and viscous effects in acoustic tubes. J. Acoust. Soc. Am. 62:S56(A).

CHAPTER 13

TECHNIQUES FOR MODELING THE HEARING AID RECEIVER AND ASSOCIATED TUBING

David P. Egolf

CONTENTS

Alteration of sound spectra by hearing aids has been the subject of numerous investigations (Cooper et al., 1975; Dalsgaard, Johansen, and Chisnall, 1966; Egolf, 1978a,b, 1979; Egolf, Tree, and Feth, 1978; Ewertsen, Ispen, and Nielsen, 1957; Franks et al., 1974; Grossman and Molloy, 1944; Lybarger, 1958, 1979; McDonald and Studebaker, 1970; Studebaker, 1974; Studebaker and Zachman, 1970).

The publications by Egolf describe a computer-assisted method for predicting, a priori, the aided sound spectrum developed at the tympanic membrane in both real and artificial ears. The material presented herein includes: 1) a review of these publications with emphasis on the mathematical principles used to develop the computer scheme, 2) the results of labora-

---

This chapter is reprinted in part from earlier publications by the author (see Egolf, 1978a, b, 1979; Egolf, Tree, and Feth, 1978). The work is based on Egolf (1976).

A major portion of the experimental work was conducted at the Ray W. Herrick Laboratories, School of Mechanical Engineering, Purdue University, West Lafayette, Ind.

This work was supported by a research grant from the National Institute of Neurological and Communicative Disorders and Stroke, National Institutes of Health.

tory experiments demonstrating the accuracy of the computer method, and 3) use of the method to predict the effects of changing several parameters on the sound spectrum in an aided ear.

MATHEMATICAL PRINCIPLES

A typical behind-the-ear hearing aid is illustrated in Figure 1. The electrical circuit of a hearing aid usually consists of a microphone, an amplifier, a battery, and a receiver. Some aids also include compression and/or filter circuitry. Sound received by the microphone is amplified and is then broadcast by the receiver into the tygon tube. The amplified sound from the receiver travels along the tygon tube, through the earmold, and into the residual portion of the ear canal, where it eventually excites the tympanic membrane.

The spectrum of received sound is drastically altered by three different parts of a conventional hearing aid (i.e., in an aid without compression or filter circuits): the microphone, the receiver, and the series of small tubes linking the receiver with the tympanic membrane. Thus, mathematical descriptions of these parts of a hearing aid must be included in any scheme designed to predict the sound spectrum at the tympanic membrane. The method described herein includes mathematical models of the receiver and the tubes between the receiver and the tympanic membrane. Mathematical descriptions of the microphone and amplifier are not included.

Receiver Model

The internal impedance of a particular source of electrical energy (e.g., a battery, generator, alternator) must be known in order to properly design

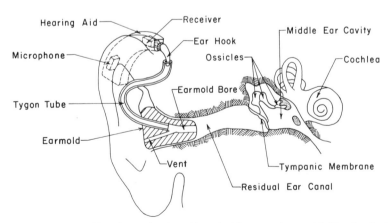

Figure 1.   Typical behind-the-ear hearing aid used with a vented earmold. (Reprinted with permission from Egolf et al., 1978.)

and/or model any electrical circuit to be attached to that source. Similarly, the internal impedance of a source of acoustic energy is an essential piece of information when attempting to design and/or model an entire acoustic system that includes that source. For example, designers of automobile mufflers require information about the internal impedance of their source of acoustic energy: the automobile engine (see Ross, 1976). Similarly, any technique for modeling an in situ hearing aid from the receiver to the tympanic membrance must include information about the internal imped- ance of the source of acoustic energy: the receiver.

Egolf and Leonard (1977) have reported the development of a computer- aided experimental technique, called the "two-load" method, for modeling internal impedance properties of an electroacoustic source, such as a hear- ing aid receiver. The two-load method is simply an electroacoustic adapta- tion of a well-known electrical principle called Thevenin's theorem (after M. L. Thevenin, French engineer, 1883). Thevenin's theorem is commonly used by electrical engineers to determine the unknown internal impedance of an electrical source. When applied to a hearing aid receiver, the two-load method is used to generate a set of coefficients, called the "four-pole" parameters (Guillemin, 1957), that are directly related to the internal im- pedance properties of the receiver.

A schematic diagram of a typical electromagnetic hearing aid receiver is given in Figure 2A. The receiver creates an acoustic response as a result of electrical stimulation of an electromagnet, which in turn excites a dia- phragm into motion. Thus there are four variables of interest: voltage and current E_i and I_i at the input terminals of the electromagnet and sound pressure and volume velocity P_o and U_o at some point in the output acoustic field. The network shown in Figure 2B is a four-pole analog model of the receiver in Figure 2A.

Circuit equations describing the network in Figure 2B are given (as in Guillemin, 1957) in matrix form as

$$\left\{ \begin{array}{c} E_i \\ I_i \end{array} \right\} = \left[\begin{array}{cc} A_R & B_R \\ C_R & D_R \end{array} \right] \left\{ \begin{array}{c} P_o \\ U_o \end{array} \right\} \tag{1}$$

where

$\left[\begin{array}{cc} A_R & B_R \\ C_R & D_R \end{array} \right]$ = matrix of unknown four-pole parameters of the receiver

E_i, I_i = complex amplitudes of voltage and current at the receiver input terminals

P_o, U_o = complex amplitudes of sound pressure and volume velocity at the receiver output terminals

The unknown four-pole parameters A_R, B_R, D_R, and C_R may be ob- tained by application of the two-load method to the receiver in Figure 2.

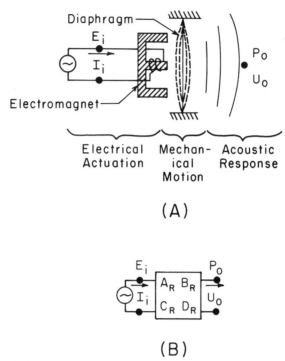

Figure 2. Typical electromagnetic hearing aid receiver. (Reprinted with permission from Egolf et al., 1978.)

Actual application of this method is a complex task involving numerous laboratory procedures and considerable mathematical manipulation, and is thus omitted from this discussion (interested readers should refer to Egolf and Leonard (1977) for details). Suffice it to say that the results of this application are four equations describing A_R, B_R, C_R, and D_R in terms of other known or experimentally derived quantities. Using these four equations, values of A_R, B_R, C_R, and D_R may then be generated with a digital computer. For example, the computed values of A_R, B_R, C_R, and D_R for a Knowles BK-1604 hearing aid receiver are given in Figure 3.

Successful application of the two-load method to another electroacoustic source—in this instance a large, high intensity sound source, such as that used to test automobile mufflers—has recently been demonstrated by D.F. Ross (personal communication).

Tube Model

Sound is transmitted between the receiver and the tympanic membrane through a series of right-circular cylindrical tubes, such as the one

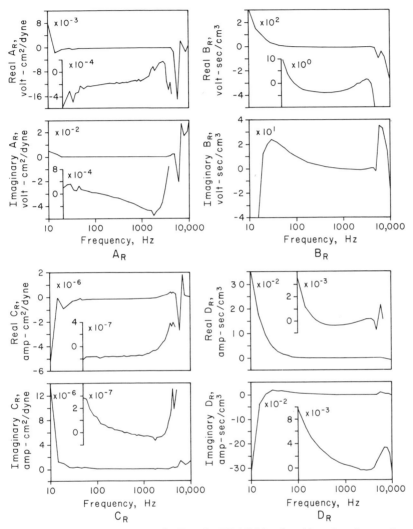

Figure 3.   Four-pole parameters of a Knowles BK-1604 hearing aid receiver (inserts show expanded scale). (Reprinted with permission from Egolf et al., 1978.)

shown schematically in Figure 4A. Using Guillemin's (1957) technique, the circuit equations describing the network in Figure 4B may be written as:

$$\left\{ \begin{array}{c} P_u \\ U_u \end{array} \right\} = \left[\begin{array}{cc} A_n & B_n \\ C_n & D_n \end{array} \right] \left\{ \begin{array}{c} P_d \\ U_d \end{array} \right\} \qquad (2)$$

where

$$\begin{bmatrix} A_n & B_n \\ C_n & D_n \end{bmatrix}$$ = matrix of four-pole parameters of tube identified by numeric subscript n

P_u, U_u = complex amplitudes of sound pressure and volume velocity at station u, and

P_d, U_d = complex amplitudes of sound pressure and volume velocity at station d

The entries in the matrix of four-pole parameters given in equation 2 are defined (Guillemin, 1957) as:

$$\begin{aligned}
A_n &= \cosh \Gamma_n l_n \\
B_n &= Z_n \sinh \Gamma_n l_n \\
C_n &= Z_n^{-1} \sinh \Gamma_n l_n \\
D_n &= \cosh \Gamma_n l_n
\end{aligned}$$ (3)

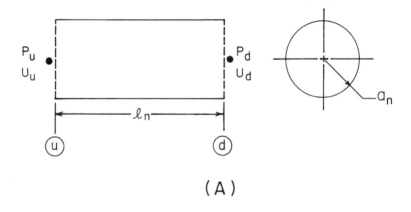

(A)

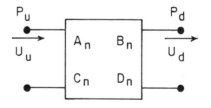

(B)

Figure 4.   Right-circular cylindrical sound transmission tube n. (Reprinted with permission from Egolf et al., 1978.)

where

Γ_n = propagation operator of tube n
Z_n = characteristic impedance of tube n
l_n = length of tube n

It is common to use approximations for the propagation operator Γ_n and the characteristic impedance Z_n given in equation 3. This is the case, for example, in textbooks on acoustics by Kinsler and Frey (1962), Morse and Ingard (1968), Rschevkin (1963), and Stephens and Bate (1966). Unfortunately, these approximations apply only to large-diameter tubes driven at low audio frequencies. They yield very poor results when applied to sound transmission through tubes smaller than about 0.5 cm in diameter if accuracy over the entire audio frequency range is desired.

The alternative to these approximations is a set of exact equations for Γ_n and Z_n developed by Iberall (1950). The exact forms for Γ_n and Z_n, as developed by Iberall, are:

$$\Gamma_n = i\,\frac{\omega}{c}\left\{ \frac{1 + 2(\gamma - 1)\,[J_1(\alpha a_n)/\alpha a_n J_0(\alpha a_n)]}{1 - 2J_1(\beta a_n)/\beta a_n J_0(\beta a_n)} \right\}^{1/2} \tag{4}$$

and

$$Z_n = \frac{\rho c}{\pi a_n^{2}}\left\{ \left[1 - \frac{2J_1(\beta a_n)}{\beta a_n J_0(\beta a_n)} \right]\left[1 + 2(\gamma - 1)\,\frac{J_1(\alpha a_n)}{\alpha a_n J_0(\alpha a_n)} \right] \right\}^{-1/2} \tag{5}$$

where

$$\alpha = \left(\frac{-i\omega\rho\sigma}{\mu} \right)^{1/2}$$

$$\beta = \left(\frac{-i\omega\rho}{\mu} \right)^{1/2}$$

c = sonic velocity in the fluid medium
γ = ratio of specific heats of the fluid medium c_p/c_v
a_n = radius of tube n
ρ = density of the fluid medium
σ = Prandtl number of the fluid medium
μ = absolute viscosity of the fluid medium
$J_0(\ ), J_1(\ )$ = Bessel functions of the first kind of orders zero and one

The use of equations 4 and 5 depends critically on several requirements, not listed herein, which were originally adopted by Iberall as fundamental assumptions in his derivation. Weston (1953) and Goodson and Leonard (1972) have developed a rationale that allows the investigator to determine just when it is appropriate to use equations 4 and 5 and when approximations for Γ_n and Z_n will suffice.

Application of equations 2–5 to the transmission of sound through small-diameter tubes has been successfully demonstrated by Egolf (1977).

COMPUTER MODEL OF A HEARING AID

Using the mathematical principles mentioned above, Egolf (1978a, 1979; Egolf et al., 1978) developed a computer model of an entire in situ hearing aid system (from the receiver electrical input to the tympanic membrane). The computer model was tested in the laboratory by comparing computer predictions of aided sound spectra with actual recordings made in one artificial ear and one real ear.

The Artificial Ear

The experimental setup for testing application of the computer model to an artificial ear is given in Figure 5A. The artificial ear employed in this test —a Brüel & Kjaer DB0260 calibration coupler—was selected because it had easily defined acoustic properties, which was a necessity for determining the validity of the computer model. Unfortunately, it and other 2-cm^3 couplers like it are rather poor acoustic replicas of the human ear (Lybarger, 1979).

A four-pole network model of the entire apparatus shown in Figure 5A is given in Figure 5B. After the application of matrix theory and considerable mathematical manipulation, circuit equations describing the network in Figure 5B reduce to an equation for sound pressure level at the plane of the microphone diaphragm (i.e., the artificial eardrum) SPL$_L$, or:

$$\text{SPL}_L = 20 \log_{10} \left| \frac{E_{ref}}{P_{ref}} \left(\frac{Z_L}{A_T Z_L + B_T} \right) \right| + 20 \log_{10} \frac{|E_i|/\sqrt{2}}{E_{ref}} \tag{6}$$

where

$$\begin{bmatrix} A_T & B_T \\ C_T & D_T \end{bmatrix} = \begin{bmatrix} A_R & B_R \\ C_R & D_R \end{bmatrix} \begin{bmatrix} A_1 & B_1 \\ C_1 & D_1 \end{bmatrix} \begin{bmatrix} A_2 & B_2 \\ C_2 & D_2 \end{bmatrix} \begin{bmatrix} A_3 & B_3 \\ C_3 & D_3 \end{bmatrix}$$
$$\times \begin{bmatrix} 1 & 0 \\ 1/Z_s & 1 \end{bmatrix} \begin{bmatrix} A_5 & B_5 \\ C_5 & D_5 \end{bmatrix} \begin{bmatrix} A_6 & B_6 \\ C_6 & D_6 \end{bmatrix} \tag{7}$$

and

$$Z_s = \frac{A_4 Z_Q + B_4}{C_4 Z_Q + D_4} \tag{8}$$

where

Z_Q = acoustic radiation impedance from the vent outlet into anechoic space

$|E_i|/\sqrt{2}$ = rms voltage at the receiver input terminals

E_{ref} = reference rms voltage

P_{ref} = reference rms sound pressure

Z_L = acoustic impedance of the microphone, derived from data given by the manufacturer (Brüel & Kjaer, 1966)

The R subscripts on the four-pole parameters in Figure 5B and equation 7 refer to their association with the receiver. The computed four-pole

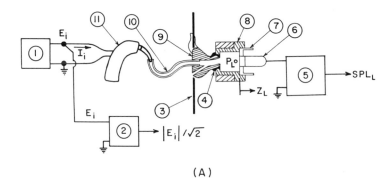

(A)

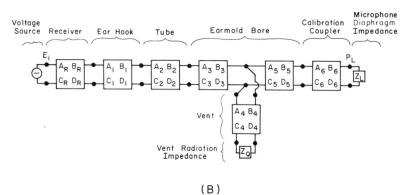

(B)

Figure 5.   Laboratory setup for measuring sound pressure levels in an artificial ear. A, Schematic (key to numbers is as follows: 1) sine wave generator, 2) rms voltmeter, 3) infinite baffle constructed from 4' × 4' poster board, 4) sealing compound, 5) B&K 2107 microphone amplifier and filter, 6) B&K 4134 microphone and 2615 cathode follower, 7) B&K DB0225 mechanical adapter, 8) B&K DB0260 calibration coupler, 9) earmold, 10) tygon tube, 11) Knowles BI receiver sealed in a Dahlberg HT 123 hearing aid case with standard internal "plumbing" and a polyethylene over-the-ear hook). B, Four-pole network model of the experimental apparatus. (In part from Egolf, 1978a,b.)

parameters for a Knowles BK-1604 receiver are presented graphically in Figure 3. The numerical subscripts on the four-pole parameters in Figure 5B and equation 7 refer to their association with the various cylindrical tubes connecting the receiver with the artificial eardrum (i.e., the microphone diaphragm).

Using equations 2–8, a digital computer program was designed to compute the sound pressure level SPL_L at the plane of the microphone diaphragm, given the rms voltage input to the receiver $|E_i|/\sqrt{2}$. The computed and experimental sound spectra derived using the receiver shown in Figure 5 with unvented, side-branch–vented, and parallel-vented earmolds are given in Figure 6.

The Real Ear

In order to test the computer scheme on a real ear, a laboratory experiment involving a live human subject with normal hearing was devised. That experiment included the recording of probe-tube measurements made in an aided real ear, as illustrated in Figure 7A. A four-pole network model of the experimental setup is given in Figure 7B. Beginning with circuit equations describing the network in Figure 7B, an equation for sound pressure level at the plane of the tympanic membrance SPL_{TM} was derived:

$$\text{SPL}_{TM} = 20 \log_{10} \left| \frac{E_{ref}}{P_{ref}} \left(\frac{Z_{TM}}{A_T Z_{TM} + B_T} \right) \right| + 20 \log_{10} \frac{|E_i|/\sqrt{2}}{E_{ref}} \tag{9}$$

where

$$\begin{bmatrix} A_T & B_T \\ C_T & D_T \end{bmatrix} = \begin{bmatrix} A_R & B_R \\ C_R & D_R \end{bmatrix} \begin{bmatrix} A_1 & B_1 \\ C_1 & D_1 \end{bmatrix} \begin{bmatrix} A_2 & B_2 \\ C_2 & D_2 \end{bmatrix} \begin{bmatrix} 1 & 0 \\ 1/Z_s & 1 \end{bmatrix}$$

$$\times \begin{bmatrix} A_7 & B_7 \\ C_7 & D_7 \end{bmatrix} \tag{10}$$

$$Z_s = (A_S Z_L + B_S)/(C_S Z_L + D_S) \tag{11}$$

$$\begin{bmatrix} A_S & B_S \\ C_S & D_S \end{bmatrix} = \begin{bmatrix} A_3 & B_3 \\ C_3 & D_3 \end{bmatrix} \begin{bmatrix} A_4 & B_4 \\ C_4 & D_4 \end{bmatrix} \begin{bmatrix} A_5 & B_5 \\ C_5 & D_5 \end{bmatrix} \begin{bmatrix} A_6 & B_6 \\ C_6 & D_6 \end{bmatrix} \tag{12}$$

and

Z_{TM} = acoustic impedance of the tympanic membrane given by Shaw (1974)

The R subscripts in Figure 7B and equation 10 refer to the receiver and the numerical subscripts refer to the cylindrical tubes linking the receiver with the tympanic membrane. According to Morton and Jones (1956), the external ear canal extending 1.5 cm laterally from the tympanic membrane may be treated as though it were a rigid-walled circular, cylindrical tube terminated by the tympanic membrane.

Equations 9–12 were then incorporated into a digital computer program designed to compute the sound pressure level SPL_{TM} at the tympanic membrane, given the rms voltage input to the receiver $|E_i|/\sqrt{2}$. Computer-predicted sound spectra and experimental measurements are presented in Figure 8. The experimental data in Figure 8 have been adjusted to the plane of the tympanic membrane by taking into account the residual ear canal volume and the probe-tube calibration curve. Sachs and Burkhard (1972) have demonstrated that the practice of making probe-tube measurements at the earmold tip and then mathematically adjusting the data to the plane of the tympanic membrane can introduce substantial error at some frequencies. Fortunately, the ex-

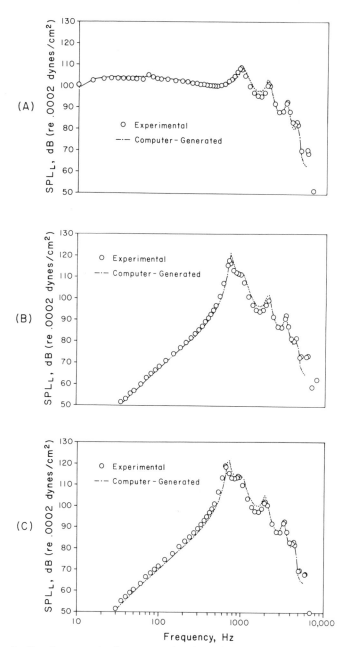

Figure 6.   Sound pressure level spectra at the plane of the microphone diaphragm in the calibration coupler of Figure 5 using unvented earmold (A), side-branch–vented earmold (B), and parallel-vented earmold (C). (Reprinted with permission from Egolf et al., 1978.)

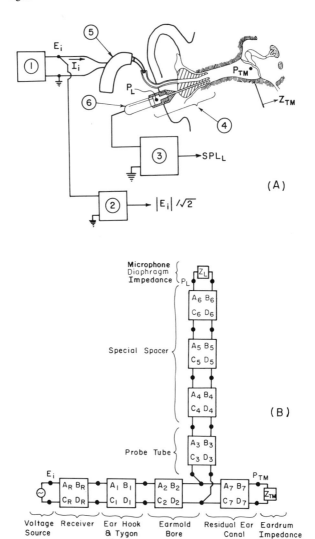

Figure 7.   Laboratory setup for measuring sound pressure levels in a real ear. A, Schematic (key to numbers is as follows: 1) sine wave generator, 2) rms voltmeter, 3) B&K 2107 microphone amplifier and filter, 4) probe tube and special spacer glued into earmold (a description of the special spacer was given by Egolf, 1977), 5) Knowles BI receiver sealed in a Dahlberg HT 123 hearing aid case with standard internal "plumbing" and a polyethylene over-the-ear hook, 6) B&K 4134 microphone and 2615 cathode follower). B, Four-pole network model of the experimental apparatus. (Reprinted with permission from Egolf et al., 1978.)

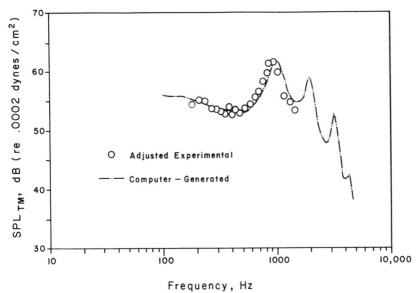

Figure 8.   Sound pressure level spectrum at the plane of the tympanic membrane in the real ear of Figure 7. (Reprinted with permission from Egolf, 1978a, Computer predictions of hearing aid sound spectra, *Hearing Aid Journal,* March.)

perimental data presented in Figure 8 are well below the frequency range where the error is significant.

EFFECTS OF HEARING AID
PARAMETERS ON AIDED SOUND SPECTRA

The data presented in Figures 6 and 8 tend to support the validity of the computer scheme as an accurate predictor of aided sound spectra in both real and artificial ears.

When a determination of aided sound spectra is desired, use of a computer technique such as this has several advantages over experimental methods. Once the computer program has been proved valid, earmold design can be altered by simply changing a statement in the computer program. Parameters such as the tygon tube length and diameter, receiver type, and residual ear canal volume can likewise be altered by changing computer statements. In addition, the computer method does not require the use of soundproof rooms, sophisticated sound measurement instruments, or live human subjects. It is easy to imagine several applications of such a scheme in the design, testing, selection, and fitting of hearing aids. Computer simulations similar to this one are used in all fields of science where experimental procedures are impossible or might prove too time consuming and costly (e.g., imagine the consequences of staging a real—rather than simulated—nuclear accident, or the expense of actually testing —rather than simulating—each of several new wing designs for a commercial jet aircraft).

In the following sections, the computer program has been used to generate sound spectra in a real, normal ear that result from commonly encountered changes in parameters, such as receiver type, tygon tube length and diameter, and vent length and diameter. The geometric parameters under investigation are illustrated in Figure 9. (During this part of the project all dimensions except those under investigation were held constant at the following values (in cm): $l_1 = 3.5$, $l_2 = 1.5$, $l_3 = 1.25$, $l_4 = 1.0$, $l_5 = 0.5$, $d_1 = 0.20$, $d_2 = 0.30$, $d_3 = 0.75$, $d_4 = 0.30$, $d_5 = 0.30$.) The computer model that was employed is essentially that shown in Figure 7B

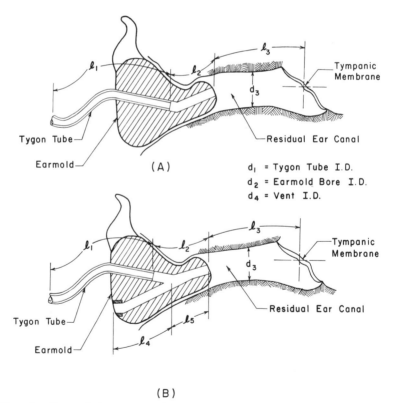

Figure 9. Geometrical parameters of in situ hearing aid under investigation. A, Unvented earmold; B, side-branch–vented earmold.

and described mathematically by equations 9–12, except that the microphone, special spacer, and probe tube have been removed.

Changing Receiver Type

The data given in Figure 10 illustrate how the aided sound spectrum (in this case the computer-predicted spectrum) in a real, normal ear can vary when two different receivers are used with the same unvented earmold/ear geometry. These curves demonstrate graphically the requirement (discussed earlier) for incorporating receiver dynamics (i.e., internal impedance properties) into the mathematical model of the entire in situ hearing aid. Lybarger (1958) stated that the receiver design had an "important influence" on the first resonance peak, an effect that is apparent in Figure 10.

Variations in Unvented Earmold Geometry

The plots given in Figure 11 demonstrate the effects of varying the length of tygon tube (i.e., dimension l_1 in Figure 9A) on the predicted sound spectrum in a real, normal ear. With regard to the fitting of hearing aids, this length is selected rather arbitrarily depending on the type of hearing aid and the user's anatomy. The results presented in Figure 10 are consistent with those described by Lybarger (1979): the first resonance peak is shifted lower in frequency as the tubing length is increased.

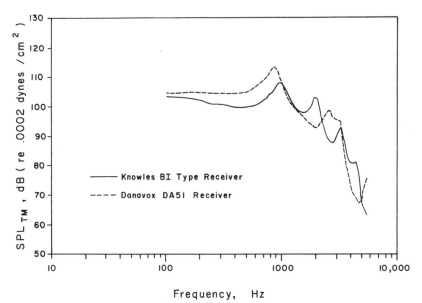

Figure 10.   Effects of changing receiver type on the computer-generated sound spectrum in an aided real, normal ear (unvented earmold).

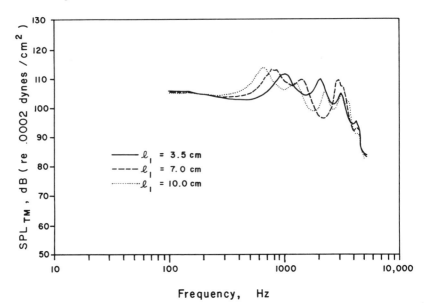

Figure 11.   Effects of varying the tygon tube length l_1 on the computer-generated sound spectrum in an aided real, normal ear (unvented earmold).

The curves plotted in Figure 12 show the effects of simultaneous changes in both the tygon tube diameter and earmold bore diameter (i.e., dimensions d_1 and d_2 in Figure 9A) on the predicted sound spectra. This type of alteration is typical of a practical situation: the outer diameter of the tygon tube usually determines the earmold bore diameter if the tube is to be cemented into the earmold. These results are consistent with data presented by Studebaker (1974) and Lybarger (1979): the first resonance peak moves higher in frequency as these diameters are increased.

With reference to Figure 9A, one commonly encountered variation in earmold geometry is the location of the point of intersection of the tygon tube with the earmold bore (i.e., dimension l_2). For example, this point of intersection can be changed by pushing the tygon tube farther down into the earmold bore before cementing it in place. Also, when fitting a particular hearing aid, the length of the acoustic transmission path between the receiver and the medial tip of the earmold (i.e., the sum of dimensions l_1 and l_2) is usually fixed. Thus, varying the point of intersection means simultaneously varying both dimension l_1 and dimension l_2, given that their sum always remains constant. The plots given in Figure 13 illustrate the effects of two extreme variations in this point of intersection. Unfortunately, this type of variation is not

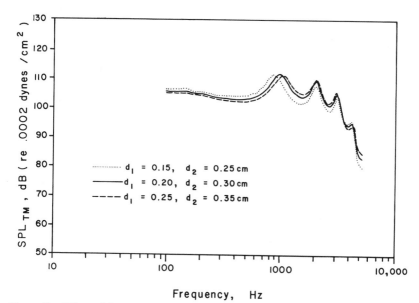

Figure 12.   Effects of the simultaneous variation of tygon tube diameter d_1 and earmold bore diameter d_2 on the computer-generated sound spectrum in an aided real, normal ear (unvented earmold).

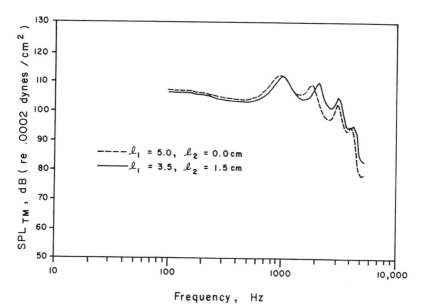

Figure 13.   Effects of varying the point of intersection of the tygon tube with the earmold bore (i.e., simultaneous variation of l_1 and l_2) on the computer-generated sound spectrum in an aided real, normal ear (unvented earmold).

reported in the literature so there are no data available for comparison.

Variations in Real-Ear Geometry

Spectral differences in the plots given in Figures 14 and 15 are the result of commonly encountered variations in the size of the residual ear canal. The ear canal lengths and diameters (i.e., dimensions l_3 and d_3 in Figure 9A) represented in Figures 14 and 15 were selected so that the volume of the residual ear canal always remained between 0.3 and 1.0 cm^3 (based on the variation between subjects of the measured residual ear canal volume lying between the earplug of an acoustic impedance bridge and the tympanic membrane, as reported by Zwislocki, 1963). According to these plots, increasing the length l_3 or diameter d_3 (and thus volume) of the residual ear canal can result in a decrease in SPL$_{TM}$ that is practically independent of frequency. Dalsgaard et al. (1966) observed similar results that occurred as a consequence of increasing the volume of the coupler in which spectral measurements were taken.

Variations in Vented Earmold Geometry

Figures 16 and 17 illustrate the effect of altering the length l_4 and diameter d_4 of the vent shown in Figure 9B. The data presented in these two figures show that the cutoff frequency, which is one characteristic of the filtering property of the vent, is increased with increases in vent diameter and with decreases in vent length. The first observation is consistent with the findings of Studebaker (1974). With regard to the second observation, it is almost intuitive that the low frequency attenuation will be greater for shorter vent lengths, because at these frequencies more acoustic energy is lost through the vent. Lybarger (1979) also states that "the length of a vent or a vent channel is much more important than has frequently been assumed." He goes on to show that increasing vent length (and earmold bore length) causes the first resonance peak to move lower in frequency, a finding consistent with that shown in Figure 16.

With reference to Figure 9B, if the vent were drilled into the earmold at a different angle, then it would intersect the earmold bore in a different location. The location of this point of intersection varied widely in the commercially produced side-branch–vented earmolds examined by the author. Changing only the point of intersection of the vent with the earmold bore amounts to changing the l_5 dimension in Figure 9B, given that l_2 and l_4 remain unchanged. The spectral results for two extreme variations of l_5 are presented in Figure 18. Effects of a variation of this type are not reported elsewhere in the literature.

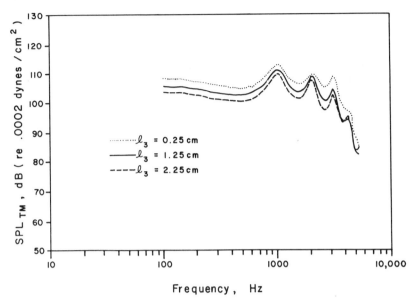

Figure 14.   Effects of varying residual ear canal length l_3 on the computer-generated sound spectrum in an aided real, normal ear (unvented earmold).

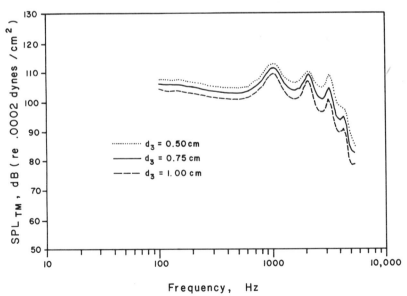

Figure 15.   Effects of varying residual ear canal diameter d_3 on the computer-generated sound spectrum in an aided real, normal ear (unvented earmold).

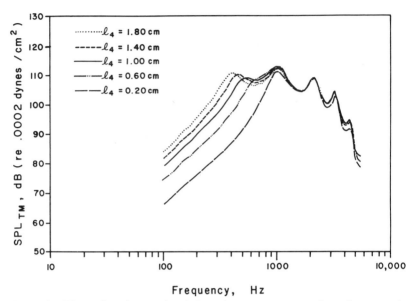

Figure 16.   Effects of varying vent length l_4 on the computer-generated sound spectrum in an aided real, normal ear (side-branch–vented earmold).

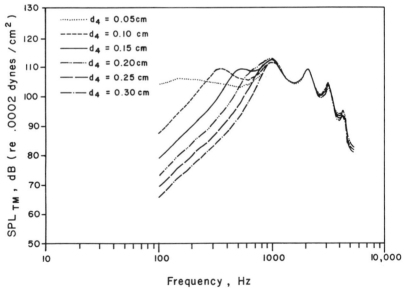

Figure 17.   Effects of varying vent diameter d_4 on the computer-generated sound spectrum in an aided real, normal ear (side-branch–vented earmold).

DISCUSSION AND CONCLUSIONS

From the data presented in Figures 6 and 8, it appears that the computer method developed herein is a reliable means for predicting aided sound spectra in real, normal ears. In addition, most of the spectral shapes given in Figures 10 through 18 (i.e., computer-generated spectra) can be verified by comparison with the findings of other investigators reported in the literature. To gain more confidence in this method, further real-ear experiments should be conducted.

The data presented in Figures 10–18 suggest that the shape of the aided spectrum in a real, normal ear is critically dependent on receiver type and internal geometry of the tubes linking the receiver with the tympanic membrane. A perusal of these figures shows that each parameter variation does, in fact, affect the spectral shape to some degree. Consequently, in a practical situation, the true spectrum shape in an aided, real, normal ear could be quite different from any curve supplied by the hearing aid manufacturer. In fact, there is little hope of determining the true shape of the sound spectrum in an aided, real, normal ear without the assistance of either an experimental technique or a computer prediction scheme, such as the one reported herein.

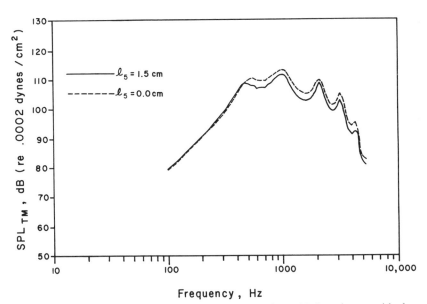

Figure 18.   Effects of varying the point of intersection of the side-branch vent with the earmold bore (i.e., dimension l_s) on the computer-generated sound spectrum in an aided real, normal ear (side-branch–vented earmold).

ACKNOWLEDGMENTS

The author wishes to acknowledge the contribution to this research work made by David R. Tree and Werner Soedel, School of Mechanical Engineering, Purdue University; Lawrence L. Feth and William A. Cooper, Department of Audiology and Speech Sciences, Purdue University; Robert G. Leonard, Mechanical Engineering Department, Virginia Polytechnic Institute; and John R. Franks, Department of Communications and Theatre, Arizona State University.

REFERENCES

Brüel & Kjaer, 1966. Instructions and Applications, Half-inch Condenser Microphones, Microphone Cartridges Type 4133/4134, Cathode Followers Type 2614/2615. Brüel & Kjaer, Copenhagen.
Cooper, W. A., J. R. Franks, R. N. McFall, and D. P. Goldstein. 1975. Variable venting valve for earmolds. Audiol. J. Commun. 14:259–267.
Dalsgaard, S. C., P. A. Johansen, and L. G. Chisnall. 1966. On the frequency response of ear-moulds. J. Audiol. Tech. 5(4):126–139.
Egolf, D. P. 1976. A mathematical scheme for predicting the electro-acoustic frequency response of hearing aid receiver-earmold-ear systems. Doctoral dissertation, Purdue University, West Lafayette, Ind.
Egolf, D. P. 1977. Mathematical modeling of a probe tube microphone. J. Acoust. Soc. Am. 61(1):200–205.
Egolf, D. P. 1978a. Computer predictions of hearing aid sound spectra. Hear. Aid J., March:6.
Egolf, D. P. 1978b. Computer predictions of earmold modifications. In P. Yanick and S. F. Freifeld (eds.), The Application of Signal Processing Concepts to Hearing Aids, pp. 61–84. Grune & Stratton, New York.
Egolf, D. P. 1979. Mathematical predictions of electroacoustic frequency response of in-situ hearing aids. In V. D. Larson, D. P. Egolf, R. L. Kirlin, and S. W. Stile (eds.), Auditory and Hearing Prosthetics Research, pp. 411–450. Grune & Stratton, New York.
Egolf, D. P., and R. G. Leonard. 1977. Experimental scheme for analyzing the dynamic behavior of electroacoustic transducers. J. Acoust. Soc. Am. 62:1013–1023.
Egolf, D. P., D. R. Tree, and L. L. Feth. 1978. Mathematical predictions of electroacoustic frequency response of in-situ hearing aids. J. Acoust. Soc. Am. 63(1):264–271.
Ewertsen, H. W., J. B. Ispen, and S. S. Nielsen. 1957. On acoustical characteristics of the earmold. Acta Otolaryngol. 47:312–317.
Franks, J. R., D. P. Goldstein, W. A. Cooper, and R. N. McFall. 1974. The effect of earmold venting: Electro-acoustic and psychoacoustic methods of evaluation. Purdue University Contributed Papers PUC4:1–4.
Goodson, R. E., and R. G. Leonard. 1972. A survey of modeling techniques for fluid line transients. J. Basic Eng. Trans. ASME 94 (ser. D):474–487.
Grossman, F. M., and C. T. Molloy. 1944. Acoustic sound filtration and hearing aids. J. Acoust. Soc. Am. 16:52–59.
Guillemin, E. A. 1957. Synthesis of Passive Networks, pp. 186–217. John Wiley & Sons, New York.
Iberall, A. S. 1950. Attenuation of oscillatory pressures in instrument lines. J. Res. Nat. Bur. Stand. 45:85–108.

Kinsler, L. F., and A. R. Frey. 1962. Fundamentals of Acoustics. 2nd. Ed. John Wiley & Sons, New York.

Lybarger, S. F. 1958. The earmold as a part of the receiver acoustic system. Radioear Voice 23(8).

Lybarger, S. F. 1979. Controlling hearing aid performance by earmold design. In V. D. Larson, D. P. Egolf, R. L. Kirlin, and S. W. Stile (eds.), Auditory and Hearing Prosthetics Research, pp. 101–132. Grune & Stratton, New York.

McDonald, F. D., and G. A. Studebaker. 1970. Earmold alteration effects as measured in the human auditory meatus. J. Acoust. Soc. Am. 22(5):329–334.

Morse, P. M., and K. U. Ingard. 1968. Theoretical Acoustics. McGraw-Hill Book Co., New York.

Morton, J. Y., and R. A. Jones. 1956. The acoustical impedance presented by some human ears to hearing-aid earphones of the insert type. Acustica 6:343.

Ross, D. F. 1976. Experimental determination of normal specific acoustic impedance of an internal combustion engine. Doctoral dissertation, Purdue University, West Lafayette, Ind.

Rschevkin, S. N. 1963. A Course of Lectures on the Theory of Sound. Moscow State University, Moscow. Translated by O. M. Blunn. Macmillan Publishing Co., New York.

Sachs, R. M., and M. D. Burkhard. 1972. On making pressure measurements in insert earphone couplers and real ears. J. Acoust. Soc. Am. 51:140(A).

Shaw, E. A. G. 1974. The external ear. In W. D. Keidel and W. D. Neff (eds.), Handbook of Sensory Physiology, Chapter 14. Springer-Verlag, Berlin.

Stephens, R. W. B., and A. E. Bate. 1966. Acoustics and Vibrational Physics. 2nd Ed. Edward Arnold, London.

Studebaker, G. A. 1974. The acoustical effect of various factors on the frequency response of a hearing-aid receiver. J. Audio Eng. Soc. 22(5):329–334.

Studebaker, G. A., and T. A. Zachman. 1970. Investigation of the acoustics of earmold vents. J. Acoust. Soc. Am. 47(4):1107–1115.

Weston, D. E. 1953. The theory of the propagation of plane sound waves in tubes. Proc. Phys. Soc. London 66(8):695–709.

Zwislocki, J. 1963. An acoustic method for clinical examination of the ear. J. Speech Hear. Res. 6:303–314.

Frequency Response
Selection Techniques

CHAPTER 14

AN EXPERIMENTAL PROTOCOL FOR THE PRESCRIPTIVE FITTING OF A WEARABLE MASTER HEARING AID

Harry Levitt and M. Jane Collins

CONTENTS

RATIONALE

A master hearing aid that is wearable is a relatively new development. Until recently, master hearing aids were heavy, relatively large pieces of equipment that served as fixtures in many research laboratories, clinics, or hearing aid dispensaries. Although much of the research on hearing aids and methods of hearing aid evaluation has been done using a master hearing aid, there has been little research on the use of a master hearing aid as a clinical tool in the prescriptive fitting of hearing aids. The primary objective of this study was the development of a protocol for the prescriptive fitting of a wearable master hearing aid (WMHA).

PROCEDURAL CONSIDERATIONS

Three versions of an experimental protocol were tried, with some favorable results. The basic structure of the experimental protocol, common to all

This chapter provides a summary of the final report of NINCDS Contract #N1H-NOI-NS-4-2323. The complete report is reprinted in CSL Research Report #11 (Levitt et al., 1978).

three versions, was that it consisted of four stages: an initial stage, where basic audiologic data on the subject were obtained; a second stage, which consisted of a small but statistically efficient experiment, the results of which yielded an initial estimate of the optimum setting of the WMHA; a third stage, in which the parameters of the WMHA were adjusted in a systematic way, to improve performance; and a fourth stage, in which the performance of the WMHA at its estimated optimum setting was evaluated using a number of different measures.

The four stages of the protocol were identified as follows: Stage I, Basic Audiometric Testing; Stage II, Fixed Test Battery; Stage III, Adaptive Sessions; Stage IV, Comparative Measurements.

It was necessary to have a speech reception test that met certain basic requirements, such as low test-retest variability and minimal learning effects. A test of this type, known as the nonsense syllable test (NST), was developed as part of this study (Levitt and Resnick, 1978; Resnick et al., 1975). Similarly, an efficient adaptive procedure that would be simple to operate and yet be reasonably reliable for inherently variable data was required. An adaptive method that is eminently well suited for such problems is the simplical method (Box, 1957); a version of this procedure, suitably modified for the needs of this study, was used during Stage III.

A crucial step in the adaptive procedure is obtaining an initial estimate of the optimum setting of the WMHA. In order to do this, a small, statistically efficient experiment was carried out with each subject during Stage II. In this experiment, known as the fixed test battery, several settings of the WMHA were chosen to form a balanced experimental design (e.g., 2×3 factorial). The NST was administered at each of these settings. The purpose of the fixed test battery was to ascertain the effect of the major variables influencing the subject's performance, and from this information to obtain an initial estimate of the optimum setting of the WMHA. With the instrumentation available, the variables that could be adjusted conveniently were slope of frequency-gain characteristic, lower cutoff frequency, upper cutoff frequency, maximum power output, form of amplitude limiting, and various parameters relating to compression amplification. Of these variables, the first two were found to be most revealing with respect to estimating the optimum setting of the WMHA for each subject. For practical reasons, the parameters relating to compression amplification were not investigated beyond a preliminary assessment.

It is important that the settings of the WMHA used in this experiment (the fixed test battery) either encompass or be close to the optimum setting of the WMHA. If a good initial estimate of the optimum WMHA setting is obtained from the fixed test battery, then relatively little adaptive adjustment may be needed beyond this first estimate. In this study, the initial estimates of the optimum WMHA setting obtained from the fixed test battery turned out to be particularly good.

The subjects participating in the study were experienced adult hearing aid users requiring monaural amplification. The subjects all had a sensorineural hearing impairment (some had an additional conductive component) acquired postlingually, but not as a result of presbycusis. The particular subject population was chosen because it minimized confounding the results with ancillary effects, such as the initial adaptation to acoustic amplification, differences in language comprehension, and central processing deficits. It was also considered wiser to restrict this first study to the simpler problem of monaural, as opposed to binaural, amplification.

DISCUSSION OF RESULTS

Major Findings

Evaluative data were obtained on three versions of the experimental protocol, the bulk of the data being obtained with the third and final version. For all three versions, consistent improvements in performance were obtained as measured by the NST, which was the basic performance measure used in the fitting procedure. The primary method of assessing the effectiveness of the experimental protocol was to compare performance at the estimated optimum setting of the WMHA to that obtained with the subject's own hearing aid. In order to check against the possibility of confounding the improvements resulting from the use of the experimental protocol with any improvements that may result from differences in the overall quality of the WMHA and the subject's own aid, additional comparisons were made using only the WMHA. Test scores obtained at the estimated optimum setting of the WMHA were compared with scores obtained at those settings of the WMHA corresponding, within reasonable limitations, to the recommendations of the Harvard Study (Davis et al., 1947). Another method of comparison was to monitor progress of the experimental protocol by obtaining a repeat measurement at each new estimate of the optimum setting of the WMHA. All of the above comparisons showed consistent improvements as a result of using the experimental protocol.

In general, the results of the study showed that the experimental protocol converged on a setting of the WMHA that yielded improved performance. The improvements were largest and most consistently revealed by the NST, which was the basic test used throughout the study. Figure 1 shows the relative improvements as measured by the NST. Scores for the subject's own aid are given on the abscissa; scores for the WMHA at its estimated optimum setting are shown on the ordinate. All data shown

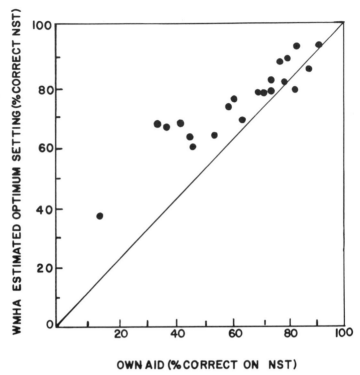

Figure 1.   Comparison of WMHA at estimated optimum setting with subject's own aid: NST scores.

are for a signal-to-noise (S/N) ratio of 20 dB. The positive diagonal represents equal performance for the two hearing aids. All points lying above this diagonal indicate better performance for the WMHA.

The data show a consistent pattern of improvement for the estimated optimum setting of the WMHA. The improvement appears to increase with decreasing score; i.e., those subjects whose performance was not as good to begin with showed the largest improvements. The four subjects with the lowest scores (as measured on the subject's own aid) showed improvements of over 20 percentage points with the WMHA. Subjects whose scores on their own hearing aids were between 50% and 80% showed a smaller average improvement on the order of 8 percentage points. Relatively small improvements, if any, were shown by subjects with scores over 80% on their own hearing aid. Three of the subjects in this region showed slightly higher scores for their own hearing aids.

It could be argued that the improvements obtained are a result of comparing a precision laboratory hearing aid with extremely smooth frequency-response curves, low internal noise, and low harmonic distortion with a commercial hearing aid (the subject's own aid) of poorer overall quality. Figures 2 and 3 show relative performance at the estimated optimum setting of the WMHA compared with the performance on the WMHA using the frequency-gain characteristics that would have been used had the recommendations of the Harvard study (Davis et al., 1947) been used. Figure 2 shows relative performance for the flat frequency-response condition and Figure 3 shows relative performance for an upward sloping frequency response of +6 dB/octave, as measured in a standard 2-cm³ coupler. The scores for the estimated optimum setting of the WMHA are shown on the ordinate, and those for the reference condition on the abscissa.

In these two diagrams, all of the scores have been obtained with the same hearing aid and thus there is no difference in quality of the instrument at the two conditions of interest. As before, consistent improvements were

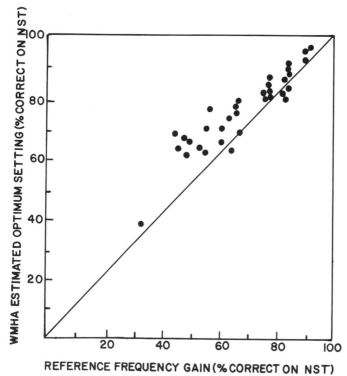

Figure 2.   Comparison of NST scores at the estimated optimum setting and at the reference frequency-gain characteristic setting of the WMHA (0 dB/octave slope and 100- to 4500-Hz bandwidth).

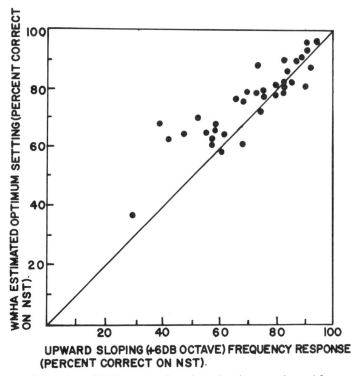

Figure 3. Comparison of NST scores at the estimated optimum setting and for an upward-sloping frequency response (+6 dB/octave as measured in a standard 2-cm³ coupler). Bandwidth was 100–4500 Hz.

obtained for the WMHA at its estimated optimum setting. The improvements were not quite as large as those obtained in comparison with the subject's own aid, but the pattern of improvement is the same, the larger improvements being obtained, on the average, for subjects with lower scores. There are more points in Figures 2 and 3 than in Figure 1, because scores at these settings were obtained during the fixed test battery for all subjects plus the additional measurements that were obtained during the adaptive sessions on a number of the subjects. There were also many more points falling close to or on the diagonal, because on several occasions the estimated optimum setting of the WMHA was identical to the reference condition.

Figures 1, 2, and 3 as a group indicate that improved performance was obtained at the estimated optimum setting of the WMHA as a result of using the experimental protocol and that although some of the improvements observed in comparison with the subject's own aid may have resulted from the differences in overall quality of the hearing aids, the relative magnitude of this factor appears to be small.

Effect of Speech Test Materials

Improvements were also obtained for several other measures of performance, such as the CID W-22 test (Hirsh et al., 1952), the CHABA sentences (Davis and Silverman, 1970), and the high frequency word test (Pascoe, 1975). The data obtained with these performance measures are shown in Figures 4, 5, and 6. As can be seen from these diagrams, the improvements were not obtained consistently across all subjects, and were more variable both across subjects and across tests. Although the majority of subjects showed improved performance and, on the average, scores for the estimated optimum setting of the WMHA were higher than those for the subject's own aid, the magnitude of the average improvement was smaller and many of the subjects showed poorer performance on individual tests.

The differences in measured performance are believed to be the result of a combination of two factors. First, the NST was designed to have a low

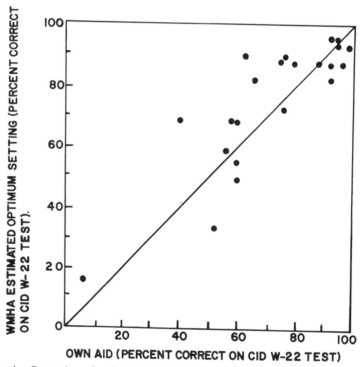

Figure 4.   Comparison of performance as measured by CID W-22 test for estimated optimum setting of the WMHA and the subject's own aid.

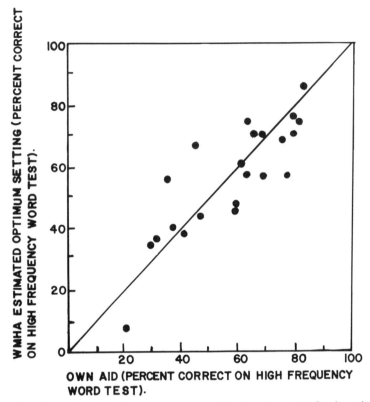

Figure 5.   Comparison of performance as measured by the Pascoe test for the estimated optimum setting of the WMHA and the subject's own aid.

test-retest variability specifically for use in hearing aid evaluation procedures requiring repeated measurement. The estimated test-retest variability under conditions of use was found to be comparable to the minimum binomial error variance for the estimate of a proportion. The differences in measured performance, particularly those that did not occur consistently for an individual subject, were believed to be partly a result of the larger test-retest variability of the other performance measures.

Second, and more important, the various performance measures are tapping different aspects of the subject's speech reception ability. This in itself is not a problem, provided that the optimum setting of the WMHA remains the same for the different types of speech material. A check on the two sets of test material that differ most in their structure, the CHABA sentences and the NST, showed some evidence that the optimum setting for the NST is not necessarily the same as that for the CHABA sentences.

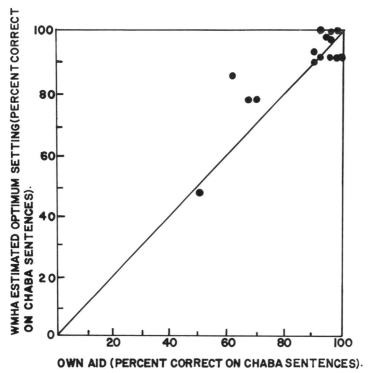

OWN AID (PERCENT CORRECT ON CHABA SENTENCES).

Figure 6. Comparison of performance as measured by the CHABA sentences for the estimated optimum setting of the WMHA and the subject's own aid.

Unfortunately, the variability of the test scores using the CHABA sentences is relatively high, and it is difficult to assess reliably the extent to which the optimum settings may differ. It is likely that the difference is small since there was considerable overlap between the two sets of measurements.

A preliminary estimate of the difference between the two optimum settings (shown graphically in the left-hand portion of Figure 7) would place the optimum slope of the frequency response for sentence material slightly lower, possibly by as much as one step size (3 dB/octave), than the optimum slope for nonsense syllables. As can be seen in the right-hand portion of Figure 7, no systematic differences were observed between the estimated optimum values for the lower cutoff frequency. The possibility that there may be differences in the optimum setting of the WMHA for test stimuli like nonsense syllables and the types of speech encountered in the everyday use of a hearing aid should not preclude the use of the test stimuli in fitting hearing aids, provided that the difference between the optimum settings is known and an appropriate adjustment is made. For example, the

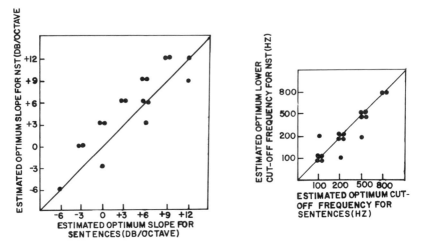

Figure 7.   Comparison of estimated optimum settings using NST and CHABA materials. The graph on the left is for the slope parameter, and the graph on the right is for lower cutoff frequency.

CHABA sentences of everyday speech may be more representative of the type of speech encountered in everyday communication than the more precise tests of limited content, such as the NST. Unfortunately, because of the limited number and the high test-retest variability of the CHABA sentence lists, this material cannot be used to find the optimum WMHA setting in a practical way. However, if it can be established that the optimum WMHA setting for the CHABA sentences differs from that for the more precise NST by a small, relatively fixed amount, then the fitting can be done using the more precise NST. Once the estimated optimum WMHA setting has been obtained for the relatively precise test material, the setting is adjusted by this known, fixed amount in order to approximate the optimum WMHA setting for everyday speech.

Effect of Speech-to-Noise Ratio

On an early version of the experimental protocol, a check was also carried out to test for differences in the estimated optimum setting of the WMHA as a function of S/N ratio. Estimated optimum settings obtained independently at two S/N ratios (+10 and +20 dB) showed no significant differences for a group of five subjects. However, the number of subjects was small and the same type of noise (cafeteria noise) was used in both cases. It is not known whether or not there would be significant differences in the optimum WMHA setting for different types of noise having distinctly different spectra.

Effect of Extended Usage

The final version of the experimental protocol included periods of usage of the aid in the test environment only (clinic usage) and in everyday situations as well as in the test environment (extended usage). The comparison between the clinic usage and extended usage phases of the protocol showed no significant differences between these two types of usage on the adaptive fitting procedure. However, it is important to bear in mind that the subjects tested were experienced hearing aid users and that the study did not involve auditory training in any way. The everyday use of the WMHA between test sessions (i.e., extended usage), and its use as part of an auditory training program, may have an important effect on new hearing aid users.

Factors Affecting Adaptive Strategy

Data on the rate of convergence toward the estimated optimum setting show that the largest improvements were obtained during the early portion of the experimental protocol (see Figure 8). By far the largest improvement in a single session was obtained for the fixed test battery. Subsequent improvements of decreasing size were obtained during the adaptive sessions that followed. An accurate starting point yielded substantial improvements in the speed with which the estimated optimum setting was reached. In this regard the recommendations of the Harvard study proved very useful in choosing parameter values for the fixed test battery, since for several subjects the estimated optimum settings were very close to or consistent with these recommendations. There were, of course, many other subjects whose estimated optimum settings differed significantly from the fixed set of parameter values recommended by the Harvard study. The recommendations of the study were used herein to establish a reference starting point. The pattern of estimated optimum settings obtained in this study indicates that the choice of parameter values for the fixed test battery could be improved in order to increase the probability of obtaining good initial estimates of the optimum setting of the WMHA.

An important factor affecting the rate at which an adaptive procedure converges on the optimum setting is the rate at which the performance measure changes with each parameter value. Figure 9 shows the change in test score as a function of the change in slope of the frequency response relative to the estimated optimum value. The change in test score is shown in arc sine units to stabilize the error variance (see Brownlee, 1965, for a definition and explanation of this transformation).

Figure 10 shows the corresponding diagram for a change in lower cutoff frequency. As can be seen from the Figures 9 and 10, a change in slope (in steps of 3 dB/octave, which is the available step size on the

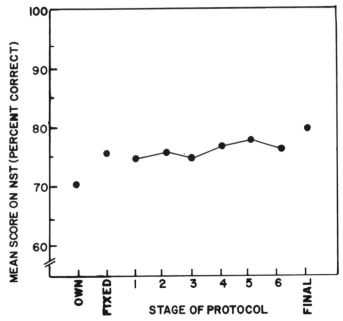

Figure 8.   Mean performance at selected stages of the protocol. Going from left to right, the first point is the score obtained for the subject's own aid. The second point shows the score at the initial estimate of the optimum setting of the WMHA as obtained from the fixed test battery. The six succeeding points show the scores at each successive estimate of the optimum setting as obtained during the Adaptive Sessions stage. The last point shows the score obtained at the final estimate of the optimum setting as measured during the Final Comparative Measurements stage. Data points represent averages across all subjects.

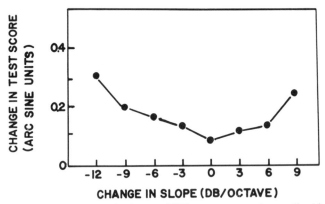

Figure 9.   Mean effect of change in slope on NST scores, averaged over all subjects.

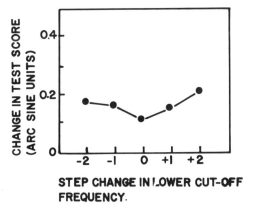

STEP CHANGE IN LOWER CUT-OFF
FREQUENCY.

Figure 10.   Mean effect of change in lower cutoff frequency on NST scores averaged over all subjects.

WMHA) leads to a comparatively larger change in average test score. Thus, a greater relative precision and a more rapid rate of convergence are to be expected on the slope dimension than that of lower cutoff frequency for the step sizes available on the WMHA.

The data of Figures 9 and 10 were obtained by averaging scores over all subjects at every WMHA setting used in the study more than twice. This tends to obscure the relative symmetry of the curves shown in the figures since the subjects tested on one side of the optimum parameter setting were not necessarily also tested on the opposite side. It is possible to extract from the available measurements a balanced set of data (i.e., for the same set of subjects at all points) over a limited range. Figure 11 shows the corresponding curve obtained in this way for three adjacent settings of the slope parameter: the estimated optimum slope and the setting on either side of

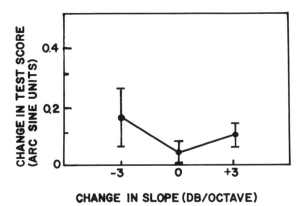

CHANGE IN SLOPE (DB/OCTAVE)

Figure 11.   Mean effect of change in slope on NST scores for selected subjects.

the estimated optimum. The value of this diagram is that it provides a more reliable picture (between-subject differences having been balanced out) of the effect of changing the slope in the vicinity of the optimum setting. Similarly, Figure 12 shows, for a common group of subjects, the effect of changing lower cutoff frequency in the vicinity of the optimum setting.

A Clinical Protocol

The last version of the experimental protocol that was tried contains the elements of a practical protocol suitable for clinical use. The data obtained on this version of the protocol show that by the end of the first of the adaptive sessions, the large majority of subjects will have reached a level of performance that is not significantly different from that obtained at the final estimated optimum setting of the WMHA. This version of the experimental protocol is thus seen to consist of three key sessions which could be modified and further reduced in size to form the basis of a practical clinical protocol.

The first key session covers that portion of the protocol in which basic audiometric data are obtained, including, in particular, loudness discomfort level measurements for noise bands in the sound field. The second key session consists of the fixed test battery, which leads to the initial estimate of the optimum setting. The third key session is the first of the adaptive sessions. It is recommended that this three-session protocol be evaluated in a clinical setting. The possibility of further reducing the protocol to two sessions should also be investigated. This could be done by incorporating portions of the first session into the subject's initial audiological workup, and using the remainder of the first session to administer the fixed test battery.

Limitations of the Study

The study has some important limitations, several of which were imposed by the particular WMHA being considered. Others were imposed by practi-

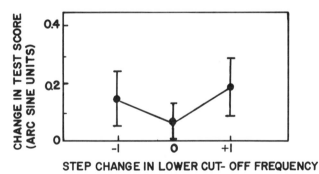

Figure 12.   Mean effect of change in lower cutoff frequency on NST scores for selected subjects.

cal considerations inherent in a study of finite size. First, the WMHA used in this study, although reasonably flexible in its adjustable parameters and range of adjustment on each parameter, was nevertheless limited to a finite number of settings. Thus the estimated optimum setting of the WMHA for a given subject should not be regarded as the optimum electroacoustic characteristics for that subject, but simply the estimated best setting from among a moderately large range of possible settings.

A major limitation of the range of settings available on the WMHA related to the control of the frequency-gain characteristic. Only three parameters of adjustment were available: slope, lower cutoff frequency, and upper cutoff frequency. Thus it was not possible to adjust for departures from a linear frequency-response curve, as measured in a standard coupler. The effect of factors such as ear canal resonance or head diffraction on the frequency-gain characteristic could only be adjusted for in terms of rough straight-line approximations.

For practical reasons, the subject population was limited (adults; experienced hearing aid users; monaural amplification; sensorineural impairment acquired postlingually, but not as a result of presbycusis). Subjects satisfying these criteria form an important subgroup of the hearing-impaired population. They are also subjects for whom the effectiveness of a prescriptive fitting protocol could be evaluated without confounding with other limiting factors (e.g., poor language and learning skills, adaptation to acoustic amplification, central processing problems). The choice of subjects was dictated primarily by a concern for obtaining reliable evaluative data on the fitting protocol itself, not on demonstrating the value of a WMHA per se. In this regard the choice of subjects represented a much more critical test of the efficacy of the fitting protocol, since it turned out that all of the subjects were reasonably satisfied with their own hearing aids. For several subjects the functional gain of the estimated optimum WMHA setting was found to be essentially the same as that of their own aids.

Another practical limitation of the study was that adjustment strategies for compression characteristics were not considered in any depth. The study began with various parameters of compression amplification considered within the first version of the experimental protocol. However, because of time constraints and serious limitations on the range of adjustment of the compression parameters on the WMHA, compression characteristics were not considered beyond the first part of the study.

CONCLUDING COMMENTS

The most important finding of this study was that improved performance could be obtained from systematic, individualized adjustment of a wearable master hearing aid. For many of the subjects, particularly those showing relatively poor speech discrimination scores with their own hearing aid, the magnitude of the improvement was large enough to warrant the time and effort required for individualized prescriptive fitting of the WMHA.

In embarking on this study it was recognized that, with time, improved instrumentation, improved testing techniques, and improved hearing aids would be developed. The underlying thrust of the study was thus to develop a fairly general approach to hearing aid evaluation that would not be limited to a specific WMHA or method of assessing performance. Thus, because of the limitations on the WMHA used in this study, the electroacoustic characteristics corresponding to the estimated optimum setting of the WMHA should not be regarded as the optimum electroacoustic character- istics for that subject, but simply the best of an available range of electroa- coustic characteristics. However, it is beleived that the method of fitting the WMHA can be generalized. That is, given a different WMHA and/or another performance measure, provided certain minimum constraints are met regarding the need for repeated testing and low test-retest variability in assessing relative performance, the experimental protocol developed in this study should still be applicable.

The essential ingredients of such a protocol are that it consists of a fixed test battery using a statistically powerful experimental design to obtain an efficient, initial estimate of the optimum setting of the WMHA, followed by a reliable, efficient adaptive fitting strategy, such as the simplical method, that can be used with data containing some degree of random variability. The choice of parameter values for the fixed test battery should be derived from previous data obtained on similar subjects, using deterministic psy- choacoustic stimuli to determine parameter values where necessary. Until such time as adequate techniques are developed for predicting speech recep- tion ability from psychoacoustic data, speech stimuli, with their greater inherent test-retest variability, should be used in the fixed test battery and subsequent adaptive fitting strategy. Although this study was concerned with developing a strategy for fitting a specific type of hearing aid, it is believed that, in its essentials, the protocol developed would be applicable to a much wider range of sensory aids.

REFERENCES

Box, J.E.P. 1957. Evolutionary operation: A method for increasing industrial pro- ductivity. Appl. Statist. 6:81–101.
Brownlee, K.A. 1965. Statistical Theory and Methodology in Science and Engineer- ing. 2nd Ed. John Wiley & Sons, Inc., New York.
Davis, H., and S.R. Silverman. (eds.). 1970. Hearing and Deafness. 3rd Ed. Holt, Rinehart & Winston, New York.
Davis, H., S.S. Stevens, R.H. Nichols, Jr., C.V. Hudgins, R.J. Marquis, G.E. Peter-

son, and D.A. Ross. 1947. Hearing Aids: An Experimental Study of Design Objectives. Harvard University Press, Cambridge.

Hirsh, I.J., H. Davis, S.R. Silverman, E. Reynolds, E. Eldert, and R.W. Benson. 1952. Development of materials for speech audiometry. J. Speech Hear. Disord. 17:321–337.

Levitt, H., M.J. Collins, J.R. Dubno, S.B. Resnick, and R.E.C. White. 1978. Development of a protocol for the prescriptive fitting of a wearable master hearing aid. Communications Sciences Laboratory Report #11, Doctoral Program in Speech and Hearing Sciences, CUNY Graduate School, New York.

Levitt, H., and S.B. Resnick. 1978. Speech reception by the hearing-impaired: Methods of testing and the development of new tests. In C. Ludvigsen and J. Barfod (eds.), Sensorineural Hearing Impairment and Hearing Aids. Scand. Audiol. (suppl. 6):107–130.

Pascoe, D.P. 1975. Frequency responses of hearing aids and their effects on the speech perception of hearing-impaired subjects. Ann. Otol. Rhinol. Laryngol. 84 (suppl. 23):1–40.

Resnick, S.B., J.R. Dubno, S. Hoffnung, and H. Levitt. 1975. Phoneme errors on a nonsense syllable test. J. Acoust. Soc. Am. 58:S114 (abstr.).

CHAPTER 15

COMPARISON OF METHODS FOR PREDICTING OPTIMUM FUNCTIONAL GAIN

M. Jane Collins and Harry Levitt

CONTENTS

The need for individualized fitting of hearing aids has been well established in the laboratory and in the clinic. A clinician faced with the problem of recommending amplification for an individual with a hearing loss needs to be able to accurately estimate optimum frequency-gain characteristics for that person and must do so in a limited amount of time. Often, unfortunately, accuracy of estimation is sacrificed to some degree for the sake of efficiency when the clinician selects the method to be used for fitting. On the other hand, laboratory methods have been developed that may prove to be highly accurate, but are often inefficient because they require more time than is feasible in a clinical setting.

One approach to minimizing the time required for testing has been to attempt to predict optimum frequency-gain characteristics on the basis of readily available or easily obtained audiometric and psychoacoustic data. However, there is not good agreement on which predictive method is the most accurate, nor is there good agreement on the psychophysical measures

This work was supported in part by NINCDS Contract #NIH-NDI-NS-4-2323.

that should serve as the basis for prediction. Among the proposed methods are those of mirroring the audiogram, bisecting the dynamic range, and paralleling the maximum comfort level. The various psychoacoustic measurements on which the predictions are based may or may not be routinely obtained in clinical practice.

In the following sections, selected representative predictive methods are described. These have been applied to data obtained on persons who served as subjects in a study by Levitt et al. (1978) and to the predicted optimum frequency-gain characteristics determined by a nonpredictive laboratory procedure. Implications for the feasibility of using predictive methods are discussed.

PREDICTIVE METHODS FOR ESTIMATION OF OPTIMUM FREQUENCY-GAIN CHARACTERISTICS

Mirroring the Audiogram

Pascoe (1975) recommended functional gain that renders aided noise-band thresholds parallel to the normal thresholds for the noise bands. This "mirroring" of the audiogram is accomplished by a subtraction of thresholds for ⅓-octave bands of noise obtained on normal-hearing subjects from individual subjects' thresholds obtained under the same conditions:

$$FG_i = NBT_i - NNBT_i$$

where FG_i is the functional gain at the ith center frequency of a ⅓-octave band of noise. NBT_i is the subject's unaided threshold for that band of noise, and $NNBT_i$ is the average, normal threshold for the same noise band. Thus, in order to apply this method, norms must be established for audibility of narrow bands of noise and individual thresholds must be established for the hearing-impaired persons. The functional gain required for optimum performance can then be readily calculated.

Skinner (1976) evaluated performance of persons with high frequency hearing loss under a variety of conditions of frequency-gain characteristics. Her best results were obtained with functional gain similar to that predicted by Pascoe's method. The optimum performance was achieved with average functional gain mirroring the audiogram between 500 Hz and 1600 Hz, but with 0-dB functional gain below 500 Hz and an average of 23 dB gain above 1600 Hz. For present purposes, this is considered a predictive method for persons who exhibit high frequency hearing loss.

Mirroring the MCL Curve

Watson and Knudsen (1940) suggested that optimum performance would be obtained with frequency-gain curves that mirror the most comfortable loudness level (MCL) curve. The formula they gave for calculating the functional gain at each frequency is:

$$A_F = HL_F - (DR_{RF} - DR_F) + K$$

where A_F is the required gain at each Fth frequency, HL_F is the hearing loss at the Fth frequency, DR_{RF} is the difference between MCL (measured in dB re normal hearing) and the hearing loss at a reference frequency, DR_F is the difference between the MCL (measured in dB re normal hearing and the hearing loss at the Fth frequency), and K is a constant. To apply this method, audiometric threshold data are required, plus an estimate of MCL at the reference frequency. Although the formula was developed on the basis of pure tone data, it is reasonable to assume it could be applied to narrowband noise data also.

Byrne and Tonnison (1976) proposed a method that has some features in common with that of Watson and Knudsen, but with important differences. Their method is based on the assumption that speech discrimination is optimal when all components of the speech spectrum are presented at equal and most comfortable loudness levels. Therefore, this is a method that requires mirroring of the MCL curve by the aided speech spectrum. They calculated real-ear (functional) gain for levels of pure tone hearing thresholds re normal to meet the assumed requirements. Initially, they determined the gain at 1000 Hz as a function of hearing loss according to measurements of preferred sensational level with a group of hearing-impaired subjects. Then, they also corrected for relative speech spectrum levels across frequencies. Thus, they provide a table that gives real-ear gain as a function of frequency for a range of hearing loss values, and the predicted optimal frequency-gain characteristic can be obtained from the table with only pure tone audiometric data.

Bisecting the Dynamic Range

Wallenfels (1967) advocates bisection of the dynamic range and recognizes that most persons with sensorineural hearing loss do better without overamplification of low frequency sounds. His method requires measurement of the thresholds of audibility and of discomfort levels for pure tones at 1000 Hz and 4000 Hz. The midpoints of the dynamic range at those two frequencies determine the straight-line frequency-gain function above 1000 Hz. If the slope of that line is greater than 10 dB/octave, the line is simply extended to the lower frequencies. If the slope is less than 10 dB/octave, the frequency-gain function below 1000 Hz is a straight line with slope equal to 10 dB/octave. If the method is applied using noise bands rather than pure tone stimuli, the functional gain required at 1000 Hz and 4000 Hz can be determined by:

$$FG_i = \frac{NBT_i + LDL_i}{2}$$

where FG_i is functional gain at the ith frequency (ith frequency is either 1000 or 4000 Hz), NBT_i is the threshold for a ⅓-octave band of noise centered at the ith frequency, and LDL_i is the loudness discomfort level for a ⅓-octave band of noise centered at the ith frequency.

A NONPREDICTIVE METHOD

Levitt et al. (1978) developed an experimental protocol for the fitting of a wearable master hearing aid (WMHA). The protocol included a two-stage procedure for estimating the optimum frequency-gain characteristics for each subject. The first stage involved measurement of speech perception skills for fixed settings that covered a wide range of slope (change in gain/ change in frequency) and bandwidth characteristics. The setting (for bandwidth and slope values) that yielded the best scores was utilized as a starting point for an adaptive stage of testing. During the adaptive stage, settings of the WMHA were changed according to preestablished rules on the basis of performance on a nonsense syllable test (NST). Thus, in contrast to the predictive methods, this type of procedure manipulates the frequency-gain characteristics, and psychoacoustic measurements play no part in the estimation of the optimum frequency-gain characteristics. A detailed description of the entire protocol and its evaluation are contained in Chapter 14 (Levitt and Collins, this volume).

After the estimation procedures were completed, functional gain was calculated on the basis of sound field measurements. Using ⅓-octave bands of noise as stimuli, subjects tracked thresholds in the unaided and aided conditions. For the aided condition, the WMHA was at the estimated optimum setting. Functional gain was also determined for the subject's own aid. The differences between the aided and unaided thresholds for each noise band were taken to be estimates of the functional gain:

$$FG_i = NBT_i - ANBT_i$$

where FG_i is the functional gain for the ith frequency, NBT_i is the unaided noise band threshold, and $ANBT_i$ is the aided noise band threshold.

COMPARISON OF PREDICTIVE METHODS

In order to predict optimum frequency-gain curves on the basis of Pascoe's, Skinner's, Watson and Knudsen's, Byrne and Tonnison's, and Wallenfels's methods, one needs the following measurements: pure tone thresholds, unaided noise band thresholds, MCLs, and loudness discomfort levels (LDLs). Pure tone thresholds as well as MCLs, LDLs, and thresholds for ⅓-octave bands of noise were available for subjects in the Levitt et al. (1978) study. Thus it was possible to generate frequency-gain curves according to each of the five predictive methods. Skinner's method was applied only for those subjects who exhibited high frequency hearing loss. When formulas required normal thresholds for a ⅓-octave band of noise, the norms published by Pascoe (1975) were utilized. The curves "predicted" were compared with the functional gain curves obtained in the Levitt et al. study for the estimated optimum setting for the WMHA as far as relative gain as a function of frequency was concerned. That is, gain at each frequency for each method was determined relative to the gain at 1000 Hz so that the comparison involved the shape of the curve, not the absolute gain as predicted or measured.

Because of the nature of the narrowband measurement in the sound field, high variability was encountered in the data, and some subjects' thresholds exceeded the maximum output levels of the system. Whenever there was excessive variability in the measurements, the subjects involved were eliminated from the present comparison. Data based on 12 subjects are presented.

RESULTS

Figure 1 shows the average predictions for 12 subjects from Pascoe's method, from Byrne and Tonnison's tables, from Watson and Knudsen's formula, and from Wallenfels's method. The average functional gain is also shown for the WMHA. The dashed lines represent coupler gain for the flat and +6 dB/octave frequency response curves (as measured in a 2-cc coupler) recommended by the Harvard report (Davis et al., 1947). As can be seen, the results for Pascoe's method, Byrne and Tonnison's table, and the Levitt et al. study using the WMHA are similar. On the average, these methods yield frequency-gain functions that provide more gain in the high frequencies than in the low frequencies, with slopes on the order of 15–20 dB/octave in the steepest portion. The curves from Watson and Knudsen's formula and from Wallenfels's method tend to be flatter than those from the other methods, yet these also show the least gain in the low frequency regions.

In order to realistically compare the recommended coupler measurements from the Harvard study to the frequency-gain curves in this figure, corrections for head diffraction, earmold, and ear canal resonance effects should be made. The curves are shown here mainly for reference and are not intended to represent functional gain.

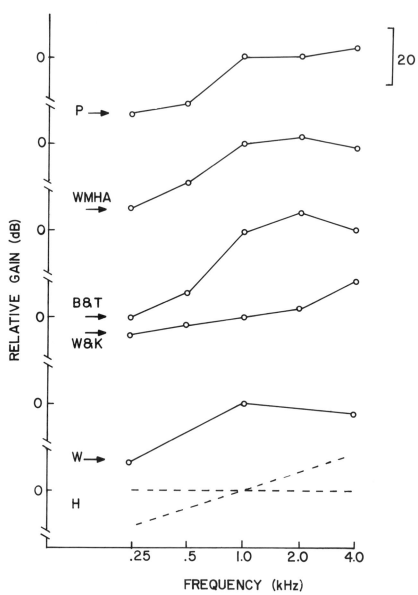

Figure 1.   Relative functional gain, predicted and measured. The data shown are average results obtained for 12 subjects according to Pascoe (P), the wearable master hearing aid study (WMHA), Byrne and Tonnison (B&T), Watson and Knudsen (W&K), and Wallenfels (W). The broken lines represent 0 dB/octave and 6 dB/octave coupler-measured gain recommended by Davis et al. (H).

The frequency-gain functions for a subject with close-to-normal hearing in the low frequencies and a severe hearing loss in the high frequencies are shown in Figure 2. (The contour of the hearing loss is, of course, reflected in the curve generated according to Pascoe's recommendations.) The differences seen between the Pascoe curve and the Skinner curve are simply related to limited gain above 1600 Hz and below 500 Hz for Skinner's optimum curve. The WMHA curve shows a similar steeply rising function below 1000 Hz, but the gain falls off slightly in the higher frequencies. Over the frequency range from 500 Hz to 2000 Hz, for this subject, the least steep slopes were predicted by Byrne and Tonnison's and Watson and Knudsen's methods. On the whole, however, the curves based on the methods shown are more similar to each other than they were for the average data shown in Figure 1.

The subject whose results are shown in Figure 3 has a relatively flat loss. Pascoe's curve predicts more relative low frequency gain than any other curve. Thus, as hearing improves in the low frequencies (relative to hearing in the high frequencies), the similarities between the Pascoe, WMHA, and Byrne and Tonnison curves tend to disappear. The Wallenfels curve changes little as the contour of the hearing loss changes since the lower portion of the curve is fixed, and therefore is independent of the shape of the audiogram.

A crucial factor in Byrne and Tonnison's computations of required gain at each frequency is the particular speech spectrum values used. The spectrum they used was for four male and four female speakers. In order to illustrate the effect of the spectral measurements on the frequency-gain function, the tabled values were corrected for a different speech spectrum. The values used are from ⅓-octave measurements by Pearsons, Bennett, and Fidell (1976) for an American male talking at a "normal" effort level. Figure 4 shows the predictions for all methods for a subject with a high frequency hearing loss. Looking at the solid lines, the results are similar to those shown in Figure 2 for another subject with high frequency hearing loss. The broken line is what would be predicted by Byrne and Tonnison's theory if Pearson's spectrum had been used. The curve would obviously provide less high frequency gain and more low frequency gain and diverge from the curves of Pascoe, Skinner, Byrne and Tonnison, and the WMHA.

It should be noted that most of the subjects in the Levitt et al. (1978) study did exhibit more loss in the high frequencies than in the lows. That is, on the average and for most subjects, the greatest similarities between the functional gain curves were found for Pascoe's method, Byrne and Tonnison's table, and the measured values with the WMHA. For more pronounced high frequency losses, such as for the subject whose results are shown in Figure 2, predictions by all methods tended to call for minimal

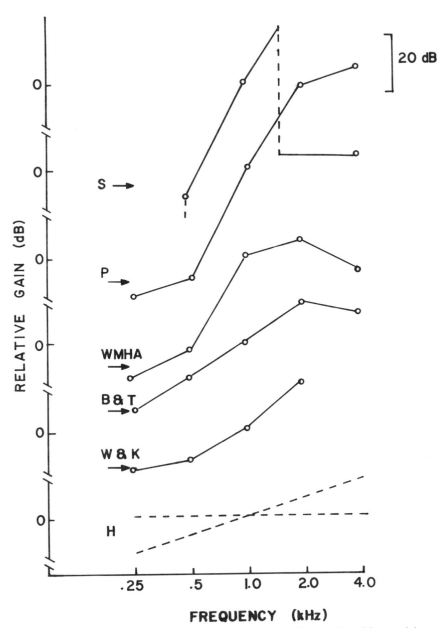

Figure 2.   Relative functional gain, predicted and measured for a subject with a precipitous high frequency hearing loss. No curve is shown for Wallenfels's method because the LDL for a ⅓-octave noise band centered at 4000 Hz was beyond the limits of the test system.

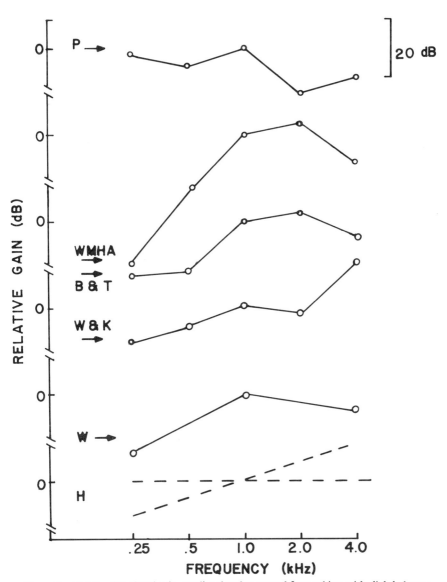

Figure 3.   Relative functional gain, predicted and measured for a subject with slightly better hearing in the high frequencies than in the low frequencies.

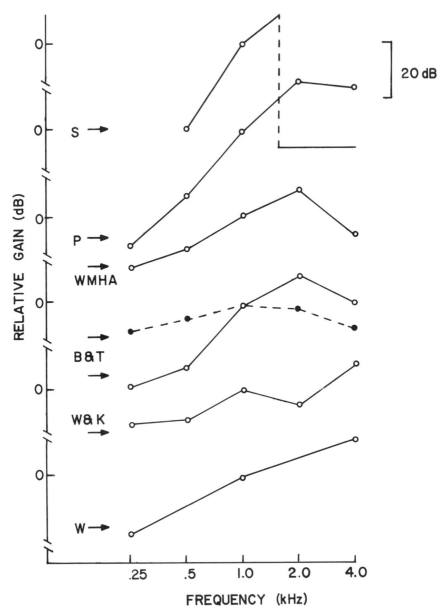

Figure 4.   Relative functional gain, predicted and measured for a subject with a high frequency hearing loss. The broken line for Byrne and Tonnison's (1976) method (B&T) was generated using Pearson et al.'s (1976) relative spectrum values.

low frequency gain and increased high frequency gain. As the hearing improved in the high frequencies relative to the low frequencies, the WMHA and the Byrne and Tonnison curves emerge as the most consistent predictors of greater gain in the high frequencies. The slope of the Pascoe curve obviously follows the contour of the loss. The shape of the Watson and Knudsen curve is not predictable according to hearing loss contour alone if the loss is anything other than a precipitous high frequency loss. The upper portion of the Wallenfels curve does not vary systematically with the contour of the loss either.

Obviously, the WMHA curves for functional gain do not mirror the audiogram, do not mirror the MCL curve, and do not bisect the dynamic range. The curves are most consistently similar to the predictions of Byrne and Tonnison using their measured spectrum, but differences are still noticeable.

A SUGGESTED PREDICTIVE RULE

Since none of the described predictive rules held for the subjects in the study by Levitt et al., the relationships between aided noise-band thresholds and audiometric or psychoacoustic data for individual subjects were further analyzed.

The most promising analysis indicated a trend for best results when the speech spectrum, after being modified by the frequency-gain function, tended to parallel the individual subject's hearing threshold curve. That is, when using the speech spectrum reported by Pearsons et al. (1976), computations were made of the levels of speech delivered to the ear within ⅓-octave bands. This was carried out for the WMHA and for the subject's own aid. An analysis of variance was performed on the resultant aided speech spectrum values and the hearing thresholds for each subject. The size of the interaction term thus served as an index of the degree to which the two curves were parallel, parallel curves yielding a negligible interaction term. A trend emerged for the interaction term to be smaller for the aid (WMHA or subject's own aid) that yielded the better speech discrimination score as measured with the NST. The greater the high frequency loss, the more obvious was this effect.

Figures 5 and 6 show data for the same two subjects used for the comparison of methods. In each figure, the threshold of audibility is shown for the ⅓-octave bands of noise (centered at the indicated frequency) presented in the sound field. Also, the aided speech spectra are shown for the WMHA and for the subject's own aid.

In Figure 5, the aided speech spectrum for the WMHA is clearly more parallel to the hearing threshold curve than is the spectrum for the subject's

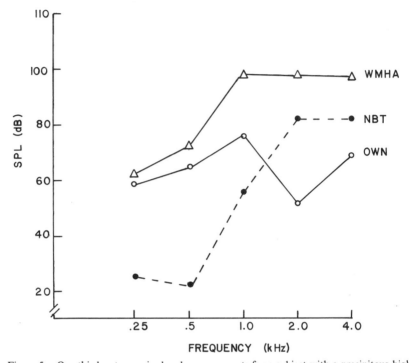

Figure 5.   One-third–octave noise band measurements for a subject with a precipitous high frequency hearing loss. Solid lines indicate measurements of threshold of audibility (NBT). The broken lines indicate estimates of the level of the speech spectrum within each ⅓-octave band when using the wearable master hearing aid (WMHA) and the subject's own hearing aid (Own).

own aid. This subject's performance on the NST with the WMHA was improved significantly. The subject whose results are shown in Figure 6 has a relatively flat hearing loss, and the WMHA was again the aid to yield the higher NST scores. At 500 Hz and above, the curve for the aided speech spectrum with the WMHA is slightly more parallel to the threshold curve than is the curve for the subject's own aid. This is representative of the trend seen in the data. However, as hearing loss in the low frequencies increased relative to hearing loss in the high frequencies, the rule did not hold as well.

IMPLICATIONS FOR THE USE OF PREDICTIVE METHODS

It appears that the "accuracy" of any of the predictive methods is dependent to some degree on the contour of the hearing loss. This is inferred from the observations that: 1) for persons with marked high frequency hearing loss, the predictive methods yield similar curves and those curves are similar to that obtained for optimum frequency-gain characteristics estimated on the

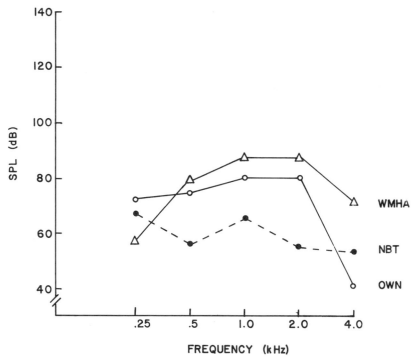

Figure 6.   One-third–octave noise band measurements for a subject with a relatively flat hearing loss. Solid lines indicate measurements of threshold of audibility (NBT). The broken lines indicate estimates of the level of the speech spectrum within each ⅓-octave band when using the wearable master hearing aid (WMHA) and the subject's own hearing aid (Own).

basis of performance, and 2) as hearing loss becomes flatter, the predictive rule derived from the data of the Levitt et al. study fit less well and greater discrepancies are noted between the other predictive methods. Perhaps the predictive rules require a weighting factor for hearing loss in differing frequency ranges.

For those methods of prediction that involve the speech spectrum, i.e., functional gain such that the aided speech spectrum parallels the threshold of audibility or the MCL curve, the shape of the frequency-gain curve is going to be highly dependent on the speaker characteristics and analysis methodology. Is there a "best" estimate of the speech spectrum to use? Might that differ depending on the communication situations a particular person encounters?

It should be emphasized that the predictive rule derived from the data of the Levitt et al. study was based on *post hoc* observations. Nonetheless, the trend seen (when hearing loss is greater in the high frequencies than in the low frequencies) for the "better" aid to provide gain such that the aided speech spectrum is systematically related to the hearing threshold curve should not be ignored. These data suggest, however, that a reasonably good first estimate of the frequency-gain characteristics for optimum performance can be obtained from simple psychoacoustic measurements (e.g., noiseband thresholds).

Further research is needed to explore the effects of contour of hearing loss on the accuracy of the predictive methods. In addition, the effects of hearing aid characteristics other than frequency gain have been ignored in the present evaluation of the predictive methods. These and other factors (such as effects of recruitment, effects of conductive components superimposed on sensorineural hearing loss, effects of etiology of hearing loss, and effects of the performance measure used to evaluate the hearing aid interacting with the subject) require further investigation in relation to predictive methods before the value of such methods in the clinic can be definitively determined.

ACKNOWLEDGMENTS

The material reported here summarizes portions of a larger project. We are indebted to all members of the research group, and to Drs. Judy Dubno, Steffi B. Resnick, and Richard E.C. White, in particular, for their help on this project.

REFERENCES

Byrne, D., and W. Tonnison. 1976. Selecting the gain of hearing aids for persons with sensorineural hearing impairments. Scand. Audiol. 5:51–59.

Davis, H., S. S. Stevens, R.H. Nichols, C.V. Hudgins, R. J. Marquis, G.E. Peterson, and D.A. Ross. 1947. Hearing Aids: An Experimental Study of Design Objectives. Harvard University Press, Cambridge.

Levitt, H., M.J. Collins, J.R. Dubno, S.B. Resnick, and R.E.C. White. 1978. Development of a protocol for the prescriptive fitting of a wearable master hearing aid. CSL Report No. 11, City University of New York, New York.

Pascoe, D.P. 1975. Frequency response of hearing aids and their effects on the speech perception of hearing-impaired subjects. Ann. Otol. Rhinol. Laryngol. 86 (suppl. 23).

Pearsons, K.S., R.L. Bennett, and S. Fidell. 1976. Speech Levels in Various Environments. Bolt, Beranek and Newman, Inc., Report No. 3281. Prepared for the Office of Resources and Development, Environmental Protection Agency.

Skinner, M.W. 1976. Speech intelligibility in noise-induced hearing loss: Effects of high frequency compensation. Unpublished doctoral thesis, Washington University, St. Louis.

Wallenfels, H.G. 1967. Hearing Aids on Prescription. Charles C Thomas Publisher, Springfield, Ill.

Watson, N.A., and V.O. Knudsen. 1940. Selective amplification in hearing aids. J. Acoust. Soc. Am. 11:406–419.

CHAPTER 16

INTEGRATION OF THE ELECTROACOUSTIC DESCRIPTION OF HEARING AIDS WITH THE AUDIOLOGIC DESCRIPTION OF CLIENTS

James D. Miller, Arthur F. Niemoeller,
David Pascoe, and Margo W. Skinner

CONTENTS

This chapter is an attempt to identify and illustrate the problems in relating the electroacoustic description of a hearing aid to the audiologic description of the patient. It is surprising that there continue to be so many barriers to elucidating these basic and simple variables. The problems arise because

The preparation of this chapter was supported in part by Research Grant NS03856 from the National Institute of Neurological and Communicative Disorders and Stroke to Central Institute for the Deaf.

many of the measurements that would provide the desired information are so difficult to make that it would be impractical to make them at this time. For example, if sounds could be accurately and practically measured at the listener's eardrum, both unaided and with earphones or earmolds in place, and if sounds could be measured accurately and easily at the microphone of the hearing aid, then many of our outstanding problems would be eliminated.

The difficulty and impracticality of the appropriate measurements have led us to a situation where the degree of confusion is limited only by our ingenuity in developing indirect methods that approximate, with noticeable error, the desired measurements.

There remain basic problems with the measurement of speech. If one takes a statistical approach, then there are the problems of sampling, filtering, and time averaging. If one chooses a phonetic approach, then problems of windowing and unending labor appear.

The description of a hearing aid is complicated by noise levels, distortions, limiting levels, and frequency responses, which are all subject to change with a variety of factors.

MEASUREMENTS ILLUSTRATED
IN THE RESEARCH OF PASCOE (1975)

Measurement of Threshold

Pascoe elected to measure the patients' thresholds for ⅓-octave bands of noise as presented in a sound field. The field measures were closely related to normal listening conditions. Also, a bandwidth of ⅓-octave seems to provide adequate selectivity in the frequency domain while reducing the extreme uncontrolled variations in level often encountered with extremely narrow bands or tones. It is also important that the same stimuli used to measure the parameters of the patient's hearing be used in the electroacoustic evaluation of the hearing aid itself. The appropriateness of these measurements is illustrated by experimental results. Figures 1 and 2 present the results of the noise-band audiometry for eight hearing-impaired patients and a normal base line. They reveal directly the patient's ability to detect acoustic energy as it is normally measured and encountered in fields.

When experimental hearing aids are worn, one can clearly determine how they change the patient's ability to detect sounds by plotting the aided and unaided thresholds in exactly comparable units, as Figure 3 illustrates. The differences between the aids, in regard to the patient's ability to detect sound fields, are apparent.

Frequency Responses of Hearing Aids

The ideal measurement of the frequency response of a hearing aid would be valid, reliable, transferable from laboratory to laboratory or clinic to clinic, and fast and inexpensive to make. Unfortunately, we have no one method of measurement that meets all of these criteria. For example, one

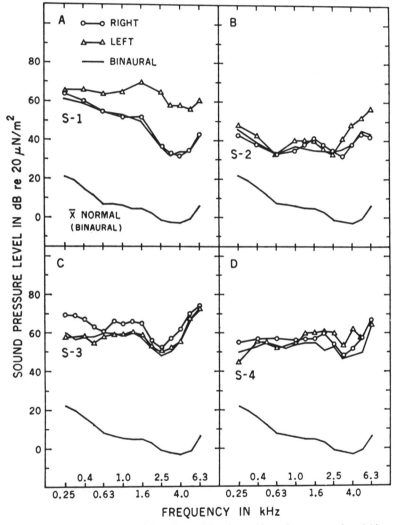

Figure 1. Individual results of four hard-of-hearing subjects for monaural and binaural minimum audible fields. The test sounds were ⅓-octave bands of noise centered at the stated frequencies. (Reprinted with permission from D.P. Pascoe, 1975, *Annals of Otology, Rhinology and Laryngology 84* (suppl. 23).)

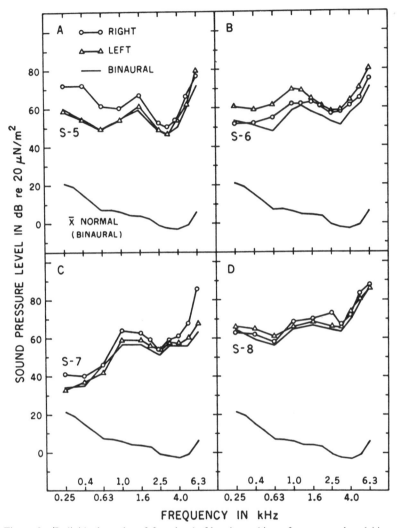

Figure 2. 'Individual results of four hard-of-hearing subjects for monaural and binaural minimum audible fields. (Reprinted with permission from D.P. Pascoe, 1975, *Annals of Otology, Rhinology and Laryngology 84* (suppl. 23).)

can attach a hearing aid to a standard 2-cc coupler and measure the difference between the sound pressure level (SPL) measured in the coupler and that measured at the position of the microphone of the hearing aid as it is held in place in a test jig. This method is reasonably reliable, transferable, and cost effective, although serious problems in each of these areas can and do arise. Unfortunately, methods of calibration of hearing aids in test jigs

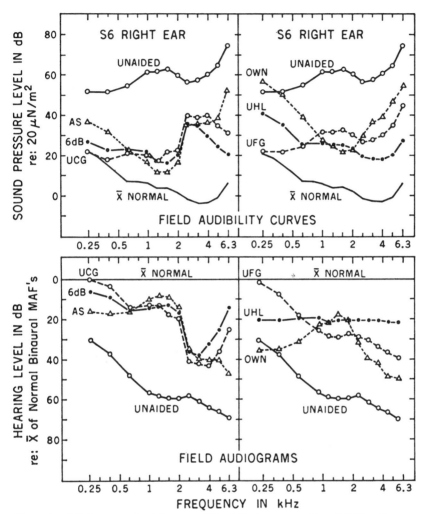

Figure 3.   Aided monaural minimum audible fields (upper panels) and field audiograms (lower panels) for one hard-of-hearing patient. The right ear was tested with each of six frequency responses: his own aid (OWN), a simulation of a commercial aid (AS), and other configurations of a master aid, which provided uniform functional gain (UFG), uniform hearing levels (UHL), or rose 6 dB/octave as measured in a 2-cc coupler (6 dB). (Reprinted with permission from D.P. Pascoe, 1975, *Annals of Otology, Rhinology and Laryngology 84* (suppl. 23).)

that utilize some form of artificial ear have consistent problems in the area of validity.

Pascoe measured the functional gain, or frequency response, by measuring the differences between the patients' aided and unaided thresholds.

The measurement, properly done, reflects the gain of the hearing aid as it is actually worn.

Differences between the two methods, that is, coupler gain vs. functional gain, can be sizable, as shown in Figure 4. Notice that differences shown in the lower right-hand panel of Figure 4 vary from about + 10 dB to − 18 dB, over a range as large as 28 dB. Clearly, the frequency response of the hearing aid as measured in a test jig that utilizes a 2-cc coupler is not a valid representation of the frequency response of the aid as it is worn.

The sizable differences between the frequency responses as they are

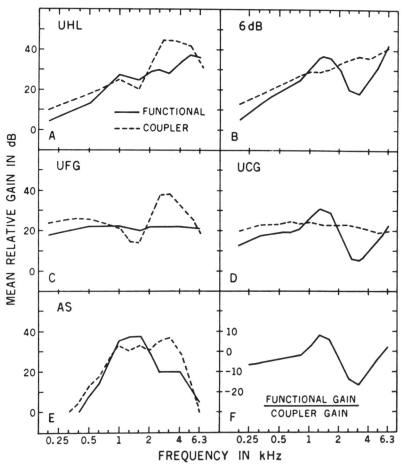

Figure 4.   Mean functional gains and associated coupler gains for each of five frequency responses. The lower right-hand panel contains the overall difference between the two kinds of measurement in dB. (Reprinted with permission from D.P. Pascoe, 1975, *Annals of Otology, Rhinology and Laryngology 84* (suppl. 23).)

measured in a test jig that utilizes a 2-cc coupler and as they are measured when actually worn can be attributed to several hard-to-measure variables. For example, there are differential effects of head diffraction in the aided and unaided conditions, but these are totally absent in the test jig. Outer-ear resonances play an important role in the unaided condition, but these are absent when the hearing aid is employed, whether in the patient's ear or in the test jig. Furthermore, the exact arrangements and volumes of tubes, earmolds, ear canals, and so on all differ between the 2-cc coupler and the aid as actually worn, as do the acoustic impedances encountered in the different situations. All of these factors indicate that while calibrations in test-jigs that utilize 2-cc couplers are useful for the electroacoustic evaluation of hearing aids and for the exchange of such evaluations, they do not reflect the results to be obtained when the aid is worn by a particular patient.

Identification of Speech Sounds

If hearing aids are to be evaluated on the basis of how well patients can identify speech sounds, then this performance needs to be measured accurately. Such measurements are not made easily. In Pascoe's work considerable care was taken in an effort to sample a variety of listening conditions, to eliminate the effects of learning and fatigue, and to achieve an adequate sample of stimuli for each listening condition.

The test materials were either the high frequency word lists described by Pascoe or the more familiar monosyllabic words of the standard phonetically balanced (PB) lists. A male and a female talker were used and measurements were made in quiet and in noise (0 or 6 dB signal-to-noise ratio). Practice tests ensured that the patients were thoroughly familiar with procedures, the talker's voices, and the possible items before the collection of data. It is extremely important to note that Pascoe ensured that the recording and playback system did not change the spectrum of the speech. Thus, high frequency energy that carried information was included by virtue of the care in recording and playback and in the selection of the words.

The tests were conducted in the field, as shown in Figure 5. The noise sources were diffuse while the speech emanated from directly in front of the listener. Thus, the speech is presented in a manner appropriate to the description of the patient's audibility curve and the measurement of the gain provided by the hearing aid.

Speech Discrimination and Frequency Response of the Hearing Aids

When such care is taken, clear and striking differences between experimental hearing aids emerge. As shown in Figure 6, there is a very orderly dependence of the percentage of words identified correctly and the frequency response of the hearing aid. In a second experiment only the extreme

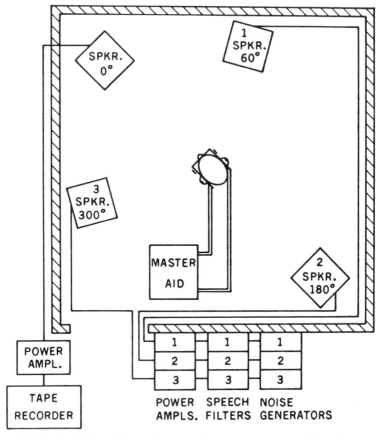

Figure 5.   Block diagram of the equipment and room arrangement used for the word discrimination tests. (Reprinted with permission from D.P. Pascoe, 1975, *Annals of Otology, Rhinology and Laryngology 84* (suppl. 23).)

frequency responses were tested and both the high frequency and the phonetically balanced word lists were used. Figure 7 shows the results. We believe that the orderly quality of the results in Figures 6 and 7 reflects not only the overall selection of procedures in terms of variables known to influence such scores but, in particular, the care taken to achieve consistency of electroacoustic and audiologic measurement.

Unaided Hearing Levels and the Discrimination of Speech

Pascoe found good correlations between the patients' unaided hearing levels and their ability to understand speech. An adjusted hearing level was defined as the mean of the hearing levels at 2.0, 2.5, 3.1, and 4.0 kHz plus

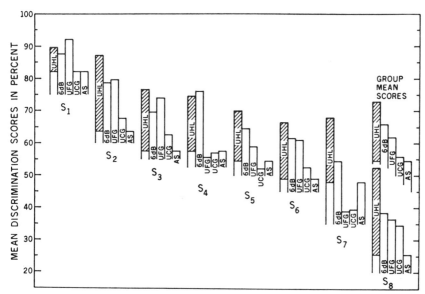

Figure 6.   Individual and group mean word discrimination scores for five frequency responses. Shaded areas represent the advantage of UHL over AS. (Reprinted with permission from D.P. Pascoe, 1975, *Annals of Otology, Rhinology and Laryngology 84* (suppl. 23).)

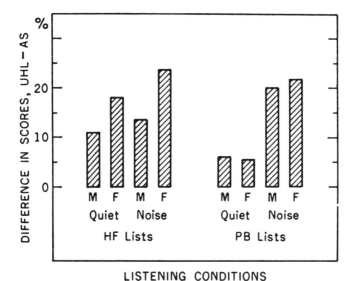

LISTENING CONDITIONS

Figure 7.   A comparison of discrimination score advantages given by UHL over AS, in the second experiment. (Reprinted with permission from D.P. Pascoe, 1975, *Annals of Otology, Rhinology and Laryngology 84* (suppl. 23).)

the difference between it and the mean of their thresholds at 0.5, 0.63, 0.8, and 1.0 kHz. The correlations between the adjusted hearing levels and the speech scores are illustrated in Figures 8 and 9. These results are truly remarkable in light of frequent claims that speech discrimination scores cannot be predicted from the audiograms. Several factors probably contribute to these excellent correlations. The patients are homogeneous with regard to slope of audiogram and other audiologic findings, including history. The measurements of threshold are highly reliable and sample appropriate spectral areas, and the speech scores were collected as described and each score is based on an adequate number of items (200/point). The results of Pascoe suggest to us that within patient groups, auditory thresholds may be very accurate predictors of speech discrimination scores if all of the relevant variables are measured properly. Of course, the measurement of other audiometric parameters, such as discomfort levels, may further serve to improve the accuracy of such predictions.

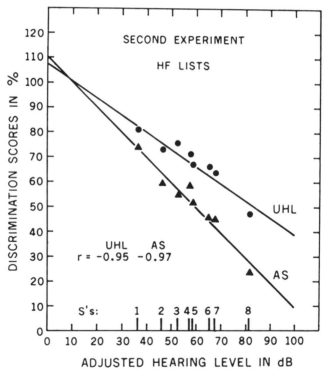

Figure 8.   Correlation between subjects' adjusted hearing levels and their discrimination scores on the UHL and AS conditions. (Reprinted with permission from D.P. Pascoe, 1975, *Annals of Otology, Rhinology and Laryngology 84* (suppl. 23).)

MEASUREMENTS ILLUSTRATED
IN THE RESEARCH OF SKINNER (1976)

Noise-Band Audiometry

Skinner used ⅓-octave bands presented in a sound field as audiometric test stimuli, as did Pascoe. In addition to finding thresholds for the detection of these, she also had the patients report the discomfort levels. To find the discomfort levels, the patients adjusted each band of noise to a level that they "would not want to listen to for a long period of time" and then reduced that level by 1 dB. Sample results are shown in Figure 10. These subjects were selected to have noise-induced hearing losses limited to the high frequency range. Note the variations in dynamic range, that is, the difference between the threshold of discomfort and the threshold of audibility and how that dynamic range varies from patient to patient and with frequency. It is important to note that thresh-

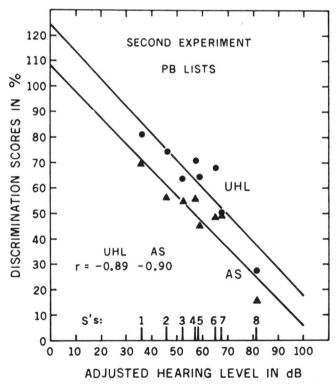

Figure 9.   Correlation between subjects' adjusted hearing levels and their discrimination scores on the UHL and AS conditions. (Reprinted with permission from D.P. Pascoe, 1975, *Annals of Otology, Rhinology and Laryngology 84* (suppl. 23).)

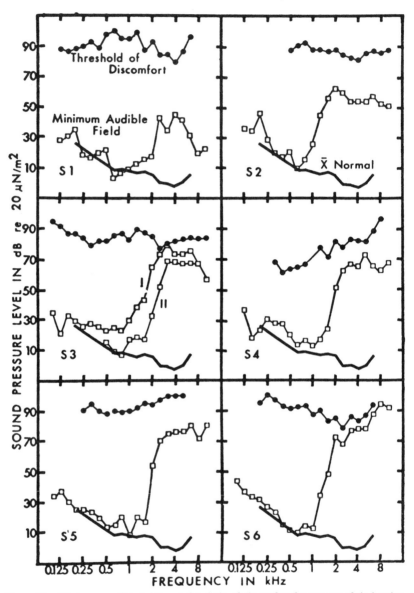

Figure 10.   Minimum audible fields and threshold of discomfort for one ear of six hearing-impaired subjects. (Stimulus: ⅓-octave bands of noise, centered at stated frequencies.) Two curves are shown for subject 3. Curve I was obtained at the beginning of the testing sessions, and Curve II, at the end. (Reprinted with permission from Skinner, 1976.)

olds for audibility and discomfort are measured in a common way and are exactly comparable.

Frequency Responses of the Hearing Aids

The experimental hearing aids were, in fact, a public address system that was assembled with laboratory quality components that included an equalizer and power amplifier that drove a loudspeaker. Since the output of this "hearing aid" was measured as the sound field at the position of the listener's head (in its absence), that output was indeed measured in terms exactly commensurate with the measures of audibility and discomfort. The selection of a frequency response was guided by the subject's audibility curve, as illustrated in Figures 11 and 12. The key idea was to provide relative amplification in the region of high frequency loss. For subject 2, whose audibility curve is shown in Figure 11, the spectral region of high frequency loss was judged to begin at 630 Hz and extend upward. The amount of gain in this region was relative to the gain below 630 Hz and was systematically varied in 11-dB steps, as shown in Figure 12.

Levels of Speech

The levels of the speech materials actually used in the speech testing were acoustically measured as follows. A word was played through the hearing aid. The output of a microphone placed in the position that the listener's head would occupy was analyzed spectrally. The very same ⅓-octave bands used for the audiometry and for the measurement of the hearing aid were then used for the analysis of the speech. The energy in each of these bands

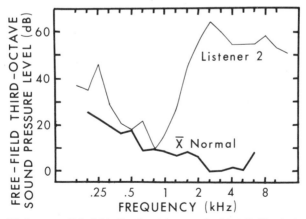

Figure 11.   Minimum audible fields for the right ear of subject 2. The test sounds were ⅓-octave bands centered at the stated frequencies. (Reprinted with permission from Skinner, 1976.)

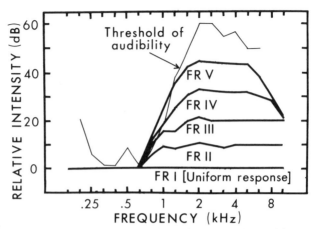

Figure 12.   Frequency response curves used for subject 2. (Reprinted with permission from Skinner, 1976.)

was measured separately for each word. Thus, for each ⅓-octave band there were 50 measures, one per word.

An example of how these measurements can be used is given in Figure 13. This shows the distribution of the speech levels in relation to the subject's discomfort and audibility thresholds. The vertical lines represent the range from the 90th to the 10th percentiles of the distribution of levels while the thin line passes through the 75th percentile. Notice that all variables are measured in exactly the same terms and that the graphic display makes the relations among thresholds, discomfort levels, and speech energy immediately obvious.

Discrimination of Speech

The Pascoe high frequency word list was used as recorded by his female talker. There were 50 words for each combination of the five overall levels of speech with the five frequency responses. Counterbalancing techniques were used to eliminate the effects of practice and fatigue. The tests were conducted in a nominally quiet environment where the signal-to-noise (S/N) ratio was 30 dB or larger in each band.

The word identification scores are shown in Figure 14. Note that in the usual range of conversational speech, responses 3 and 4 result in the highest scores.

The relations of the actual levels of speech to the audiometric measures suggest certain rules for predicting the intelligibility of the speech. The example graphs for subject 2 serve to illustrate these points. In Figure 15

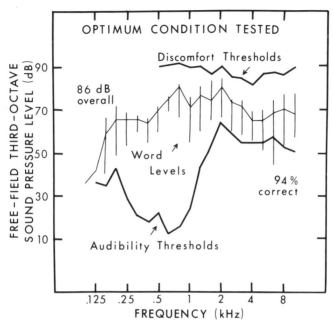

Figure 13.   Discomfort levels, thresholds, and speech levels for optimal frequency response for subject 2. (Reprinted with permission from Skinner, 1979, in P. Yanik (ed.), *Rehabilitation Strategies for Sensorineural Hearing Loss,* Grune & Stratton, New York.)

one can see that as more and more of the speech energy exceeds the subject's threshold, the percentage of words identified correctly also increases. When the 75th percentile of the speech energy measurements begins to approach the subjects' thresholds of discomfort, however, this trend is reversed and the scores decline. The balance between the low and high frequency levels is also important, even when the energy in all bands is well positioned between threshold and discomfort, as is illustrated in Figure 16. Presumably, the excess energy in the low frequency region results in the lower discrimination score.

THE DEVELOPMENT OF THE LIMITING MASTER HEARING AID

Flowing from the work of Pascoe and Skinner are three main substantive conclusions: 1) In general, the more speech energy between about 200 and 6000 Hz that is audible, the better the ability to understand speech. 2) When speech energy in any band of approximately ⅓ octave exceeds the threshold of discomfort, then the ability to understand speech will decline or the subject will turn down the gain so that the discomfort level is no longer

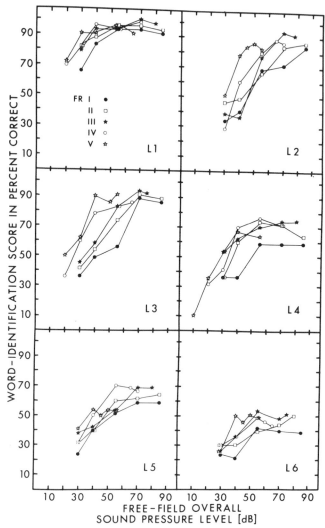

Figure 14. Word identification scores of six hearing-impaired subjects as a function of intensity level for the frequency responses in experiment 1. (Stimulus: Pascoe High Frequency Word List spoken by the female talker.) (Reprinted with permission from Skinner, 1976.)

exceeded or the subject will refuse to use the aid. 3) The relative levels of speech between 500 and 1000 Hz and between 2000 and 4000 Hz must be kept within certain, as yet unspecified, limits. Naturally, we were intrigued by the possibility suggested by Skinner's results, as shown in Figure 14, that with appropriate limiting, the subjects with high tone losses might show additional benefit from greater gain in the high frequencies.

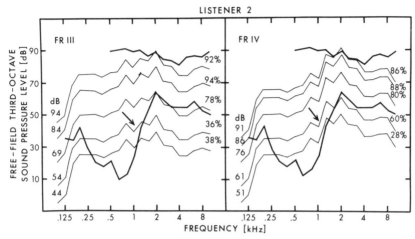

Figure 15.   For subject 2, word identification improves as more of the speech exceeds threshold up to the point where the word levels (75th percentile) begin to exceed the discomfort levels. (Reprinted with permission from Skinner, 1979, in P. Yanik (ed.), *Rehabilitation Strategies for Sensorineural Hearing Loss,* Grune & Stratton, New York.)

The concept of the limiting master hearing aid (LMA) was suggested by one of our colleagues, A. Maynard Engebretson. By this scheme, the spectrum is divided into channels by a bank of bandpass filters. In each channel the gain is controlled by an amplifier and the limit is established by appropriately ganging an amplifier, a nonlinear element, and a second amplifier. The two amplifiers are related so that the net gain through the limiter remains constant. The distortion products introduced by the nonlinear element are removed by a subsequent filter. The limited outputs of the channels, which are nearly distortion free, are then mixed, amplified, and presented to the listener via a transducer.

The LMA has 12 channels, the upper 10 of which are spaced at ½-octave intervals, (322–7242 Hz) each with bandwidths of ½ octave. The lowest channels (100 and 211 Hz) are about an octave apart with bandwidths of about ³⁄₁₀ octave; a block diagram is given in Figure 17.

It is convenient to describe the performance of the LMA and the subjects' audiologic findings in terms of the calibration of the earphone, which is used with both. The important point is that one may describe these parameters in terms of sound fields, as did Pascoe and Skinner, or in terms of a calibration on an artificial device, such as KEMAR, as long as *all* of the relevant parameters are described in directly comparable terms. In the case of the LMA, we have often used earphones mounted in circumaural cushions. Any reliable measure of the output of these phones can serve as the needed common measure as long as noise levels, thresholds, discomfort

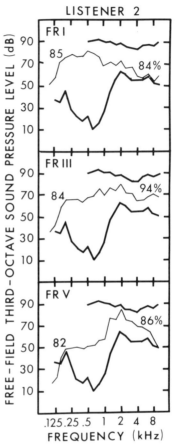

Figure 16.   For subject 2, word identification depends on the choice of frequency response within the area of comfortable audibility. (Reprinted with permission from Skinner, 1976.)

levels, speech levels, and so on are measured in precisely the same way.

The electroacoustic performance of the LMA has been carefully measured, and over 5300 parameters are stored in a computer program that calculates, for each channel, the output noise floor, the maximum output, and the average output of conversational speech (Pearsons, Bennett, and Fidell, 1976) and displays these in relation to the patient's thresholds of audibility, most comfortable listening levels, and his thresholds for discomfort. This is done for ⅓-octave bands of noise centered on the passbands of each channel of the LMA. In this way the performance of the aid can be compared quickly to the client's hearing.

A typical procedure would be to start with the aid adjusted to initial values, as shown in Figure 18. The fit is unacceptable. The maxi-

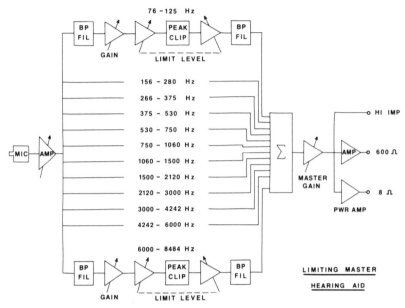

Figure 17.   Block diagram of the limiting master hearing aid (LMA). (Reprinted with permission from Skinner, 1979, in P. Yanik (ed.), *Rehabilitation Strategies for Sensorineural Hearing Loss,* Grune & Stratton, New York.)

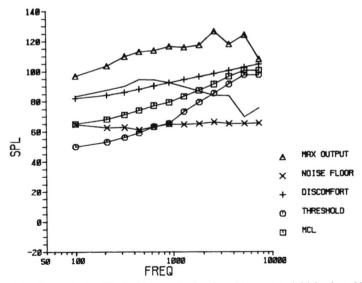

Figure 18.   The controls of the limiting master hearing aid are set to initial values. Notice the poor fit of the performance of the aid to the parameters of the hypothetical client's hearing.

mum outputs, for the most part, fall 20 dB above the patient's discomfort levels. The average spectrum of speech, shown by the thin line, is above the discomfort levels at the low frequency end of the spectrum and falls below threshold at the high frequency end of the spectrum. Similarly, the noise floor is above threshold for the lower bands. As shown in panels A and B of Figure 19 for channel 7, the controls of the aid can be adjusted so that the limiting level (L) or maximum output can be set precisely to the discomfort level (D). The gain can be adjusted to bring the average level of speech (S) precisely to the most comfortable listening level (M) and the noise floor can be set below threshold (T). This procedure is followed for each of the bands and the result is shown in Figure 20. The aid has been adjusted so that lowest band is turned off and the speech spectrum is gently rolled off at the low and high frequency ends.

The process illustrated in Figures 18, 19, and 20 is for a hypothetical patient. Similar fits for two of our subjects are shown in Figures 21 and 22. Our preliminary results suggest that such a method of precisely fitting the electroacoustic parameters of a hearing aid to quantitative measures of the client's hearing can provide very satisfactory results in a laboratory setting. That is, the hearing aid is comfortable, no matter how strong the input, and the user can understand speech as well as or better than he could with competing aids. Of course, many problems remain in proving the value of this approach and in the eventual development of it as a useful clinical tool.

SUMMARY

We are moving toward a technology that will allow a very complete description of the relations between the electroacoustic properties of the hearing aid and the important audiometric parameters such as thresholds and discomfort levels. There remain many problems to be solved with regard to improving the accuracy of such description. The output side of the graphs is often unreliably measured as the coupling between the output and the patient changes, or coupler measures may be irrelevant to particular situations. Also, audiometric and hearing aid data are often not commensurate, being referred to different standards. The measurement of speech in terms that are accurate and useful in this regard has not yet been solved.

As the accuracy of measurement and the ability to depict the relations between the significant variables improve, so will the quality and relevance of the hypotheses on what is an "optimal" hearing aid for each class of patient.

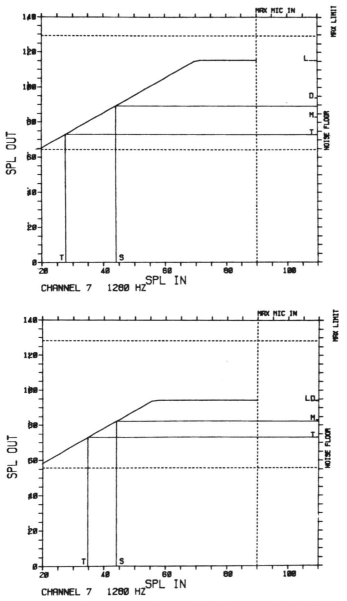

Figure 19.   Input-output functions of the channel of the LMA centered at 1280 Hz. The upper panel shows results before fitting, and the lower panel shows results after appropriate adjustments of the controls of the limiting master aid. L, Limiting level or maximum output; D, threshold of discomfort; M, most comfortable listening level; T, threshold of hearing; S, average level of conversational speech.

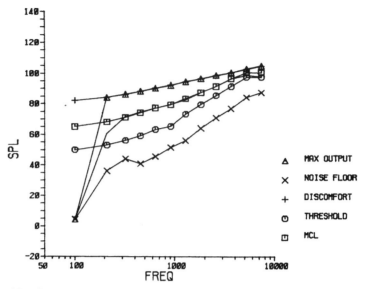

Figure 20.   Correspondence between performance of aid and hypothetical client's hearing. The thin line is the average spectrum of conversational speech after amplification.

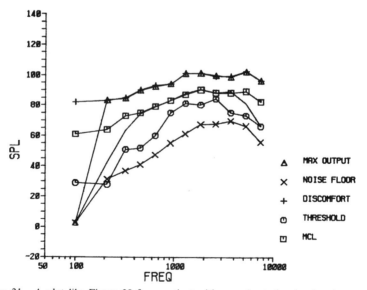

Figure 21.   A plot like Figure 20 for a patient with a moderate hearing loss but narrow dynamic range.

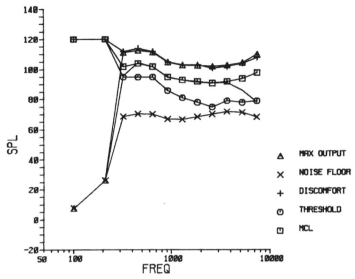

Figure 22.   A plot like Figure 20 for a patient with a severe hearing loss and a narrow dynamic range.

REFERENCES

Pascoe, D.P. 1975. Frequency responses of hearing aids and their effects on the speech perception of hearing-impaired subjects. Ann. Otol. Rhinol. and Laryngol. 84 (suppl. 23):1–40.

Pearsons, K.S., R.L. Bennett, and S. Fidell. 1976. Speech Levels in Various Environments. BBN Report No. 3281. Bolt, Beranek, and Newman, Inc., Cambridge, Mass.

Skinner, M.W. 1976. Speech intelligibility in noise-induced hearing loss: Effects of high frequency compensation. Doctoral dissertation, Washington University, St. Louis.

Skinner, M.W. 1979. Audibility and intelligibility of speech for listeners with sensorineural hearing losses. In P. Yanik (ed.), Rehabilitation Strategies for Sensorineural Hearing Loss, pp. 159–184. Grune & Stratton, New York.

CHAPTER 17

IMPLICATIONS OF PREVIOUS RESEARCH FOR THE SELECTION OF FREQUENCY-GAIN CHARACTERISTICS

Raymond L. Dugal, Louis D. Braida, and Nathaniel I. Durlach

CONTENTS

The selection of the frequency-gain characteristics for hearing aids used by listeners with sensorineural hearing losses has received a great deal of attention from researchers and clinicians. Although this problem has been studied for at least half a century (e.g., see the review by Braida et al., 1979), the solution continues to be elusive. This is evident in the inconsistencies among the results of previous studies, the lack of general agreement about the value of individual fitting of hearing aids or about the most appropriate techniques for such fitting, and the existence of a number of ongoing research projects concerned with this problem.

To a certain extent the failure to solve this problem reflects both its inherent complexity and our inadequate understanding of speech perception and auditory processes. The optimum choice of frequency-gain characteristic is likely to depend on interactions between properties of the sound source (e.g., the speech of men, women, and children typically have different spectra and amplitude distributions), the transmission channel (which may be noisy and reverberant and which may introduce nonlinear distortions), and the receiver (i.e., the details of the hearing loss and the preferred listening level). Also, speech perception is not thoroughly understood even

379

for listeners with normal hearing, and the perceptual effects of sensorineural impairments are only imperfectly known. In light of these difficulties, it is not surprising that most studies concerned with the choice of frequency-gain characteristic have been empirical. Nevertheless, the large number of experimental variables and the difficulty of obtaining precise measurements of speech reception make it extremely unlikely that the problem can be solved by purely experimental approaches.

Two considerations underscore the difficulty of this problem. First, current understanding of hearing impairments is sufficiently limited so that it is very difficult to predict performance for a given impaired listener from data from other listeners, even when audiometric configurations are similar. Second, speech testing procedures have limited power to discriminate between different frequency-gain characteristics for a given listener because of test-retest variability and the need to control for differential practice effects. In consequence, empirical studies are faced with the need to trade off experimental time between the number of characteristics studied and the accuracy with which performance on a given characteristic is determined. Although the use of adaptive procedures can alleviate this problem, it cannot eliminate the conflict entirely. Even if these considerations did not exist, however, it would ultimately be necessary to formulate probabilistic models of the set of situations in which hearing aids are used in order to determine the optimum characteristic for everyday use in the real world because of the wide variety of signals to be understood and the wide range of environments in which communication occurs.

Many of the issues that complicate the selection of the frequency-gain characteristic for hearing aids are similar to those encountered in the design of voice communication systems. Such systems, and in particular the frequency-gain characteristics of such systems, are generally designed according to the predictions of articulation theory (e.g., Fletcher and Galt, 1950; French and Steinberg, 1947; Kryter, 1962), which incorporate the results of a wide variety of measurements of speech transmission systems. At least for listeners with normal hearing, the effect of varying the frequency-gain characteristic can generally be predicted as well by articulation theory as it can be measured by speech tests.

There have been a number of attempts to apply articulation theory to experiments involving listeners with impaired hearing. In conjunction with the MEDRESCO master hearing aid study (Radley et al., 1947), a method was developed for determining the optimum characteristic for hearing aids designed to achieve maximum intelligibility when total delivered power is limited. Fletcher (1952) analyzed the results of the Harvard master hearing aid study (Davis et al., 1947) in terms of articulation theory and derived a characteristic for listeners with losses that could be divided into conductive and sensorineural components. These two studies are considered in more detail below. Wilber (1964) obtained mixed results when using a highly simplified form of articulation theory to analyze the performance of impaired listeners: relatively good predictions were found for listeners with normal hearing or conductive losses, but not for those with sensorineural impairments. More recently, Macrae and Brigden (1973) reported that the reception of sentences by listeners with sensorineural hearing loss was predicted better by means of articulation theory than by the three-frequency average hearing loss. Aniansson (1974), in a study concerned with the effect of high frequency hearing loss on speech perception in a variety of realistic environments, concluded that articulation theory could be used to predict average intelligibility scores for groups of individuals with roughly similar audiograms, but generally not for individuals within these groups.

It is somewhat surprising that there has been no study of the extent to which articulation theory can be used to select the frequency-gain characteristic of hearing aids, or to analyze the results of experiments in which the dependence of intelligibility on the characteristic was determined for impaired listeners. Three factors seem to have been responsible for this failure. First, most studies of the effect of varying the frequency-gain characteristic have not taken adequate account of relevant acoustic effects, so that the characteristics were not specified completely. In order to calculate the performance of an amplifying system, it is necessary to specify the functional gain precisely with reference to the orthotelephonic condition (French and Steinberg, 1947; Inglis, 1938; Richards, 1973). Second, in many cases the results clearly reflect an interaction between the frequency-gain characteristic and other (poorly specified) properties of the amplifying systems used, such as nonlinear distortion and internal noise. In general, the effects of these interactions are difficult to analyze in terms of articulation theory because the relevant properties are inadequately described and because the theory is known to have only limited power to handle certain of these factors. Third, in most studies the frequency-gain characteristics have not been evaluated at a sufficiently wide range of presentation levels for each listener. This is in contrast to cases in which articulation theory has been applied to listeners with normal hearing (e.g., Fletcher and Galt, 1950; French and Steinberg, 1947), and the failure makes it extremely difficult to use the theory to predict the performance of impaired listeners. Fortunately, Skinner (1976) controlled the necessary parameters of the amplifying systems, reported the test conditions in sufficient detail, and tested a relatively homogeneous set of highly trained impaired listeners at a large number of levels.

DETAILED THEORY

The theory presented here is based on the form of articulation theory developed by Kryter (1962). Since extensive discussions of this formulation are available elsewhere (e.g., ANSI, 1969), this presentation is brief except where significant alterations have been made.

Articulation theory uses a measure called the articulation index (AI), which is a weighted average over frequency of the proportion of the speech signal that is available to convey information to a given listener. Performance on a given test of speech reception (in terms of intelligibility-percentage of items correctly identified) is related to the AI by a monotonic increasing function, the intelligibility-articulation function, which reflects the effects of constraints placed upon the message set by the structure of the test. Intelligibility-articulation functions for test vocabularies consisting of sets of 32, 256, and 1000 monosyllabic words have been published (ANSI, 1969) and are shown in Figure 1. The function for a test vocabulary of 50 words has been estimated from these data. Note that when the test vocabulary is highly restricted, test scores initially rise rapidly as the AI increases initially, but grow at only a much reduced rate as the AI approaches unity.

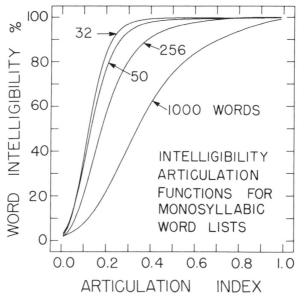

Figure 1.   Dependence of intelligibility on articulation index for test vocabularies consisting of 32, 50, 256, and 1000 monosyllabic words. The curves for 32, 256, and 1000 words are from ANSI (1969); the curve for 50 words is interpolated.

The AI itself is computed from estimates or (preferably) measurements of the spectrum and short-term level distribution of the speech materials used, the effective masking spectrum of the interfering sounds present, and the audibility and discomfort thresholds of the listeners tested. This computation is carried out on a band-by-band basis throughout the speech spectrum. In the form of the computation used in this chapter, $15\frac{1}{3}$-octave bands with center frequencies of 200–5000 Hz are used. The AI is computed as the weighted average band efficiency, where the weighting coefficients are chosen to reflect relative band importance. More formally, one has:

$$AI = P \cdot \sum_{i=1}^{15} BI(i) \cdot BE(i)$$

where P = proficiency factor, $BI(i)$ = importance of band i, $BE(i)$ = efficiency of band i, and i = band index. The importance weights are chosen to represent the relative contribution of the different frequency bands to speech transmission under ideal acoustic reception conditions. According to the reports of Fletcher (1952), French and Steinberg (1947), Black (1959), and Radley et al. (1947), the importance per cycle is roughly constant below 2000 Hz and decreases at roughly 3 dB per octave above 2000 Hz. The weights used in our formulation, which reflect both the importance per cycle and the progressively increasing widths of the $\frac{1}{3}$-octave bands, are those given by Kryter (1962), which are shown in Figure 2.

Band efficiency measures the proportion of the speech signal in a given band that is above the listener's masked threshold and below the listener's discomfort level, as shown in Figure 3. The speech level distribution for the band is assumed to be uniform on a logarithmic scale, ranging 12 dB above and 18 dB below the long-term rms level for the band. The effect of changing the functional gain for a given band is to alter the relation between the speech-level distribution and the listener's threshold and discomfort levels.

The proficiency factor *(P)* reflects such elements as the degree to which the listener is practiced in listening to the talker and also the precision with which the test materials are enunciated. This constant, which is independent of the listening condition, is typically treated as a fitting parameter and estimated on the basis of experimental results rather than from theoretical considerations (e.g., Fletcher, 1952; Fletcher and Galt, 1950). In principle, the proficiency factor can be determined from a single data point (intelligibility score for a single test condition), but is often chosen to provide the best overall fit to the listener's performance. Note that although the proficiency factor serves to determine the absolute level of performance in a given listening condition, it does not affect the relative levels of performance

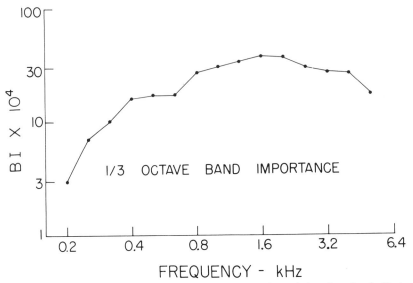

Figure 2.   Band importance function for ⅓-octave bands of speech, based on data in Kryter (1962).

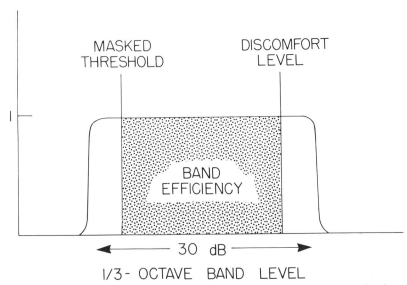

Figure 3.   Determination of band efficiency. The band efficiency, represented by the shaded area, is the range of speech levels (in dB) above the listener's threshold level and below his discomfort level.

for different listening conditions, since the factor has the same value for all listening conditions.

The method used to compute the masked threshold in this chapter is a modified form of the procedures suggested by French and Steinberg (1947) and Kryter (1962). As is discussed in a later section, precisely the same method will be used for listeners with normal hearing and for listeners with hearing loss. In the method to be used, the masked threshold in a given band is assumed to reflect the quiet threshold in the band, the masking effects of in- and out-of-band external noise, and the spread of masking from speech elements in other bands. The in-band self-masking of speech is ignored (e.g., French and Steinberg, 1947). The quiet threshold is treated as if it were caused by an internal noise. The masking effect of external noise is assumed to spread in frequency and the interband masking is calculated by the method developed by Kryter (1962). The effective external noise in a given band is computed as the maximum of the external noise in the band and the internal noise in the band required to produce the same masking as that resulting from external noise in other bands. The upward spread of masking produced by the speech itself is modeled as an effective noise and is computed by the method given by French and Steinberg (1947). The downward spread of masking produced by the speech is computed to reflect the results of Bilger and Hirsh (1956). According to this study, the downward spread of masking produced by a band of noise is roughly the same at all frequencies below the band, is proportional to the sensation level of the noise, and is less severe for high frequency noise than for low frequency noise. These results are accounted for by assuming that speech in a given band produces an effective masking noise in lower bands that is at the same sensation level at all frequencies. The ratio *(D)* of the speech band sensation level *(X)* to the effective masking noise sensation level in lower bands *(Y)* is shown below:

F (Hz)	D (dB)
515	40
820	40
1200	45
1650	45
2160	50
2765	55
4500	65

The masked threshold in a given band is determined by computing the total equivalent power of the internal noise required to account for the quiet threshold, the effective external noise, and the equivalent noise representing the masking caused by speech in other bands. The threshold is then taken to be equal to the total equivalent noise power.

EXPERIMENTS

Although there have been many experimental studies of the performance of hearing-impaired listeners under conditions in which the frequency-gain characteristic, or filtering condition, was varied, nearly all of these are unsuitable for detailed tests of the theory. In many cases significant properties of the amplification system used—in particular the actual frequency-gain characteristic—have not been described adequately. Also, important properties of the speech material (e.g., the presentation level, spectrum, and level distribution) and of the individual listeners (e.g., detection and discomfort thresholds) have not been reported. It should be evident from the previous discussion that it is essential to specify such parameters with substantial precision in order to predict speech intelligibility based on the AI.

In the recent study at the Central Institute of the Deaf conducted by Skinner (1976), considerable effort was expended to ensure that the relevant parameters of the speech transmission system were carefully measured and specified. For example, tests were conducted in a sound-treated room with the listener facing the sound source (a loudspeaker) from 1.5 m, thus ensuring that the reference gain characteristic was functionally flat. The study is also noteworthy for its use of an impaired listener population that, although small, was relatively homogeneous in etiology and audiometric configuration, although not in chronologic age or duration of impairment. In addition, the listeners were trained extensively on the test materials used. Since an extensive report of the experiments is available elsewhere, only a brief summary of relevant details is presented here.

Six young adults with normal hearing and six impaired listeners were included in the tests. Detection and discomfort thresholds for three of the impaired listeners, measured monaurally in the sound field using 1/3-octave bands of noise, together with representative values for listeners with normal hearing (obtained from the study of Pascoe, 1975), are presented in Figure 4. The measured discomfort thresholds are substantially below the 120–140 dB SPL values used when articulation theory is applied to listeners with normal thresholds (e.g., Kryter, 1962). Detection thresholds for one of the impaired listeners measured after speech testing was completed were significantly lower than those measured before testing. Since it is not possible to determine which set of measurements is more appropriate for the actual test conditions, this listener is treated as two listeners (I3 and IX) in the work that follows. All impaired listeners had nearly normal thresholds at low frequencies, but sharply sloping losses above roughly 1000 Hz. It is possible that the small low frequency losses observed for some of the impaired listeners reflect masking by low frequency background noise rather than

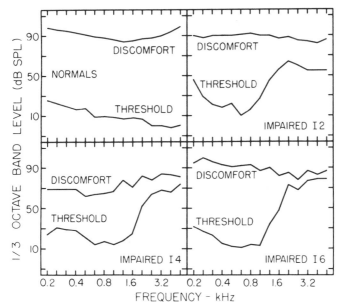

Figure 4.   Threshold and discomfort levels for impaired listeners I 2, I 4, and I 6 tested by Skinner (1976), plus results for normal listeners obtained by Pascoe (1975).

hearing loss. Also, the measured discomfort levels were relatively low (70–90 dB SPL) and relatively independent of frequency. Most of the subjects reported an aversion to loud sounds.

The main set of speech tests utilized the 50-item high frequency word list devised by Pascoe (1975) and spoken by a female. Although the short-term level distribution for this material was not measured, the long-term average spectrum for the materials was computed on a word-by-word basis for ⅓-octave bands. The peak speech levels (exceeded by only 1% of band level measurements made in ⅛-sec intervals) shown in Figure 5 were assumed to be 10 dB above the 25th percentile level measurements reported by Skinner. These estimates correspond very well to the idealized peak level spectrum suggested by Kryter (1962), and are also roughly consistent with the data of Dunn and White (1940).

Measurements of the background noise spectrum were also reported. This noise consisted of two principal components: low frequency building noise concentrated primarily below 200 Hz and noise generated by the electronic reproduction and playback system, which produced a relatively flat spectrum above 200 Hz. Unlike the building noise, which was relatively constant, the electronic noise reaching the listener was a function of the gain of the amplification system. An estimate of the resultant noise spectrum is

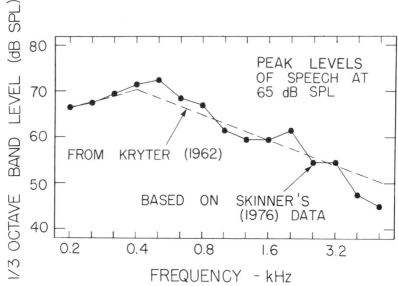

Figure 5.   Estimated peak speech levels for the word list used by Skinner. The curve suggested by Kryter (1962) is for male voices.

shown in Figure 6 for the orthotelephonic presentation condition at a speech level of 85 dB SPL. As the gain in a given band was reduced relative to this condition, the noise level in the band decreased linearly relative to that shown. For the first three bands, however, the resultant noise level never decreased below 20–22 dB SPL because of the presence of building noise.

In the main set of speech tests, the frequency-gain characteristics, which were varied by filtering the electrical signal fed to the loudspeaker, included a functionally flat (orthotelephonic) characteristic (denoted FR 1) and four characteristics with high frequency emphasis (denoted FR 2, FR 3, FR 4, FR 5). Each of these characteristics mirrored the audiogram, except that the maximum high frequency emphasis was limited to 11, 22, 33, and 44 dB, respectively. In supplementary experiments, the 22-dB characteristic (FR 3) was compared with three characteristics (FR 6, FR 7, and FR 8) derived from it by the addition or deletion of low or high frequency gain. A representative set of the characteristics tested is shown in Figure 7. Intelligibility tests were administered at a variety of levels (reported in terms of the overall level reaching the position of the listener's head for the orthotelephonic condition) for each of the characteristics. Throughout the tests, a listener with normal hearing was paired with each impaired listener and tested under precisely the same conditions.

THEORETICAL PREDICTIONS AND COMPARISONS WITH DATA

In this section we apply the theory outlined previously to the data obtained by Skinner on the performance of normal and impaired listeners by computing the performance that would have been predicted by the theory. In making these predictions, we have used the same set of band importance weights, and the same procedures for computing band efficiency factors, for all listeners. The only distinction made between listeners concerned the measured values of detection and discomfort thresholds. In particular, no "abnormal spread of masking" was assumed in computing the masked thresholds for the listeners with impaired hearing. Although this assumption may be somewhat naïve, we believe that it is appropriate for initial work. Measurements of the spread of masking in impaired listeners are relatively sparse and somewhat contradictory. Furthermore, predictions based on the assumption of normal spread of masking should be of interest. If they can be shown to be inadequate, it will be important to revise the assumption in future work.

An intelligibility-articulation function for the 50-item lists used in this experiment was derived by interpolating logarithmically between the func-

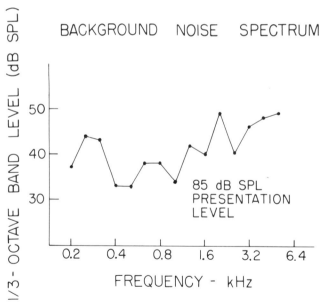

Figure 6.   Estimated background noise spectrum in Skinner's (1976) tests for the 85-dB SPL test condition. At lower presentation levels the spectrum was attenuated proportionally, but the levels in the lowest three bands were never less than 20–22 dB SPL. (Reprinted with permission from Skinner, 1976.)

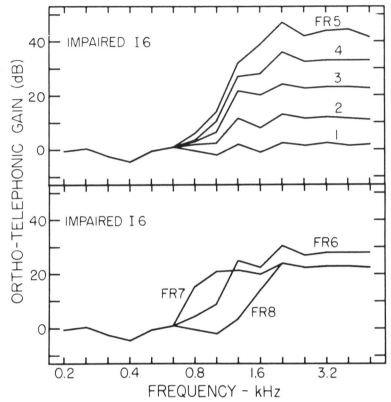

Figure 7.   Frequency-gain characteristics tested by Skinner (1976) for impaired listener I 6. Characteristic FR 1 is functionally flat (orthotelephonic), characteristics FR 2, FR 3, FR 4, and FR 5 mirror the listener's audiogram with maximum high frequency emphases of 11, 22, 33, and 44 dB, respectively. Characteristics FR 6, FR 7, and FR 8 are derived from FR 3 by the addition or deletion of gain in the low or high frequencies. (Reprinted with permission from Skinner, 1976.)

tions for the 32- and 256-word vocabularies presented in Figure 1. Unfortunately it is not possible to verify the validity of this function from independent measurements. The listeners with normal hearing in Skinner's study were not tested under conditions sufficiently severe to determine the shape of this function empirically.

Articulation indices were computed for each subject, each frequency-gain characteristic, and each presentation level used in Skinner's experiment. Separate indices were computed for I 3 and I X corresponding to the different threshold measurements. In these calculations, the proficiency factors for the listeners with normal hearing were set equal to unity. For each impaired listener a single proficiency factor was chosen to give the best

Table 1.   Proficiency factors and fits for Skinner's listeners

Listener	P	Fit in 2σ bounds
I 1	0.44	90%
I 2	0.36	74%
I 3	0.44	74%
I X	0.30	92%
I 4	0.32	86%
I 5	0.23	94%
I 6	0.23	84%
Normals	1.00	25%

overall fit to the results. These factors are listed in Table 1. The extent to which the calculated articulation indices and the observed intelligibility scores agree with the assumed form of the intelligibility-articulation function is shown in Figures 8 and 9. The error bounds in these figures represent only the 2σ dispersion of test scores expected from binomial sampling

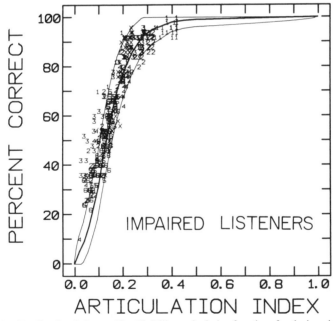

Figure 8.   Predicted and observed intelligibility-articulation functions for the impaired listeners tested by Skinner (1976). The computed AI and observed test score for a given listening condition and listener is represented by a character coded for each listener. The heavy curve is the predicted intelligibility-articulation function for a 50-word test vocabulary. The light curves represent the 2σ bounds expected for the variability of the test scores because of sampling variance.

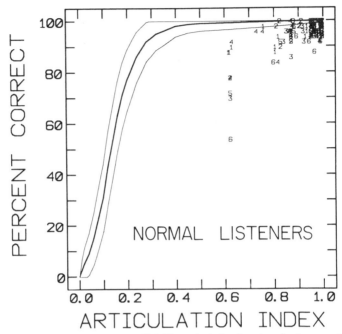

Figure 9.   Predicted and observed intelligibility-articulation functions for the normal-hearing listeners tested by Skinner (1976). The computed AI and observed test score for a given listening condition and listener is represented by a character coded for each listener. The heavy curve is the predicted intelligibility-articulation function for a 50-word test vocabulary. The light curves represent the 2σ bounds expected for the variability of the test scores because of sampling variance.

variance for 50-item lists. The variation in articulation indices associated with errors in measurement of the detection and discomfort thresholds is not included in these bounds. Unlike the statistical fluctuations in the test scores, which are likely to be independent from data point to data point, errors in the determination of thresholds are likely to have correlated effects for all of the data points of a given listener. Although we have not made a careful model of the effects of such errors, it appears that under the worst conditions (small values of AI) random errors of 5 dB in threshold measurements could lead to errors of 0.05 in the AI or to errors of 20 percentage points in the predicted word score.

The fit for the impaired listeners appears to be relatively good: roughly 85% of the data points fall within the 2σ error bounds for the proficiency factors used. The fit is improved to 87% if the data for listener I 3 are excluded but those for I X are retained. Parenthetically, Skinner felt that the hearing loss for this listener was better represented by the posttesting measurements, which we have included as listener I X. The fits for individual listeners are also relatively good, with results ranging from 74% to 94%. There appears to be little correlation between the quality of the individual fit and the proficiency factor.

For the listeners with normal hearing, on the other hand, the fits are generally much poorer: only 25% of the data points fall within the error bounds. Several aspects of this failure are noteworthy. First, roughly half of the data points fall in the region where the articulation index is high (AI > 0.85) and intelligibility scores are high (> 93%). In this region the data points tend to fall below the curve, which predicts intelligibility scores of 100%. This failure may reflect an error in the assumed form of the intelligibility-articulation function. If this function took on the value 97% instead of 100% for large values of AI, the fit would improve to include over 75% of the data points for the normal listeners. Second, there is great variance among the results for the normal listeners for the more adverse test conditions. For example, under the orthotelephonic condition at the 30 dB SPL presentation level, scores ranged from 54% to 88%, although all listeners were predicted to have the same AI (0.62). This variance presumably reflects the fact that the listeners did not all have the same thresholds, as was assumed. Third, although we attempted to fit the results for the normal listeners by assuming that the proficiency factors were all equal to 1.00, it would have been possible to obtain a much better fit by choosing proficiency factors less than unity for these listeners.

Figure 10 illustrates the observed and predicted dependence of word scores for the impaired listeners on presentation level for the various frequency-gain characteristics. The predicted scores generally tend to increase from a low value at low presentation levels to a plateau region where scores are relatively independent of level, and then to decrease or roll over at higher levels. Nevertheless, there is appreciable variation in the predicted curves both across listeners and across characteristics. The curves for listener I 1, for example, have very nearly the same shape and differ primarily by translations along the presentation level axis. By contrast, the curves for listener I 2 differ both with respect to initial slope and general shape. For this listener, the curve for characteristic FR 5 (with the greatest high frequency emphasis) rises more steeply than the others and has only a relatively narrow plateau region. In general, the curves for the characteristics with greater amounts of high frequency emphasis exhibit greater tendency to roll over at high presentation levels. Since the predicted curves for listeners with small dynamic ranges (e.g., I 3, I 4, and I 6) exhibit large rollovers for certain characteristics, it is likely that the rollovers reflect the fact that amplifying speech components above the discomfort level causes a decrease in band efficiency. Since Skinner avoided testing at levels where discomfort thresholds were likely to be exceeded, it is not surprising that

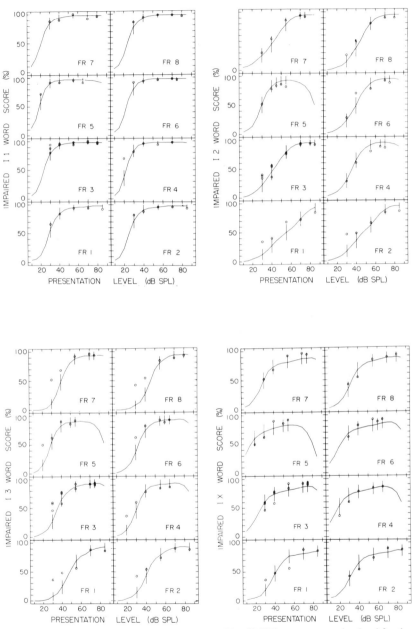

Figure 10.   Predicted and observed dependence of intelligibility on presentation level for the impaired listeners tested by Skinner (1976). The curves represent the predicted functions, the circles indicate measured intelligibility scores, and the vertical bars indicate the 2σ error bounds associated with the sampling variance of the test scores.

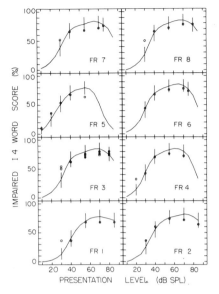

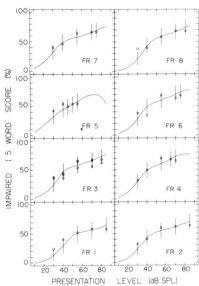

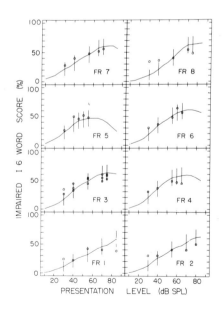

the data points exhibit less rollover than the predicted curves. Finally, it is interesting to note that there is relatively little variation in the maximum predicted scores for a given listener as the characteristic is varied. This implies that there would be little basis for choosing among the characteristics if the sole criterion were the maximum predicted score, since very nearly the same maximum score can be achieved with each of the characteristics if the appropriate presentation level is used.

The error bounds in Figure 10 represent the 2σ dispersion of the test scores expected on the basis of sampling variance, and correspond precisely to the error bounds shown in Figure 9. In particular, averaged over characteristics, levels, and listeners, 85% of the observed data points fall within these bounds. Since the proficiency factors (the only fitting constants used) were chosen on the basis of overall results for a given listener, and since only a single proficiency factor was used for each listener, variations in the degree of fit between the predictions and the data with respect to presentation level and frequency-gain characteristic are of interest. Averaged over listeners and characteristics, the predicted scores agree more closely with the observed data for the three intermediate presentation levels (94% within the 2σ bounds) than for the lowest (56%) or highest (87%) levels used. In general, the predictions for the lowest levels tend to be below the observed scores. This discrepancy is particularly striking for listener I 3: all of the data points at the lowest presentation level fall above the 2σ bounds around the predicted values. Presumably this reflects the fact, noted by Skinner, that the reported detection thresholds for this listener are inappropriately high. In contrast to the wide variation in degrees of fit with presentation level, there is relatively little variation for the different characteristics. The best fits (90%) are obtained for characteristics FR 2, FR 5, FR 6, and FR 7, which have intermediate amounts of high frequency emphasis, and the poorest (69%) for FR 1, which had a functionally flat response.

In summary, it appears that the performance of the impaired listeners in Skinner's experiments can be predicted by the theory if an appropriate proficiency factor is used for each listener. Furthermore, the variation in the performance-intensity functions with listener and frequency-gain characteristic is relatively well described by the theory, particularly at levels well above threshold. These results suggest that articulation theory can provide an analytical basis for selecting the characteristic that would permit a given impaired listener to achieve best performance in a given listening situation.

OPTIMUM FREQUENCY-GAIN CHARACTERISTICS

Whether accomplished empirically or analytically, the determination of the frequency-gain characteristic that is most appropriate for a given listener in a given listening condition requires careful specification of the criterion that defines optimal performance. A number of such criteria have been used in previous investigations, e.g., maximum intelligibility and maximum range of input levels over which intelligibility exceeds a threshold value. Given such a definition, the determination of the optimum characteristic requires the selection of the characteristic that yields maximum performance from the set of possible characteristics. Although this set could, in principle, be unconstrained, in most previous work it has been restricted as the result of two factors. First, certain characteristics are often automatically excluded from consideration, independent of performance, because they could lead to discomfort on the part of the listener. Second, the selection process generally considers only a small subset of likely characteristics that samples the set of reasonable alternatives in order to minimize testing or search time. In general these considerations affect empirical and analytical approaches in different ways. In empirical studies, it is often straightforward to determine whether a given characteristic leads to discomfort, but it is extremely difficult to examine performance reliably with a large number of characteristics. Analytical approaches permit a much larger number of characteristics to be considered, but there is no satisfactory means of detecting the characteristics that are likely to cause discomfort. In this section we focus on the application of articulation theory to the problem of determining the optimum frequency-gain characteristic. In all the work considered, the same optimal criterion, maximum intelligibility, has been used, but different approaches have been taken toward the problem of listener discomfort.

Within the context of articulation theory the problem of determining the characteristic that yields maximum intelligibility for a given listening condition is logically equivalent to determining the characteristic that maximizes the value of AI for that listening condition. The proficiency factor of the listener and the intelligibility-articulation function of the test materials determine the level of performance that is achieved, but do not affect the choice of optimum characteristics. The optimum characteristic thus depends only on the listener and the acoustic properties of the test materials, and is specified as that characteristic with the highest weighted average band efficiency. In general, the dependence of band efficiency on functional gain is complicated: increasing the gain in a given band increases the efficiency in that band until the tolerance level is exceeded, but it also reduces the efficiency in other bands because of masking effects. These interactions are difficult to represent in terms of closed-form analytic expressions. Thus, the more elaborate forms of articulation theory, which may well be the more accurate ones, have not proved tractable. This difficulty has led to the use of simpler formulations by some early investigators.

There have been two noteworthy previous attempts to determine the optimum frequency-gain characteristic analytically. Both of these were carried out in conjunction with the empirical master hearing aid studies conducted in Great Britain and the United States. These two approaches employed distinct theoretical formulations and treated the problem of listener discomfort in different fashions. Although current versions of articulation theory are likely to be more accurate than those employed in these studies, these attempts continue to be used as models for the analysis of this problem, and we thus review these attempts in some detail before presenting our own approach.

An analytic method was developed for determining the optimum frequency-gain characteristic based on a simplified form of articulation theory. This method was presented in an appendix to the MEDRESCO report, *Hearing Aids and Audiometers* (Radley et al., 1947). In this formulation, interband masking of speech is ignored. Furthermore, band discomfort levels are taken to be arbitrarily high, but the listener is assumed to be unwilling to tolerate sounds whose average power exceeds a certain level. The formal problem, summarized in Figure 11, is essentially equivalent to maximizing the value of the articulation index subject to a constraint on total delivered power. Using the calculus of variations, it is possible to show that the optimum characteristic has the property that the rate of change of band articulation with respect to band power gain, dA/dG, must be proportional to the average speech spectrum (S) divided by the importance density

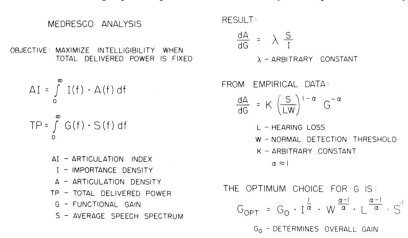

MEDRESCO ANALYSIS

OBJECTIVE: MAXIMIZE INTELLIGIBILITY WHEN
TOTAL DELIVERED POWER IS FIXED

$$AI = \int_0^\infty I(f) \cdot A(f) \, df$$

$$TP = \int_0^\infty G(f) \cdot S(f) \, df$$

AI — ARTICULATION INDEX
I — IMPORTANCE DENSITY
A — ARTICULATION DENSITY
TP — TOTAL DELIVERED POWER
G — FUNCTIONAL GAIN
S — AVERAGE SPEECH SPECTRUM

RESULT:

$$\frac{dA}{dG} = \lambda \frac{S}{I}$$

λ — ARBITRARY CONSTANT

FROM EMPIRICAL DATA:

$$\frac{dA}{dG} = K \left(\frac{S}{LW}\right)^{1-\alpha} G^{-\alpha}$$

L — HEARING LOSS
W — NORMAL DETECTION THRESHOLD
K — ARBITRARY CONSTANT
$\alpha \approx 1$

THE OPTIMUM CHOICE FOR G IS:

$$G_{OPT} = G_0 \cdot I^{\frac{1}{\alpha}} \cdot W^{\frac{\alpha-1}{\alpha}} \cdot L^{\frac{\alpha-1}{\alpha}} \cdot S^{-1}$$

G_0 — DETERMINES OVERALL GAIN

Figure 11. Theoretical analysis of the choice of frequency-gain characteristic developed as part of the MEDRESCO master hearing aid study (Radley et al., 1947). The analysis derives an expression for the power gain of the characteristic yielding maximum articulation in terms of the importance density of speech, the detection threshold for normal listeners, the hearing loss, and the speech power spectrum, under the assumption that total delivered power is constrained.

(I). In order to determine the optimum characteristic, it is necessary to specify dA/dG as a function of G. The form of this dependence suggested by Ackroyd is given in Figure 11. According to the MEDRESCO report, the parameter $\alpha = ¾$, but the data considered are reasonably well fit by various choices in the range $¾ \leq \alpha \leq ⁵⁄₄$. In the form of the theory presented in this chapter, $\alpha = 1$ over roughly the entire range of speech levels. Note that in order for the function $A(G)$ to converge to a finite limit, it is necessary that $\alpha > 1$, at least for large values of G. Figure 11 also presents the predicted form of the optimum characteristic that results from this analysis. For this characteristic, the power gain (G) is predicted to increase with the loss for $\alpha > 1$, to be independent of the loss for $\alpha = 1$, and to decrease with the loss for $\alpha < 1$. In general, the band importance function is a more significant determinant of G than is the hearing loss when α is near unity. The authors of the MEDRESCO report interpreted these predictions as being generally in agreement with the results of their experimental studies.

Fletcher (1952) analyzed the results of the Harvard master hearing aid study (Davis et al., 1947) in terms of a version of articulation theory developed by Fletcher and Galt (1950). Fletcher modeled the conductive component of the hearing loss as a frequency-dependent attenuation and the sensorineural component in terms of the masking effect of a fictitious internal noise. Since the aid used in the Harvard study employed limiting to prevent discomfort at high levels, Fletcher did not address the problem of discomfort. With these assumptions, Fletcher was able to predict the dependence of intelligibility on presentation level for the different frequency-gain characteristics tested in the Harvard study with an accuracy that he believed to be comparable to that of the experimental results. It is noteworthy that Fletcher found it necessary to use proficiency factors substantially smaller than unity to achieve this agreement. The proficiency factors required for the listeners with purely sensorineural losses (0.6 and 0.8) were somewhat smaller than those for the conductive or mixed cases (0.8 and 1.0). Based on the success of such predictions, Fletcher suggested, without providing derivation or proof, that the optimum characteristic should have the following form:

$$G = LC \cdot (LN)^R$$

where LC = conductive loss, LN = sensorineural loss, and $0.2 \leq R \leq 0.4$ Such a characteristic would compensate for the entire conductive component of the hearing loss, but for only a fraction of the sensorineural component. To the extent that R is smaller than unity, the predicted optimum characteristic is only weakly dependent on the sensorineural component of the loss.

Based on the success of these two early efforts, we have attempted to use the same formulation of articulation theory that we applied to the analysis of Skinner's data to the problem of specifying the optimum frequency-gain characteristic. Specifically, we have sought to determine the characteristic that yields the largest articulation index for a listener with a hearing loss typical of the listeners in Skinner's study. In order to represent Skinner's deliberate avoidance of testing conditions likely to result in listener discomfort, we have excluded from consideration those characteristics and presentation levels for which the speech peaks exceed the discomfort level in any band. Since we have not developed a closed form expression that relates the articulation index to functional gain, it is necessary to search for the optimum characteristic computationally. Furthermore, since the number of independent parameters is relatively large (e.g., 15 values of functional gain for the ⅓-octave bands), efficient techniques are needed to keep the computational requirements reasonable. In one algorithm that has proved useful for such optimization problems (Powell, 1964), roughly 1000–2000 iterations are required to determine the maximum of a function of 15 variables.

To evaluate these theoretical predictions, we have computed optimal frequency-gain characteristics for a hypothetical listener corresponding roughly to the average of Skinner's impaired listeners. The discomfort thresholds of this listener were assumed to equal 90 dB SPL in all bands. The detection thresholds for this listener, the gain of Skinner's FR 3 characteristic (which gave the best results), and the theoretically optimum characteristics (with gain normalized at 500 Hz) are shown in Table 2.

Several aspects of these data are noteworthy. First, the characteristic suggested by Fletcher and that computed by the method described in this chapter are similar to one another and to the FR 3 characteristic in that they are flat at low and high frequencies, and provide 20–22 dB of high frequency emphasis. These similarities are not unexpected. Fletcher's prediction is based on the same considerations as those that underlie the articulation theory analysis developed in this chapter. Also, to the extent that articulation theory provides a good fit to Skinner's results, agreement between the predicted and observed optimum characteristics merely indicates that both procedures evaluated the same range of characteristics. The MEDRESCO characteristic for $\alpha = 1$ provides a greater and more uniform high frequency emphasis throughout the frequency range, whereas that for $\alpha = ¾$ is peaked with maximum gain at 1000 Hz and differs sharply from FR 3. Second, both the characteristic suggested by Fletcher and that predicted by this analysis introduce the emphasis at significantly higher frequencies (by roughly ½ octave) than FR 3. According to Skinner's results, such modification of the FR 3 characteristic would be slightly deleterious for most listeners. The reasons for this discrepancy are not yet understood.

Table 2. Predicted optimal frequency-gain characteristics

Band	Center frequency (Hz)	Loss (dB)	Skinner (FR 3)	MEDRESCO ($\alpha=1 = \frac{3}{4}$)	Fletcher	Dugal et al.
1	200	6	0	−9	−11	2	0
2	250	4	0	−7	−8	1	0
3	315	6	0	−6	−8	2	0
4	400	4	0	−4	−6	0	0
5	500	1	0	0	0	0	0
6	625	2	0	6	5	0	0
7	800	4	0	12	11	1	0
8	1000	6	6	14	12	2	0
9	1250	18	18	17	11	6	5
10	1600	22	22	18	11	7	10
11	2000	42	22	19	5	14	15
12	2500	60	22	20	0	20	16
13	3150	63	22	21	0	21	17
14	4000	67	22	21	0	22	19
15	5000	66	22	20	−2	22	20

Finally, although not indicated in Table 2, the method described in this chapter leads to an optimum characteristic corresponding to a presentation level condition of roughly 80 dB SPL. While a number of impaired listeners were not tested at this level, presumably because it was not comfortable, the listeners who were tested at this level obtained roughly the same scores as at lower levels (e.g., 70 dB SPL). In general we regard the agreement between the predicted optimum characteristic and the findings of Skinner to be highly encouraging.

CONCLUSIONS

The predictions of articulation theory have proved to be in relatively good agreement with the data obtained in Skinner's study.

An important feature of the analysis presented in this chapter has been the use of the proficiency factor to describe the performance of the impaired listeners. Although this factor was used merely as a fitting constant, there are ample precedents for introducing such a factor to account for differences between listeners. It is possible that some of the negative results obtained by previous attempts to apply articulation theory to impaired listeners have resulted from a failure to consider the proficiency factors of the listeners studied.

Although we regard the results obtained thus far as encouraging, it is clear that further work is required. Additional computations are needed to obtain predictions of performance under a wider variety of conditions, including different types and degrees of losses, different interference conditions, and different speech tests. It will also be important to consider modifications of the theory to account for properties of the impairment other than those concerned with detection and discomfort thresholds (e.g., abnormal spread of masking, poor temporal resolution, etc.). Also the theory will have to be revised to deal with preferred listening levels, since there is at present no assurance that listeners will utilize optimum characteristics at appropriate levels. Finally, it is important to develop a generative model for the proficiency factor, so that actual intelligibility scores can be predicted.

It would also appear worthwhile to conduct additional experimental tests of the theory. Carefully controlled experiments should be conducted to test the predictions of the theory directly for a wider variety of listeners and listening situations. Experiments are also needed to analyze the components of the theory (e.g., the method used to estimate the masked threshold) and to provide insight into the factors that determine the listening levels chosen by impaired listeners.

Although the research reported in this chapter is largely theoretical, it has implications for future empirical studies of the effectiveness of linear amplification systems. For example, articulation theory provides a basis for deciding on the trade-off between reliability of test results and number of characteristics studied. In addition, the intelligibility-articulation function provides a means for selecting test materials so that testing time is used efficiently. Furthermore, the value of including listeners with normal hearing would probably be greater if they were tested under conditions sufficiently adverse that the intelligibility-articulation function could be determined independently from their results. It is hoped that the results we have obtained will encourage others to consider the implications of articulation theory when designing their experiments and analyzing their results.

REFERENCES

American National Standards Institute. 1969. Methods for the Calculation of the Articulation Index, ANSI-S3.5–1969. American National Standards Institute, New York.

Aniansson, G. 1974. Methods for assessing high frequency hearing loss in everyday listening situations. Acta Oto-Laryngol. (suppl. 320).

Bilger, R.C., and I. Hirsh. 1956. Masking of tones by bands of noise. J. Acoust. Soc. Am. 28:623–630.

Black, J.W. 1959. Equally contributing frequency bands in intelligibility testing. J. Speech Hear. Res. 2:81–83.

Braida, L.D., N.I. Durlach, R.P. Lippmann, B.L. Hicks, W.M. Rabinowitz, and

C.M. Reed. 1979. Hearing aids—A review of past research on linear amplification, amplitude compression and frequency lowering. ASHA Monogr. 19.

Davis, H., S.S. Stevens, R.H. Nichols, C.V. Hudgins, G. Peterson, R.J. Marquis, and D.A. Ross. 1947. Hearing Aids—An Experimental Study of Design Objectives. Harvard University Press, Cambridge, Mass.

Dunn, H.K., and S.D. White. 1940. Statistical measurements on conversational speech. J. Acoust. Soc. Am. 11:278–288.

Fletcher, H. 1952. The perception of speech sounds by deafened persons. J. Acoust. Soc. Am. 24:490–497.

Fletcher, H., and R.H. Galt. 1950. Perception of speech and its relation to telephony. J. Acoust. Soc. Am. 22:89–151.

French, N.R., and J.C. Steinberg. 1947. Factors governing the intelligibility of speech sounds. J. Acoust. Soc. Am. 19:90–119.

Inglis, H.A. 1938. Transmission features of new telephone sets. Bell Syst. Tech. J. 17:358.

Kryter, K.D. 1962. Methods for the calculation and use of the articulation index. J. Acoust. Soc. Am. 34:1689–1697.

Macrae, J.H., and D.N. Brigden. 1973. Auditory threshold impairment and everyday speech reception. Audiology 13:272–290.

Pascoe, D.P. 1975. Frequency responses of hearing aids and their effects on the speech perception of hearing-impaired subjects. Ann. Otol. Rhinol. Laryngol. 84(suppl. 23).

Powell, M.J.D. 1964. An efficient method for finding the minimum of a function of several variables without calculating derivatives. Computer J. 7:155–162.

Radley, W.G., W.L. Bragg, R.S. Dadson, C.S. Hallpike, D. McMillan, L.C. Pocock, and T.S. Littler. 1947. Hearing Aids and Audiometers. Report of the Committee on Electroacoustics. His Majesty's Stationery Office, London.

Richards, D.L. 1973. Telecommunication by Speech. Butterworth & Co., London.

Skinner, M.W. 1976. Speech intelligibility in noise-induced hearing loss: Effects of high-frequency compensation. Doctoral dissertation, Washington University, St. Louis.

Wilber, L.A. 1964. The application of the articulation index concept for prediction of speech discrimination scores. Doctoral dissertation, Northwestern University, Evanston, Ill.

CHAPTER 18

Discussion Summaries

CONTENTS

Panel discussions took place in five scheduled sessions. One session followed each of the four major sections of the conference and an additional session was scheduled midway through the papers on the factors affecting the hearing aid's output signal.

The discussions were substantially edited for readability and continuity. Some questions and answers were deleted because they involved relatively trivial matters or material dealt with more fully in the chapters. The primary goal of the editors was to present as accurately and clearly as possible the apparent intent of the questioner and of the respondent. To help ensure this, the edited versions of the discussions were circulated among the panel members with the near-verbatim transcripts of the discussion for their review, correction, and comment. Their corrections were used by the editors to modify the edited discussions and to produce the final versions of the discussion synopses.

SESSION I
(Chapters 1–5)

Acoustic Factors Affecting the Hearing Aid's Input Signal

Moderator

George F. Kuhn

Discussants

D.A. Berkley, N. Erber, B. Johansson,
A. Nábělek, R. Plomp, E.A.G. Shaw, G.A. Studebaker

Shaw was asked how an in-the-ear (ITE) hearing aid would affect the pinna effects since most, if not all, ITE aids plug the concha. Shaw responded that one could derive substantial benefit from a hearing aid microphone placed at the base of the concha provided that most of the concha is free of obstruction. With current aids the concha gain is lost because the concha is filled. The presence of the pinna flange increases the response at frequencies greater than 2 kHz by approximately 2 dB in the frontal sector and substantially reduces the response from the rear (see Teranishi and Shaw, 1968).

Kuhn agreed with Shaw's response and corroborated this by citing an article by Gardner and Gardner (1973) of Bell Laboratories, who studied the effects on localization by occluding various parts of the pinna. Kuhn stated that he had repeated a few of these experiments by measuring the pressure transformation to the coupler microphone under various occluded conditions. He concluded that "if you occlude the entire concha, it will not do anything for you. If you occlude various parts of the pinna, such as the fossa or the cymba, you will get some response, which in some cases has some relationship in terms of directivity to what you get with the unoccluded pinna." Kuhn added that the response with ITE hearing aids depends on the size of the aid, the size and location of the occlusion, and the operational frequency range of the hearing aid.

The next question was why a hearing aid that featured a pinna microphone, where the entry port of the microphone was directed down to the meatus, had so quickly disappeared from the market. If only a canal mold was used (rather than a shell or full mold), some of the wave guide effects of the pinna should remain. Johansson was familiar with this type of aid since the Danish industry presented two models of it after investigation in his laboratory. He encountered success with this aid in certain cases, but widespread usage did not continue because of a feedback problem.

A member of the audience then asked about the effect of open molds on hearing aid performance. Shaw began the discussion with an explanation of the effects of open molds on the output of the hearing aid, stating that the response of the external ear is changed when the entrance to the ear canal is closed off, but a resonance of lower frequency remains if it is at all open. An extreme case is the use of a "no-mold fitting" in which only the sound tube is placed in the ear canal. If the tube is inserted deeply in the ear canal, then there is good coupling at high frequencies, but very little coupling at low frequencies. All open molds do this to some extent, but with a "no-mold" arrangement, the gain rises with the ear canal resonance and full gain is only attained at 2 to 2 ½ kHz. There is very little gain at 1 kHz. This characteristic would be most valuable for people with high frequency loss.

Shaw then explained the effects of open molds on the input to the hearing aid. If the open mold has a fairly wide opening, the ear will collect sound quite normally at low frequencies. However, once appreciable gain is provided by the hearing aid, the directionality of the amplified signal will depend on the microphone. Thus, the direct pickup of signal by the open ear is only useful at low frequencies where there is little directionality apart from that caused by head diffraction.

Berkley was asked if the microphones from the dereverberation processor he described can be effectively placed in a wearable hearing aid to enhance speech intelligibility. Berkley responded by explaining that since speech intelligibility was not the measurement they studied, he does not know if speech intelligibility was improved with the dereverberation processor. The normal-hearing subjects generally had about an equal preference for the full binaural presentation and the output of the dereverberation processor. The question remains whether there is a relationship between speech intelligibility and preference. Berkley emphasized that the processor itself is a large-scale, rather slow computer. There is no real-time processor to do this. It probably can be done in real time, but not with wearable equipment. Thus, it is not feasible at present to connect the processor with a wearable hearing aid.

Pursuing an earlier question concerning the open mold, the next questioner asked Studebaker whether he observed the problem of phase shift in the hearing aid amplifier producing destructive interference. Studebaker stated that he had observed destructive interference of an amplified signal by the direct signal that enters the ear canal through the open mold. In a specific instance, it decreased the level by 4–5 dB over a frequency range of approximately 800 to 2500 Hz. There are also other kinds of effects of interactions between the signals, which pass both inward and outward through open molds, causing both constructive and destructive interference effects. (Some of these are discussed in Chapter 9.)

A question from a member of the audience was addressed to Shaw citing an investigation by Batteau (1967). Batteau did detailed studies of the pinna, but instead of using steady-state stimuli in the frequency domain, he used transient stimuli analyzing in the time domain. It was asked whether the ear is designed to process the time cues for transients and if Shaw's findings are a result of this processing, as suggested by Batteau, or if the reverse is true.

Shaw replied that Batteau's hypothesis (the existence of a monaural time domain processor) would be difficult to prove because there appeared to be no psychologic or physiologic evidence to support it. Essentially, Batteau was concerned about the interference between direct and reflected waves from parts of the pinna. If one could explain the acoustic behavior of the ear in terms of a *single* reflected wave, this could be a fairly powerful idea. For example, the discrimination between front and back around 4 to 5 kHz can be ascribed to a "reflected" wave from the edge of the pinna. However, Batteau was mainly interested in the concha, where there are multiple reflections. In Shaw's view the mode description is the correct one to use and the one that will produce powerful results when dealing with a cavity that is only a half-wavelength wide.

Berkley felt that, in general, whether to use the time domain or the frequency domain mathematical description is largely a question of aesthetics. However, in the case of the very short time delays encountered in the outer ear, *perceptual* effects are entirely in the frequency domain and Berkley concluded that there is no question that the normal mode description (or some type of modal description) is mathematically more appropriate for these complex situations.

Killion returned briefly to the problem of an occluded concha by providing a short overview of microphone placements over the years. He noted that at one time most hearing aid microphones were located behind the ear. In the mid-1960s the microphone was moved from *behind* the pinna to *over* the pinna, producing an improvement of several decibels in the front-to-back ratio (Berland and Nielsen, 1969; Lybarger and Barron, 1965). Moving the microphone from over the ear down to in front of the pinna flange results in a further enhancement. The entire effect of the concha-pinna combination is not realized, "but neither is it true to say that if you move it [the microphone] to in front of an occluded concha, it has no useful effect," Killion stated.

The next question, directed to Nábělek, was why two rooms with the same reverberation time but different volumes would have a different effect on intelligibility. Nábělek explained that time distribution of reflections differs in rooms of different volumes. How the differences in pattern of reflection influence intelligibility is not well known. Because most investigators have described test rooms only in terms of reverberation time, she compared intelligibility in rooms of different volume but similar reverberation time.

Plomp injected that if investigators would instead do their measurements at distances expressed relative to the "critical distance" (i.e., the distance where the direct and reflected signal levels are equal), results would be comparable.

Kuhn suggested that in order to define the parameters across rooms, a measure of the diffuse field spectrum relative to the progressive wave spectrum would probably be appropriate.

Berkley asked Nábělek if, in her study, she always measured at the same distance in a small and a large room. Nábělek responded that she did not because in both rooms she chose a distance that was greater than the critical distance for that particular room.

Berkley commented that even if one is working outside the critical distance, significant problems may arise. Using a chamber of 13 m³, one is below a critical size (for a 1-sec reverberation time), and the modal structure of the room (below about 1 kHz) no longer overlaps, the modes no longer behave statistically, and linear decay on a log scale may not be obtained. In short, reverberation time becomes meaningless. If one wishes to use reverberation time as a measure, the room should be in excess of 400 m³ (for a 1-sec reverberation time decay). Decay, above approximately 200 Hz, may then be reasonably represented as a reverberation time.[1]

Plomp stated that, using metric measures, $\frac{1}{16}$ of the square root of the volume divided by reverberation time equals the critical distance:

$$r_c \sim \frac{1}{16} \sqrt{\frac{V}{T}}$$

where r_c = critical distance.

Erber commented that it is an important practical consideration to develop specifications for maximum allowable talker-microphone distance based on room reverberation effects, so that teachers of the hearing impaired could be informed about how far they can place themselves from the children (or the microphone) without adversely affecting speech intelligibility.

Shaw noted that a young scientist had suggested that there was only one real solution for persons with severe hearing losses: a hearing aid with

---

[1]The formula for critical frequency is given by:

$$f_c > 4000 \sqrt{\frac{T}{V}}$$

where V = room volume in m³, T = reverberation time in sec, and f_c = frequency statistical modal behavior in Hz.

a microphone that can be placed close to the talker in order to greatly reduce the reverberant sound level relative to that received directly from the source. The disadvantages of this system are obvious.

A member of the audience noted that in architectural acoustics reverberation time is typically calculated with a broadband stimulus (e.g., a gunshot or clap of the hands). There is a resulting decay of the rms amplitude of the total signal. However, there is more than one way, "frequency-wise," to generate the same reverberation time, e.g., a room of large volume that has a reverberation time dominated by low frequency components as opposed to a room of small volume with high frequency components dominating the decay. Both may have the same reverberation time, but because of the frequency content of what remains during the decay process, there may be more speech masking with one stimulus than with the other.

Kuhn summed up this comment by saying that reverberation time should be measured as a function of frequency, reiterating that the diffuse field spectrum vs. the progressive wave spectrum should be measured.

Nábēleck concurred and added that most of her studies were done in rooms in which reverberation time vs. frequency was fairly flat. The mean values that she cited were fairly close to values at all frequencies.

Shaw was asked in what circumstances he would prefer the diffuse field response to the response at a particular angle of incidence.

Shaw stated that one circumstance in which the diffuse field response might be useful is in evaluating noise exposure in factories, i.e., in determining what signals are reaching the eardrum and how to measure the noise (since most noisy places are reverberant). On the other hand, if more directionality could be obtained in the middle frequencies (between 1 and 5 KHz), this would be optimal for the hearing impaired in crowded situations.

Berkley discussed misconceptions associated with the precedence effect. He explained that when a signal and its echo occur close in time and one is distinctly lower in amplitude, it strongly affects frequency response and produces what is known as coloration. However, the signal and echo are not heard as two independent signals. Thus, precedence does not *prevent* the perception of two signals vs. one signal, and "the diffuse field measurement is . . . a good representation of the way in which we perceive the frequency structure of the received sound . . . in the diffuse field."

A member of the audience commented that for some tests of the hard-of-hearing person one may want a directional signal. It is important, therefore, that the user understand the difference between the random response curve and the direct incidence response.

SESSION II
(Chapters 6–11)

Acoustic Factors Affecting the Hearing Aid's Output Signal

Moderator
James F. Jerger

Discussants
B. Johansson, M.C. Killion, H.S. Knowles, G.A. Studebaker

Knowles began the second discussion session with some historical perspectives on hearing aid research in general and on the development of the KEMAR manikin in particular. He commented first on the unfortunate dichotomy between investigations in the behavioral and physical sciences and the need for broadly trained persons in hearing research. He noted that for nearly three decades, research on the signal processing required by different types of hearing loss and hearing tasks was largely neglected because hearing aid research was held in low esteem by those in the behavioral field. As a result, research on the physical devices (such as transducers) has outstripped fundamental behavioral research so that hearing aid designers have had few validated design objectives. Most hearing aids on the market today are not the results of fundamental behavioral research but of market research and market acceptance.

Knowles observed that there was already concern over the differences between "laboratory" and "in situ" measurements in the 1940s and 1950s. In 1959 the hearing aid industry held a series of workshops concerned primarily with the "occluded ear effect" portion of the insertion gain problem. None of the national research laboratories was interested in thoroughly researching an anthropomorphic manikin suitable for obtaining estimates of the in situ performance of hearing aids. Thus the Knowles group undertook this task, with the objective of providing a manikin that met the immediate needs of those involved in fundamental hearing aid research.

Knowles expressed concern that people now are attempting to use the KEMAR manikin for purposes for which it was not intended. He recalled that when Committee Z24.1 accepted the 2-cc coupler, it was thought that it would last only 5 years. That was around 1950. He felt that we may be falling into the same trap with KEMAR because it is being used in ways that were not intended. He stated that perhaps there is a "need for a short talk . . . about what KEMAR does not do, cannot do, and was not intended to do."

The first question was addressed to Knowles. It concerned how the characteristics of the transducer (receiver) are affected by the complex impedance of the load into which it is driven. Knowles responded that the effects are not large: The impedance seen looking back into the receiver as a sound source is generally high compared with the impedance seen looking down into the load comprised of the tubing and ear (or ear simulator), so that under most circumstances the volume velocity out of the receiver is largely independent of the load.

Studebaker agreed and added that he had observed virtually no interactions at all between receivers and earmold venting effects (which change the load impedance) using receivers of the type used in ear-level hearing aids. He suggested that an explanation for this might be that all of these receivers work through a tubing system that has a relatively high impedance and, therefore, acts as a constant isolating impedance between the receiver and the ear canal.

Killion noted, however, that there is quite a variation among hearing aid earphones in acoustic source impedance at very high frequencies. If one compares such an earphone's response connected directly into a Zwislocki or 2-cc coupler to the response it will produce through a "horn" coupling system designed to extend the high frequency response, one type may produce a 10-dB improvement with the coupling system, whereas another type may produce only a 3- or 4-dB improvement.

The next question from the audience was whether or not the impedance of the load will have an effect on the resonance peak in the transducer response. Studebaker had observed an influence of 2- to 5-dB difference between the magnitude of the peak when observed with the Zwislocki coupler as a load, compared with the same resonance with a 2-cm^3 coupler as a load. This may vary somewhat with different receivers and it was limited to the "primary" resonance zone (i.e., the lowest frequency resonance).

Knowles cautioned that a description of the receiver response is generally much more complex than a description of a resistor because of the influence of the specific load on the transfer characteristics of the receiver. When a "receiver response has a peak in it you can specify the location of that peak and its size only by assuming certain impedance values in the load with which it interacts." (In this regard, it should be noted again that the load represented by the ear or coupler, per se, is often effectively isolated from the receiver by the impedance of the coupling tube(s). Thus large changes in coupler impedance may have little effect on a system response peak whereas changes in the coupling tube may produce large changes in the system response peaks.)

Killion was asked about effects of a higher frequency resonance introduced into the ear canal shortened by an earmold. He responded that although one might superficially expect to see an 8-kHz quarter-wave resonance in the shortened canal, both the eardrum and the earmold outlet appear as relatively high impedances (so that the first important resonance mode is the half-wave resonance at roughly 16 kHz). You might see some effect of the quarter-wave resonance, but you are really dealing with a distributed system: the shortened portion of the ear canal is just the final section of a multiply discontinuous transmission line coupling the receiver to the ear. It is the overall system response that is most important.

It was asked what effect the different dimensions of the child's anatomy has on the responses of the systems designed for adults.

Studebaker emphasized the importance of more research in this area. Using available data on acoustic impedance measurements on children, a distinction should be made between the eardrum impedance (which is of interest for diagnostic work) and the impedance of the ear canal plus the eardrum (which is important for hearing aid evaluations). It is clear that the volume of the ear canal of a small child is significantly smaller than the volume of the ear canal of an adult. This has implications in that the maximum output level of a hearing aid can be greater by as much as 4 to 6 dB in the ear of a child than revealed by measurements in standard couplers or in the ear of an adult.

Tonndorf added that the problem exists only for children under 1 year of age who are fitted with a hearing aid; their ear canal is mainly cartilaginous (the bony canal is only a few millimeters long). Since the difference in canal size between a 3-year-old child and an adult is relatively small, and since hearing aids are not normally fitted before the age of 3, Tonndorf believed that this is not much of a problem.

An audience member posed two questions on Zwislocki's paper: 1) Why does the Zwislocki coupler show a hole representing the leak between the ear cap and pinna, but no low frequency resonance is seen that would correspond to that leak? and 2) What is the future of the complex Zwislocki coupler? What advantages does it have over the 2-cc coupler? He suggested that the Zwislocki coupler was too complex and had the disadvantage that it required factory calibration and adjustment.

In answer to the first question, Studebaker explained that the leak built into the coupler is effective for use under a supraaural earphone cushion, not under an earmold tip, and therefore data derived with hearing aid receivers would not show the effect of this leak.

In answer to the second question, Knowles pointed out that many years had gone by during which standards groups had struggled unsuccessfully with the coupler problem. The Knowles group searched the literature for an acceptable coupler and felt that the Zwislocki coupler was the preferred one to start with. The decision to manufacture the Zwislocki coupler reversed the normal procedure where the setting of standards and definitions precedes manufacture.

Knowles agreed that it was a good point to question the need for four meshes. He observed that several two-mesh coupler designs have been proposed that provide a satisfactory approximation to the ear for many purposes. Another manufacturer has developed a one-mesh system, but it is based on measurements on only four ears of two people. That seemed to Knowles to be a precipitous swing of the pendulum from one end to the other. A final point made by Knowles was that Zwislocki coupler calibration has been sufficiently simplified that it can now be conducted in almost any acoustic laboratory.

Johansson added that the two couplers serve two different functions. A 2-cc coupler is usually used for a control purpose. It is very rigid and you can rely on it. An ear simulator, however, is a laboratory instrument that requires more control because it is adjustable.

A member of the audience opined that current researchers are overly concerned with fine details and median values, because there are spreads of 25 to 30 dB across individuals in insertion gain and another 25 dB spread in functional gain, as some research has indicated.

Killion responded that some of the large individual differences that have been obtained can be tracked down to some of the same experimental artifacts that caused the "missing 6 dB" problem to remain in the literature for several decades. (See Chapter 8 for further discussion of some of the measurement problems.)

Studebaker suggested that each possible source of individual differences be examined separately. Some of the possible sources may be attributable to the location of the individual's head in the room, others to differences in ear canal and eardrum impedances. Some of these factors might be controlled but others cannot be.

A member of the audience requested that Johansson elaborate on attack and release time for AGC hearing aids.

Johansson replied that in order to avoid a discomfort reaction in the patient, the attack time should be of the order of a few milliseconds (optimally below 5 msec). In order to keep intelligibility at a maximum level, release time should be under 100 msec. If there exists a fairly large dynamic range and a good signal-to-noise (S/N) ratio, the release time can be made longer.

SESSION III
(Chapters 6–11)

Acoustic Factors Affecting the Hearing Aid's Output Signal

Moderator
James F. Jerger

Discussants
R.M. Cox, D.P. Egolf, S.F. Lybarger, M.C. Killion

Jerger asked Killion and Cox to comment on the clinical implications of their papers.

Killion stated that with currently available technology and earmold modification techniques almost any parameter of hearing aid performance can be modified to suit very specific needs. If what is desired is not available from some manufacturer, it is probably only because he has not been convinced that enough people would purchase the desired aid to justify the cost of bringing it to the market. Similarly, it is possible to make earmold systems "jump through a hoop" through the use of vents, damping, and horn-like construction, but such sophisticated earmolds may or may not be practical in the field. Many dispensers do not feel comfortable with simple venting techniques; if methods to vary the high frequencies are added, the whole system might bog down.

Cox said that soon clinicians will be in the position to perform hearing aid evaluations with recorded samples of speech through various hearing aids, should that appear to be desirable for other reasons.

Jerger asked Lybarger if one of the clinical implications of his talk was a suggestion to use closed earmolds and electronic modifications as much as possible and to use vents and skeleton molds only as a last resort.

Lybarger indicated that he did not intend to convey that impression. There are many instances when open ear canal fittings should be used (e.g., ski slope losses). Lybarger said that what he intended to convey was that in higher gain aids, unnecessary venting can cause feedback problems. It is vital to know the responses of various vent lengths and diameters and the closed response before the addition of the vent to the system. He added that data to be published would give the clinician a good idea of what he would get with a particular vent.

A member of the audience noted that the proper fit of a hearing aid depends on the manufacturer, the dispenser, and the earmold laboratory, all three of whom are dependent on one another. Would it, therefore, be more beneficial to supply a device with the earmold bore as part of the system?

Lybarger agreed and noted that such a system would increase the applicability and usefulness of the type of tubing systems described by Killion.

Killion noted that, with ITE hearing aids, the earmold system is already a part of the hearing aid. He believed that if there were a demand for earmolds of a particular type, there would be no difficulty convincing the earmold laboratories to supply them. The initial cost of the mold would be greater, but battery drain would be reduced compared with systems that produce the same high frequency output using a conventional earmold.

An audience member asked why, if there has been information regarding earmold acoustics for some time, has it not been implemented?

Killion responded that perhaps it has not been implemented because poor earmold acoustics has not been the biggest problem until recently. Designers like Mas Harada have been pointing out for years that the hearing aid sound quality problem was not primarily a frequency-response problem but an amplifier overload problem. Another reason is that only recently has the sound *quality* in hearing aids received much research attention (notably at Bertil Johansson's laboratory). Indeed, we really know disgracefully little about what characteristics are most important to the hearing aid user: we have seen an enormous amount of research regarding speech discrimination with hearing aids, but very little research indicating that it was really the most important problem for the user. A large amount of anecdotal evidence from the marketplace indicates that when the price of a 15%–20% improvement in speech discrimination in noise (the usual clinical measure) is a harsh or tinny sound, the user would rather structure his environment to get another few dB improvement in S/N ratio (which is all the improved discrimination usually represents; see Carhart and Tillman (1970), for example) than cope all the time with a harsh sound quality. (The multiple adjustments offered by many modern hearing aids may circumvent that dilemma by allowing the user to switch between maximum intelligibility and best quality.)

Egolf mentioned that current investigations are being carried out that will make it possible to have programmable digital filters built right into the hearing aid. We should consider standardizing some kind of a vent and then tailoring the frequency response electrically with a programmable filter rather than acoustically by altering the vent. As a result, Egolf concluded, the work of audiologists and acoustic engineers might be reduced. An audience member commented that such digital filtering would not diminish the work of the audiologist. Tests would still be required with different settings of the aid but the changes could be made very quickly.

Jerger recalled that in the late 1950s a receiver was introduced on the market that had a very broad frequency response, but the sound quality was poor and consequently it was not a commercial success. He wondered why high frequency hearing aids would be any better received now.

Killion responded that there are now more people with mild to moderate hearing losses (who attend symphonies, operas) who wear hearing aids. Also, with the early broadband aids, there may have been a tendency to keep the 2-cc coupler response high throughout the frequency range. That would result in a sharply rising real-ear (insertion gain) response above 3 kHz (see Chapter 8), which might cause a tinny sound.

Lybarger concurred with Killion regarding the reason for the current popularity of high frequency hearing aids, pointing out that many more people with 30- to 40-dB hearing losses are wearing hearing aids now than was true of the time referred to by Jerger.

Johansson related that he found a positive correlation between peaked high frequency reproduction and detailed dimension disturbances. He argued that the threshold of detectability for both nonlinear and linear distortion should be ascertained and an attempt be made to maintain the reproduction below the threshold of detectability for these distortions for the different groups of hearing-impaired people (perhaps with AGC circuits). He felt that nonlinear distortion is important in contrast to some earlier comments that had been made.

A member of the audience suggested that it would be beneficial for hearing-impaired children in public schools to have an ear-level FM reception hearing aid.

Egolf stated that with current technology, this would pose no problem.

Lybarger asked Killion what he meant by the term *maximum undistorted output* on the receivers mentioned in his papers.

Killion explained that most manufacturers have shown their earphone's response to be obtained with a very high electrical source impedance, with no restrictions on the AC voltage developed across the earphone terminals. In a typical hearing aid, however, only 1.5 V are available, so that at high frequency voltage clipping occurs long before earphone overload. In order to obtain large, undistorted, high frequency output, one must use a receiver with low electrical impedance (and accept the higher battery drain) or use special earmold construction.

A member of the audience asked what the maximum low frequency harmonic distortion permissible without interfering with speech intelligibility is.

Johansson said that 50% second-order harmonic distortion will still result in nearly a 90% speech discrimination score. He noted that even clipping to a square wave can result in good discrimination, as demonstrated in the 1940s.

Killion referred to a study by Peters and Burkhard (1968) that compared four easily generated nonlinearities and found that one system, which had a 40% total harmonic distortion, had an intelligibility rating of about 98%, whereas another, which had a 20% harmonic distortion, resulted in a 40% to 50% intelligibility rating. Thus, it is important to specify the nature of the distorting mechanism.

Jerger asked the panel whether the clinician need concern himself with the effects of minute deviations in tube length and diameter or harmonic distortions.

Johansson replied that if S/N ratio is high, these considerations do not affect speech intelligibility to any extent, just sound quality.

A member of the audience remarked that one effect of an aid that has frequency response peaks is to cause the user to reduce the gain below the optimal level. He asked Cox to elaborate on her comments that positive feedback increased the peakness of the hearing aid's response.

Cox replied that it was not unknown for a clinician to select a hearing aid with a vented earmold and then to increase the gain as much as possible. In other words, it would be turned up to just a little less than would produce audible feedback. This practice increases the irregularity of the frequency response by a great deal in most hearing aids, even though audible feedback is absent. Most clinicians believe that this is satisfactory as long as there is no audible feedback, but this is not so.

Lybarger concurred and commented that by increasing the gain by only 2–3 dB, a small peak may be transformed into a sharp peak. The hearing aid is sensitive just below oscillation, and it only requires 4 to 5 dB to produce the peak.

SESSION IV
(Chapters 12–13)

Modeling the Acoustic System

Moderator
James F. Jerger

Discussants
D.P. Egolf, M.C. Killion, H. Levitt,
S.F. Lybarger, D. Mook, G.A. Studebaker, R.E.C. White

Levitt was asked to comment on the clinical implications of modeling the hearing aid acoustic system.

Levitt stated that the original objective was to design an acoustic coupling system to meet certain frequency response requirements in fitting a hearing aid. The relevance of this work is that it is possible to obtain great insight mathematically into the various parameters of the hearing aid. Once programmed, a computer can solve the mathematical equations. The clinician will then be able to utilize these solutions for different specific problems. Also this work will lead to better understanding of the acoustic system through the computer solving these rather complex equations.

Studebaker elaborated further on some of the implications of current research on modeling. A necessary preliminary to clinical applications of modeling techniques is a knowledge of all the variables in hearing aid systems that influence the signal at the eardrum. In the research laboratory modeling helps identify areas where additional information is needed. Such information will ultimately permit the clinician to control frequency response exactly and to know exactly what is delivered to the individual patient's ear.

One specific application of models is in the design and utilization of hearing aid playback systems in the hearing aid evaluation. A possible system is one wherein hearing aid-processed signals are stored in a computer or on tape. These stored signals would then be presented to the subject instead of using real hearing aids. During the reproduction of the stored speech signals the computer could be used to introduce frequency-response effects, which would simulate the effect of various earmold modifications, tone control settings, output limitation methods, etc. Thus, the result at the eardrum could be adjusted by the clinician to specific values quickly and easily. Also, the clinician could receive from the computer instructions on what combinations of hearing aid, tubing, earmolds, and vents would best produce the desired frequency-response result. As a part of this approach it might be desirable for the clinician to obtain impedance information about the individual patient's ear. The computer program would use this information as part of the calculation scheme, thereby taking into account even the interactions between an individual's ear and a specific hearing aid-earmold tubing system. However, he noted, the possibility of such capability emphasizes the enormous lack of information on what frequency response is in fact needed at the eardrum. There is a great need for research in this area.

Lybarger remarked that modeling techniques have been used in the transducer industry for 30 to 40 years, but is just now being appreciated by other disciplines. He noted that it got its big start in 1926 when Maxfield and Harrison published data on a recorder cutting head. He believes that all transducers made for hearing aid use take advantage of analog modeling.

Levitt noted that there still exists a large gap between our understanding of acoustic systems and the clinical application of this knowledge. The situation is analogous to what occurred in the case of the acoustic bridge and the Zwislocki coupler. Acoustic impedance was understood long ago, but only in the 1950s was the clinical value of acoustic bridges recognized. Similarly, the Zwislocki coupler is an application of straightforward acoustic principles, but its clinical utilization was realized only recently.

Egolf commented on the desirability of using a two-part modeling technique on the computer (if the model was proved to be accurate) to assist the clinician in the selection and fitting of a hearing aid. He pointed out that a clinican who had little or no knowledge of the mathematical principles involved could feed patient data (e.g., eardrum impedance) to the computer and have the computer select the hearing aid configuration that best suited the patient's needs. In this way the choice of microphone, receiver, vent diameter, vent length, and other variables would be based on principles of mathematics. Egolf stated that technology is currently available to implement such a scheme. He also added that a computer method of this type should not be viewed as a substitute for common sense in clinical practice.

A member of the audience asked Egolf about his technique of two-port modeling and whether it would provide an efficient means of measuring eardrum impedance from the actual human ear by reversing the paradigm and using a calibrated sound source.

Egolf replied that he had attempted to do this, but that it had not yielded favorable results. It is a marginal technique when used in that way. The mathematics work out, but there are many difficulties, physically. The major problem Egolf encountered was the great impedance mismatch through the transducer. A "soft" transducer (i.e., one in which the acoustic load has a large effect on the electrical side of the model) would produce more favorable results in this application. Egolf stated that the underlying question is how to build such a load-sensitive transducer.

A member of the audience asked, If two sets of identical impedance information obtained from two patients were fed into the computer, would the computer recommend two different aids for better speech perception?

Levitt replied that if the two patients exhibit different loudness discomfort levels, obviously less gain would be required for the one who shows less tolerance, and therefore two different aids should be recommended. The computer cannot substitute for a clinician's common sense. Also, one must bear in mind that the computer gives only approximations, and audiometric measurements are only an approximation of the status of the residual hearing mechanism.

White emphasized how vital it is to keep open communication channels among the various disciplines involved in order to be aware of the limitations of these systems and to ascertain how to apply these techniques sensibly in practice.

Lybarger was asked whether he thought hearing aids should be designed with greater "variability" in performance characteristics to better fit the individual with specific problems, to do away with the "assumptions" that Levitt had mentioned.

Lybarger responded that the crux of the situation, as Studebaker emphasized, is knowing what to deliver. If one has variability in a hearing aid, he continued, the result is more flexibility, but the important issue is what the relationship is between what is delivered to the ear and what is actually heard.

A question was posed concerning the need to determine the impedances of individual ears over the frequency ranges in which hearing aids are used, rather than the medians that have been used in establishing these systems. With more open earmolds the resonance of the ear canal is within the hearing aid operating range, requiring knowledge of the canal length unoccluded portion. Where do we find the information that we need?

Levitt replied that information about impedances at the higher frequencies is not well known. The approximations made in the use of conventional bridges are unreliable. Only recently have more reliable impedance measurements become available.

Levitt emphasized that there is a need for volume velocity measurements or, alternatively, two pressure measurements in close proximity in order to solve the necessary equations for impedance.

Killion noted that it is easy to get carried away with the engineering aspects. With the computer programs now available, it is possible to predict, within a few percentage points accuracy up to 20 kHz, the coupler response of entire hearing aid systems. However, in dealing with people Killion favored the psychoacoustic approaches used by Fournier in 1965 and Pascoe in 1975 for obtaining the functional gain of the hearing aid. There are potential errors in these procedures, but they are all solvable. Most of the preceding discussion concerned matters that would reduce 2- to 3-dB errors to 1-dB errors, whereas routine psychoacoustic measurements on people may be accurate only to about 5 dB.

A member of the audience pointed out that models are applicable to perceptive distortions (e.g., recruitment, lack of frequency selectivity) as well. Such models are useful in discriminating between the more viable hypotheses and ones that are not likely to be useful.

Mook stated that there are two types of models. The first is a very precise, complex model that reliably translates results from one situation to another and helps to avoid redundant measurements. However, because of its complexity, it is difficult to obtain any insight from this type of model. The simpler model (e.g., the lumped-parameter model) helps one to understand a system and to redesign a system in order to change its performance.

Braida asked the panel if any applications of modeling theory can be used to reduce feedback effects, which can limit the functional gain of a hearing aid to the order of 50 dB.

Egolf replied that the next logical step in the mathematical model he has created is to consider the feedback problem. He said it was simply a matter of reversing the role played by the vent in his computer scheme. The vent would be treated as a source of sound rather than as a vehicle by which sound energy is lost. He added that the sound entering through the vent would necessarily be different in amplitude and phase from that entering by way of the hearing aid microphone.

Levitt replied that if the conditions are described explicitly, the occurrence of feedback can be minimized. Techniques have been developed for minimizing acoustic feedback in public address systems. They require sophisticated processing. By using some of these procedures from other fields the problem may be overcome using a microprocessor hearing aid.

Kuhn returned to the question of accuracy and the range of measurements on the impedance of the eardrum. He suggested the possibility of avoiding acoustic measurements entirely, and instead using fiber optics (specifically, an optical interferometer) to observe displacement at the eardrum. The pressure could be measured at the canal entrance and predictions could be made statistically on what the pressure would be at the eardrum. It would still be difficult, however, to measure impedance at very high frequencies.

Egolf returned to the question posed earlier by Levitt of how to measure acoustic volume velocity in order to determine eardrum impedance by direct measurement. He noted that the interferometry technique mentioned by Kuhn is one way to circumvent the requirement for measuring volume velocity in the ear canal by measuring velocity of the eardrum instead. The inability to measure volume velocity directly has forced acoustic engineers to make measurements under two sets of conditions rather than one (as electrical engineers do) when determining impedance. The two-load method is a good example of this, Egolf concluded.

Mook agreed that the use of a flexible hearing aid would avoid the problems of using a sophisticated modeling system. With a flexible digital hearing aid, the tests could be conducted through the hearing aid on the subject without need for a model.

Lybarger noted that the simplest way to eliminate feedback is to observe the response of the hearing aid relative to the leakage signal escaping from a vent or earmold. Generally, this leakage occurs around 2–3 kHz, even with a vent that only functions in the low frequencies. One solution is to use a notch filter in the hearing aid to reduce the performance at the frequency where feedback radiation is high.

Kuhn noted that White had measured the open-circuit transfer function of the diaphragm and that this had resulted in some phase effects that could not be accounted for. Kuhn suggested that one possibility was that there were some losses in the coil mechanism or losses in the magnetic material. He suggested the possibility of modeling the transduction mechanism by a complex transformer ratio.

White agreed that that was one possibility. He had made the point that this type of transducer mechanism is unusual because when variations in the static magnetic field are introduced in parts of the circuit, energy is lost because of hysteresis. Other losses are caused by eddy currents and by mechanical resistances to movement, but were eliminated from consideration in the model presented.

SESSION V
(Chapters 14–17)

Hearing Aid Response Characteristic Selection Strategies

Moderator
N. Erber

Discussants
L.D. Braida, M.J. Collins, R.L. Dugal, J.F. Jerger,
H. Levitt, J.D. Miller, M.W. Skinner, E. Villchur

Erber pointed out an important factor that had not been mentioned in the formal presentations: in most situations hearing-impaired individuals use their vision (speechreading) to supplement auditory cues (Erber, 1975). The use of vision during speech perception is overlooked by many researchers because it can be a confounding variable that makes many acoustic assumptions difficult to support.

With respect to the role of visual cues, Levitt agreed with Erber that it is an important factor to be considered, and that visual cues need to be studied in isolation and in conjunction with acoustic cues. The influence of visual cues on auditory perception is highly interactive. Most of the clinical tests that have been standardized for evaluation of auditory perception are based on purely acoustic input. Very large interlist differences can occur when a simultaneous visual presentation is included.

Levitt was asked why he had not used commercially available taped speech stimuli, such as the synthetic sentence index (SSI). He replied that since the SSI is a closed response set test, using sentences, a major cue for identification is sentence prosody. These cues are primarily in the low frequencies, whereas high frequency cues help to differentiate between hearing aids. Thus, it was decided to use a test with a fair number of consonants, controlled in a particular fashion (e.g., preceding or following certain vowels) to concentrate on those cues that are sensitive to differences between hearing aids.

In response to a question Levitt indicated that the noise he used was generated by a broadband noise source passed through a ⅓-octave band multifilter. Levitt was then asked whether only narrowband noise was used. He replied that ⅓-octave bands of noise were used throughout for sound field measurements. He noted that there were special problems with this stimulus. For example, a person with a steeply falling loss might appear to detect a band of noise at a certain frequency, when in fact the noise in the skirt of a neighboring filter passband is being detected. The reasons for

424

choosing ⅓-octave bands of noise as the narrowband stimulus were that they were standardized stimuli and the necessary instrumentation was commercially available. He went on to note that the threshold norms they used were those developed by Pascoe (1975).

If the speech spectrum was crucial to Levitt's results, a questioner wondered, then why not obtain most comfortable listening level (MCL) and loudness discomfort level (LDL) with speech stimuli? Levitt answered that the speech spectrum was not crucial to the results but was a factor considered in post-hoc interpretation. The use of actual speech through ⅓-octave band filters is a very good idea because it does contain the same temporal characteristics of real speech.

Levitt then commented on the use of the articulation index (AI) as a predictive tool. He felt that although it is a potentially valuable procedure, it may be premature to draw strong conclusions. There are studies (Flanagan and Levitt, 1969) that reveal weaknesses in AI procedures with normals. The AI procedure works very well for linear distortions for low-pass or high-pass filtering or additive noise. However, if the noise is of the comb-filter type (i.e., noise with gaps in the frequency spectrum), or intermittent, or if the distortions are nonlinear (e.g., harmonic), or if one is combining additive noise with various distortions (i.e., noise plus reverberation), then this procedure is not a good predictor. The hearing-impaired listener has a nonlinear form of distortion, heavily superimposed on whatever frequency-filtering distortion may exist.

Braida indicated that the form of articulation theory that was used in his investigations did indeed incorporate some nonlinear distortions that are known to exist in hearing impairments (e.g., reduction in dynamic range and the spread of masking). Since the data on spread of masking in abnormals are sparse, Braida used what he considered crude approximations. This method was not used to supplant more traditional methods, but it should not be ignored. Braida maintained that studies in this area have not been sufficiently systematic in their variation of parameters (e.g., variation of level and frequency-gain characteristic) to permit even a crude analysis to be applied to the results.

A member of the audience asked whether there is any justification for drawing conclusions about what will happen with dynamic signals (speech) based on static measurements and conditions.

Skinner said that we should use a variety of narrowband stimuli other than ⅓-octave bands of noise to obtain the MCL, uncomfortable loudness level (UCL), and threshold of audibility of hearing-impaired listeners. The findings will be important for comparisons between predicted and observed optimum frequency responses of amplification systems.

Villchur responded that there is a danger in combining steady-state information with transient phenomena (e.g., discomfort levels may differ). However, most investigations have ignored the dynamic range; therefore, one has had to rely on approximations based on steady-state data and attempts at refinining these approximations by allowing the subject to make readjustments with the transient stimuli.

A member of the audience voiced disagreement with the repeated favorable references to the Harvard study (Davis et al., 1947) because of what he felt were the misleading conclusions that have been derived from it and because of its poor experimental design.

Levitt responded that it was a good study. Although, in retrospect, certain aspects of the study could be faulted, the basic design was good. Unfortunately, the results of the study have been overgeneralized. The majority of subjects in that study (and in the British Medical Research Council study) had conductive impairments. Today, the majority of hearing-impaired persons who are considered candidates for amplification have primarily sensorineural impairments. Also, many persons being fitted with hearing aids today have severe losses that would not have been considered aidable at the time of the Harvard study.

Braida was asked to elaborate on his use of the "proficiency factor."

Braida explained that the proficiency factor measures the training of the talkers and listeners. Theoretically, it is derived from the performance of the subject under the best possible test conditions—from a single data point, possibly. In reality, since he did not know the shape of the articulation function, he estimated the proficiency factor from all the test points.

Fletcher used proficiency factors of 0.8 and 0.6, Braida continued, whereas he (Braida) needed proficiency factors from 0.2 to 0.4. It is possible that one would want proficiency factors that are functions of frequency and that might reflect degraded frequency resolutions. The accuracy of the data, however, is such that it is not necessary to have an elaborate set of proficiency factors to get a good statistical fit of the data. Subjects with the same audiogram listening to essentially the same frequency gain characteristics get very different performance scores, as shown in Miller's presentation of Pascoe's results (see Chapter 16).

Miller noted that the point he was making was that if you knew the audibility curve, you could predict the subjects' speech discrimination scores very well. Therefore, two subjects with similar audiograms and similar frequency responses would get similar scores.

"But," Braida interposed, "in reality they do not."

Miller indicated that at least in the Pascoe study they did.

A member of the audience asked whether noise stimuli should be used in determining functional gain in a sound field situation, instead of pure tones, warble tones, or pulsing tones, which are used currently in hearing aid evaluations. He wanted to know why there was such concern with changing the signals used for functional gain measurements to narrow bands of noise.

Levitt responded that he used ⅓-octave bands of noise as stimuli because there is a standardized set of filters to produce them. There are certain inherent disadvantages in the use of pure tones in a real room (because of standing waves) or warble or pulsing tones (because in many cases they are not properly defined). These are important considerations because it is important that other laboratories be able to replicate the experiment. He also indicated that they were not pushing for any particular stimulus as the one to use.

An audience member added that one obtains different MCL measures if one uses ⅓-octave band levels of noise as opposed to pure tones, warble tones, or pulsed tones.

Levitt stated the opinion that LDL (rather than MCL) should be used because of its improved test-retest repeatability and because it provided information on the ear's dynamic range.

Braida agreed with Levitt that each individual researcher should decide on the signal to be used based on the requirements of the measurements he is making. They can then be translated into other stimuli that are more appropriate for clinical practice.

Miller was asked how he dealt with some of the problems inherent in using hearing aids in sound field measurements, e.g., head diffraction effects.

Miller responded that it was not an actual hearing aid that was used but a filter set, an amplifier, and a loudspeaker. It was called a hearing aid because it was possible to alter the frequency response.

Skinner added that a uniform response was used as the baseline. A change in the uniform response in the field was produced using the attenuators on the multiband filter. The frequency responses were measured by using pink noise through the whole amplification system. By comparing the ⅓-octave band analysis of the output from the new attenuation settings on the multifilter with that for the uniform system, the actual change produced by the high frequency emphasis was obtained. This difference in dB between the compensatory frequency responses and the uniform response for each ⅓-octave band between 100 and 10,000 Hz defined the functional gain.

One cannot equate the intensity levels shown on the graphs to what the individual would get through a hearing aid, Skinner remarked, because the sound pressure level at the eardrum was not measured. The transfer function from the input to the hearing aid to the eardrum is still under investigation. However, Miller said that if a commercial aid were built that had the same functional gain, it would then presumably produce the same results.

Lybarger pointed out that in all of Skinner's curves, the LDLs were almost uniformly 90 dB and wondered whether this value was consistent enough that it might be assumed in most instances.

Skinner remarked that it is important to know how the stimuli were measured to compare LDLs (across studies). The noise bands at the Central Institute for the Deaf (CID) were measured with an rms slow meter. Measuring these stimuli with a peak reading on a fast scale would result in higher levels. Ninety dB seems a common setting for everyday use of a hearing aid. Also, Skinner added, speech-like stimuli, rather than noise, would give a better estimate of LDL.

A member of the audience asked Skinner how audiologists at CID perform hearing aid evaluations.

Skinner replied that for clients with mild to severe hearing losses, ⅓-octave bands of noise centered at the audiometric frequencies are used to determine their unaided thresholds of audibility and discomfort as well as MCL (field measurements). If the intensity level is insufficient to reach the client's MCL and LDL, they are obtained under earphones and an appropriate conversion to field measurement is made. The desired frequency response of a hearing aid is determined by calculating the difference in dB between Pascoe's perceived spectrum of summed speech (Pascoe, 1978) and: 1) the mean of the MCL range at 500 Hz and 2000 Hz, 2) 5 dB less than the mean of the MCL range at 1000 Hz, 3) approximately 10 dB less than the mean of the MCL range at 250 Hz, and 4) positive sensation level at 3 to 6 kHz, if possible. This frequency response represents the desired functional gain. Then the client's aided threshold of audibility is obtained with his own earmold and one of the CID clinic hearing aids, which has a known coupler response and approximates the projected requirements. The aided thresholds are subtracted from the unaided thresholds to determine the actual functional gain, and these are compared with the coupler gains to obtain a functional vs. coupler correction. Aided LDL measurements are also made to define the maximum power output (MPO) requirements. From this information we are able to specify approximately the aid characteristics for the dealer. This procedure would be more exact if the input voltage at the hearing aid, as suggested by Miller, could be monitored (couplers vs. real ear), and the same voltage repeated for the two conditions. Basically, hearing aid evaluations in the CID clinic follow Gengel's procedure, Skinner concluded (Gengel, Pascoe, and Shore, 1971). Erber added that a modification of the Gengel procedure (Erber, 1973) was the one used with the children in the CID school, all of whom have severe to profound hearing losses.

Braida was asked what would happen to his results with flat or rising audiometric configurations. He replied that he did not know since he had not tried it, but that on the basis of general principles and some preliminary work it would probably be successful and should be pursued.

Levitt commented that to maximize the AI, one should raise speech to the highest level possible in every ⅓-octave band or octave band plus an empirical correction for upward spread of masking. A procedure based on this approach has shown relatively good results (Abramovitz, 1979).

Skinner mentioned that she has done work in this area that takes into account the spread of masking of speech on itself. By putting 5–7 dB deemphasis in the region between 1000 and 2000 Hz, slightly better intelligibility scores were obtained.

Miller stated that the reason for the notch at 1.6 kHz in the frequency response, which divides the speech spectrum into halves and the basilar membrane into halves, should be investigated. This notch suggests that the importance function should be modified, and could be caused by the way midfrequency transitions are segmented both in time and frequency domains.

Braida disagreed with Miller's concern over a particular notch since communication must occur with different talkers and in different room acoustic situations. One would thus expect the predicted frequency-gain characteristics to be smooth. It is unlikely that sharp notches at discrete frequencies would show dramatic results that are consistent across speakers.

Collins asked Braida whether, with different types of hearing loss, particularly those that involve distortion of the signal, the band importance function term in the equation might need modification depending on the degree of distortion introduced by the ear in specific frequency ranges, as well as changes in critical bandwidth in the various frequency ranges.

Braida responded that when they first began getting large amounts of scatter in the predictions between scores and AI, they considered modifying the band importance function, which would have resulted in 15 parameters per subject and an improved fit. However, Dugal did not do this but was able to obtain reasonable fits without it.

Collins suggested that this was perhaps because they had subjects with similar etiologies and similar thresholds.

Braida replied that not having to modify the band importance values for this study does not indicate that the function used is correct. A great problem is the lack of sufficient data with various frequency-gain characteristics to give one the confidence to make the theory more elaborate. To improve matters, we could either use subjects with different audiometric configurations, or establish these audiometric configurations more solidly, or test the subjects more thoroughly. Without an interplay between the collection of data and theory development, the field is not going to progress, Braida concluded.

REFERENCES

Abramovitz, A. 1979. Frequency shaping and multiband compression in hearing aids. J. Acoust. Soc. Am. 65(suppl. 1):S136(A).

Batteau, D.W. 1967. The role of the pinna in human localization. Proc. Roy. Soc. B168:58–180.

Berland, O., and E. Nielsen. 1969. Sound pressure generated in the human external ear by a free sound field. Audecibel, Summer:103–109.

Carhart, R., and T.W. Tillman. 1970. Interaction of competing speech with hearing losses. Arch. Otolaryngol. 91:273–279.

Davis, H., S.S. Stevens, R.H. Nichols, C.V. Hudgins, G. Peterson, R.J. Marquis, and D.A. Ross. 1947. Hearing Aids—An Experimental Study of Design Objectives. Harvard University Press, Cambridge.

Erber, N.P. 1973. Body-baffle and real-ear effects in the selection of hearing aids for deaf children. J. Speech Hear. Disord. 38:224–231.

Erber, N.P. 1975. Auditory-visual perception of speech. J. Speech Hear. Disord. 40:481–492.

Flanagan, J., and J. Levitt. 1969. Speech interference from community noise. ASHA 4:167–174.

Gardner, M.B., and R.S. Gardner. 1973. Problem of localization in the median plane: Effect of pinna cavity occlusion. J. Acoust. Soc. Am. 53:400–408.

Gengel, R.W., D. Pascoe, and I. Shore. 1971. A frequency response procedure for evaluation and selecting hearing aids for severely hearing-impaired children. J. Speech Hear. Disord. 36:341–353.

Lybarger, S.F., and F.E. Barron. 1965. Head-baffle effects for different hearing aid microphone locations. J. Acoust. Soc. Am. 38:922(A).

Maxfield, J.P., and H.C. Harrison. 1926. Methods of high quality recording and reproducing of music and speech based on telephone research. Bell Syst. Tech. J. 5:493–523.

Pascoe, D.P. 1975. Frequency response of hearing aids and their effects on the speech perception of hearing impaired subjects. Ann. Otol. Rhinol. Laryngol. 86(suppl. 23):5–40.

Pascoe, D.P. 1978. An approach to hearing aid selection. Hear. Instr. 29:12–16,36.

Peters, R.W., and M.P. Burkhard. 1968. On Noise Distortion and Harmonic Distortion Measurements. Industrial Research Products Report #10350–1, Knowles Electronics, Franklin Park, Ill.

Teranishi, R., and E.A.G. Shaw. 1968. External ear acoustic models with simple geometry. J. Acoust. Soc. Am. 44:257–263.

Author Index

Subject Index